Periodic Pattern Mining

R. Uday Kiran · Philippe Fournier-Viger ·
Jose M. Luna · Jerry Chun-Wei Lin ·
Anirban Mondal

Editors

Periodic Pattern Mining

Theory, Algorithms, and Applications

 Springer

Editors
R. Uday Kiran
Division of Information Systems
University of Aizu
Aizu-Wakamatsu, Fukushima, Japan

Jose M. Luna
Department of Computer Science
and Numerical Analysis
University of Córdoba
Córdoba, Spain

Anirban Mondal
Department of Computer Science
Ashoka University
Sonepat, Haryana, India

Philippe Fournier-Viger
College of Computer Science and Software
Engineering
Shenzhen University
Shenzhen, Guangdong, China

Jerry Chun-Wei Lin
Department of Computer Science,
Electrical Engineering, and Mathematical
Sciences
Western Norway University of Applied
Sciences
Bergen, Norway

ISBN 978-981-16-3966-1   ISBN 978-981-16-3964-7   (eBook)
https://doi.org/10.1007/978-981-16-3964-7

This Springer imprint is published by the registered company Springer Nature Singapore Pte Ltd.
The registered company address is: 152 Beach Road, #21-01/04 Gateway East, Singapore 189721,
Singapore

Preface

Technological advances in the field of Information and Communication Technologies (ICTs) have facilitated organizations to collect, store, and process massive amounts of data. Useful information that can empower the end-users to achieve socio-economic development lies in this data. However, finding interesting information in data can be very challenging due to the sheer scale of the data. In the last decades, researchers from the field of data mining have aimed at tackling this challenge by proposing various techniques to discover knowledge hidden in voluminous real-world data. Over the years, data mining has received more and more attention from both industry and academia. Pattern mining is one of the fundamental knowledge discovery techniques used in data mining. It involves discovering all user interest-based patterns in a database. Much of the past research on pattern mining has focused on utilizing the *frequency-based* measures to discover different types of interesting patterns such as frequent patterns, correlated patterns, top-k frequent patterns, maximal frequent patterns, closed frequent patterns, rare patterns, coverage patterns, high utility patterns, and emerging patterns.

Although discovering frequent patterns in a database is beneficial for many applications, *frequency* may not always be enough to find user interest-based patterns, especially if the data contains temporal information. For example, the user may consider an irregularly occurring frequent pattern to be less interesting over a regularly occurring infrequency (or rare) pattern in the data. Based on this observation, efforts have been put forth in the literature to discover periodically occurring patterns (or periodic patterns) in a temporal database. Since real-world data often contain temporal information, finding periodic patterns has received a great deal of attention. Furthermore, periodic pattern mining has been extended to consider other forms of data, such as quantitative temporal databases and sequences.

From a research perspective, discovering periodic patterns is more challenging than frequent pattern mining. It is because of two main reasons: (i) we need to explore new measures to determine the interestingness of a pattern in the *time* dimension, and (ii) we need to investigate new data structures to effectively record the temporal occurrence information of a pattern in the database. Thus, traditional frequent pattern mining techniques cannot be directly used for finding periodic patterns. In the last decade, this has led to the proposal of many new interestingness measures and novel data structures to discover periodic patterns.

The main motivation for writing this book is that the research on periodic pattern mining has become quite mature. There is thus a need to provide an up-to-date introduction, overview of current techniques, and recent advances in periodic pattern mining. The book is a collection of chapters written by experienced researchers who published several papers on the related topic in top conferences and major journals. The chapters were selected to ensure that the key topics and techniques in periodic pattern mining are discussed. Several of the chapters are written as survey papers to give a broad overview of current work in periodic pattern mining, while other chapters present techniques and applications in more detail. The book is designed so that it can be used both by researchers and people who are new to the field. Selected chapters from this book could be used to teach an advanced undergraduate or graduate course on pattern mining. Besides, the book provides enough details about state-of-the-art algorithms so that it could be used by industry practitioners who want to implement periodic pattern mining techniques in commercial software, to analyze temporal database. Several of the algorithms discussed in this book are implemented in the open-source PAttern MIning (PAMI) software, which is available at `https://git hub.com/udayRage/PAMI`. Anyone can download this software through "pip install pami".

Aizu-Wakamatsu, Fukushima, Japan   R. Uday Kiran
June 2021

Contents

Introduction to Data Mining

Jose M. Luna

Abstract This chapter introduces data mining, also known as knowledge discovery from data, as a process of discovering useful, interesting and previously unknown patterns from data. Some techniques and domains related to data mining are described, explaining their similarities and differences. Some data types are then analysed since data on multiple data inputs might be considered due to the natural evolution of information technology. Data processing approaches are also described, stating how to transform raw data into a readable and useful form and presenting different data representations. Finally, general data mining techniques are outlined. Mining frequent patterns and associations; predictive analysis; supervised descriptive analysis; cluster analysis; and outliers analysis, to list a few.

1 Introduction

Data in the twenty-first century is considered as the new oil. Like oil three centuries ago, learning to extract and use data's value produces huge rewards [54]. Data is, therefore, an essential resource that powers the information economy, also known as the knowledge economy, where the amount of data that a company or individual has is proportional to the knowledge they have to make the right decisions. Nevertheless, unlike oil, data availability seems infinite, and it is a cumulative resource, existing multiple ways of representing and handling data.

The computerization of our society together with the development of data collection techniques and the reduced costs of storage tools have given rise to an explosive growth of available data volumes. Almost any business, regardless of its type, generates enormous datasets every day. It estimates that 2.5 quintillion bytes are created by an average person every day in 2020 [39]. Based on that, 463 exabytes of data will be generated each day by humans as of 2025. According to the Google search statistics, over 40,000 search queries are processed every second on average, which translates

J. M. Luna (✉)
Department of Computer Science and Numerical Analysis, Andalusian Research Institute in Data Science and Computational Intelligence (DaSCI), University of Cordoba, Córdoba 14071, Spain
e-mail: jmluna@uco.es

© The Author(s), under exclusive license to Springer Nature Singapore Pte Ltd. 2021   1
R. Uday Kiran et al. (eds.), *Periodic Pattern Mining*,
https://doi.org/10.1007/978-981-16-3964-7_1

to over 3.5 billion searches per day and 1.2 trillion searches per year worldwide [39]. When Google was founded in September 1998, 10,000 search queries were served per day. Eight years later, by the end of 2006, the same amount served in a single second.

This growing data availability and usage are what truly makes our time the data age. Powerful tools are needed to gather, handle and transform tons of data into valuable information, also known as knowledge. This necessity led to the birth of data mining [19], which is a young, dynamic and promising field of research. Data mining is the result of considering critical procedures on databases, including data gathering and management and advanced data analysis. Until recently, important decisions were often made based on intuition since collected data were so diverse and huge that no proper tools were available to reach a conclusion/decision in a few seconds [5]. Important efforts have been made in recent years to develop expert systems with no biases and reducing errors as much as possible. Hence, data mining [51] is considered as an interdisciplinary subject related to the concept of *mining knowledge from data.*

Despite the data mining concept is widely used and accepted for the research community, it is still a controversial issue. It is generally defined as a synonym of knowledge discovery from data, while others consider it as a major step within the process of knowledge discovery (data cleaning, data integration, data selection, data transformation, data mining, pattern evaluation and knowledge presentation). Confusion about the data mining term was even higher when the Big Data [8] concept appeared in 2005. Since that, different companies, researchers and institutions wrongly took Big Data as the new data mining concept [60]. The truth is that Big Data [37] states for the source, variety, volume of data and how to store and process this amount of data. It can be said that data mining does not need to rely on Big Data [45], as it can be done on a small or large amount of data, but Big Data surely does rely on data mining because if we cannot find the value of a large amount of data, then that data will not have been useful. In this book, we adopt the data mining concept as the process of discovering interesting patterns and knowledge from large amounts of data considering different data sources (either small or huge ones).

Han et al. [19] rightly defined techniques and domains that fall under the umbrella of data mining: databases, statistics, machine learning. Fig. 1 aims to illustrate how these terms are related. At this point, it is important to highlight that authors assume that the figure illustrates some important concepts or techniques due to the overwhelming number of techniques that can be found nowadays would make it hardly understandable. Machine learning appears as a subset of Artificial Intelligence, existing Artificial Intelligence techniques that do not require learning (e.g. path planners [44], traditional expert systems [57], etc.). Continuing with the analysis, and from the author's point of view, it is the interdisciplinary nature of data mining research and development what eases its success. Let us start analysing the connection with databases, which is the required input for any data mining process. Without a database, it is impossible to perform the analysis and extraction of useful knowledge. At this point, it is required to consider good and high-performance query languages, query processing methods, data storage systems and data access processes so scalability is guaranteed regardless of the data size and type. As for statistics, they are sets

Fig. 1 Data mining related terms

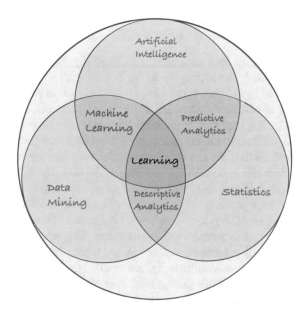

of mathematical functions that describe the behaviour of the items/events/objects in terms of their associated probability distributions. Statistics are usually considered by data mining to build models or just identify outliers, missing values or whatever targeted variable. They are also considered by data mining to represent any type of homogeneity and regularity in data, extracting good descriptors of intrinsic and important properties of data. Statistics are not only useful from the mining point of view but also to verify the outputs through statistical hypothesis testing to discard that the results were obtained by chance. Nevertheless, a major drawback of considering statistics as a part of the data mining process is the difficulty to be scaled up, which is a major problem when truly Big Data are analysed. It turns into a bottleneck, especially for some specific statistical methods that are time-consuming by nature. Finally, as for machine learning, it takes data input to learn to recognize patterns and take right decisions. Even when both machine learning and data mining are used interchangeably [58], some differences are in their purpose. Data mining is designed to extract useful, interesting, unknown patterns from data. Machine learning, on the contrary, trains a model to perform complex tasks based on data and experience. Machine learning is a well-known discipline that includes many classic problems: supervised learning (it builds a classification model to recognize/categorize future patterns); unsupervised learning (a synonym for clustering, that is, grouping patterns according to some criteria); semi-supervised learning (it makes use of both labelled and unlabelled examples when learning a model. The first ones are used to build a model, whereas the second ones are used to refine the boundaries of labels); active learning (it lets users play an active role in the learning process, asking users to label specific examples, to optimize the model quality).

2 Data Types

The key to data mining [19] is its ability to consider any kind of data input. As previously stated, we are living in a data era what implies data being handled and stored in multiple ways. Despite the multiple existing data types, four are the basic forms of data considered here: databases, transactional data, data warehouse and data lake. It is important to highlight that some of these concepts are used interchangeably by different authors, and some of them might be gathered under the same term. We include the four terms on purpose since it is our understanding that these data forms are useful to be known regardless of the way they are then processed.

A database system also called a database management system [46], is a software package properly designed and implemented to define, manage and retrieve data in a database. The system consists of a collection of interrelated data and a set of software programs to manage and retrieve data. The interrelated data is usually known as a relational database and it comprises a set of tables each of which is denoted by a unique name. Additionally, each table consists of a set of attributes/columns and stores data records (rows). Each data record denotes an object identified by a unique key and described by a set of attribute values. Fig. 2 illustrates a toy relational database comprising four tables, one of them being used as a fact table (sales) including one attribute (quantity) and foreign keys to dimensional data—the rest of tables— where descriptive information is stored. The interrelated data is semantically defined through what is known as an entity relationship data model, which represents the database as a set of entities and their relationships. For data accesses, a database

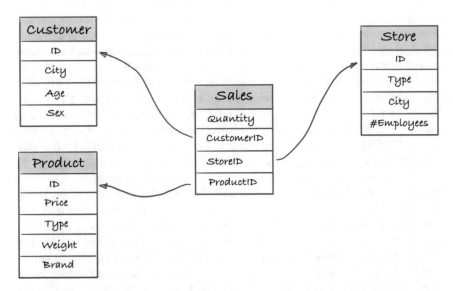

Fig. 2 Sample relational database in a star schema

Table 1 Sample qualitative (binary) tabular data representation

tid	Item1	Item2	Item3	Item4	Item5	Item6	Item7
#1	1	0	1	0	0	1	0
#2	0	1	1	0	0	0	0
#3	0	1	1	1	1	0	0
#4	0	0	0	1	0	1	1
#5	1	1	0	0	1	1	1

Table 2 Sample quantitative (non-binary) tabular data representation

tid	Item1	Item2	Item3	Item4	Item5	Item6	Item7
#1	2.3	0	6.1	0	0	6.5	0
#2	0	1.1	4.6	0	0	0	0
#3	0	1.4	4.5	1.8	1.9	0	0
#4	0	0	0	4.3	0	2.6	6.4
#5	3.0	0.9	0	0	2.1	8.7	6.2

management system makes use of a relational query language [61] including multiple relational operations such as join, selection and projection, as well as aggregate functions such as sum, average, count, etc.

Transactional databases are different types of data organizations in which information is kept in single tables stored in flat files. Columns represent attributes, whereas rows state for data records. Information is therefore recorded into transactions, which are sequences of information that are treated as a unit. A transaction typically includes a unique identifier together with a list of the items that form the transaction. There are different types of transactional databases depending on their organization. The most simple one is the binary table format (see Table 1) in which each column represents a variable or feature and each row comprises binary values for every variable. If the variable is satisfied by a transaction, then it is represented by 1, and 0 otherwise. For some specific problems, it is not enough to consider whether the variable or item appear in data but its associated quantity (profit, for example). Table 2 illustrates a tabular data representation including quantitative values. Finally, one of the most used data formats in pattern mining tasks is the transactional database. Unlike previous data organizations, each row includes a list of items (attributes or features) considered by such transaction. In this data representation, rows are variable in size. Table 3 illustrates a sample transactional database where the first transaction includes the items *Item1*, *Item3* and *Item6*. As it is shown, the number of items satisfied by each transaction varies (three items for transaction #1 and four items for transaction #3). Any of these data representations can be defined through the well-known comma-separated values file, also known as CSV, which allows data to be saved in a tabular format and it is supported by almost all spreadsheets and database management systems. ARFF (Attribute-Relation File Format) file is another file format to

Table 3 Sample
transactional database

tid	Sets of items
#1	Item1, Item3, Item6
#2	Item2, Item3
#3	Item2, Item3, Item4, Item5
#4	Item4, Item6, Item7
#5	Item1, Item2, Item5, Item6, Item7

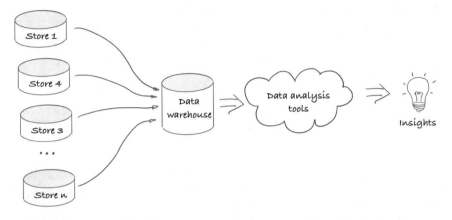

Fig. 3 Sample data warehouse database gathering information from multiple stores

keep, in ASCII text file, a list of instances sharing a set of attributes. ARFF files were developed by the Machine Learning Project at the Department of Computer Science of the University of Waikato for use with the Weka machine learning software [22].

On the contrary, a data warehouse [53] is usually defined as a central repository where information comes from multiple and heterogeneous data sources. Data are provided and managed from both transactional data and relational databases and different data types might be stored (see Fig. 3): structured, unstructured and semi-structured. Structured data is usually represented in a clearly defined way, being easily understandable for the search engine. Unstructured data, on the contrary, is not easily searchable and includes multiple data formats like audio, image and video. Finally, semi-structured data is a mix of the above, maintaining internal tags that identify separate data elements so information can be grouped and sorted into a hierarchy easily. Examples of semi-structured data are more and more common nowadays. Mark-up language XML [30] is an example of the semi-structured document language. This is a set of document encoding rules that simplifies data exchanging among systems containing data in incompatible formats. Open standard JSON [41] is another semi-structured data interchange format. Its main feature is that it is a language-independent data format. Thanks to the data warehouse, which merges information coming from different sources into one comprehensive database, any institution/organization ensures to have all the information available in a central repository from which meaningful business insights can be extracted.

Finally, it is important to talk about data lake [38], which is a centralized repository that allows you to store all your structured and unstructured data at any scale. These data repositories are of great interest nowadays mainly due to the interest in Big Data. A data lake can store the data as is, that is, raw data that is required to be processed later. The structure of the data is not defined when data is gathered so it is possible to keep all of your data without the need to know what they will be used for in the future. This is the main difference concerning a data warehouse, where data are structured, and a well-defined schema is considered in advance.

3 Data Processing Approaches

Data processing can be considered as the process that transforms raw data into meaningful information to be considered to any aim. Data processing starts with data in its raw form and translates it into a more readable and useful form (databases, graphs, streams, etc.). Thus, data is manipulated to produce results through a series of steps that begin with collecting data from trustworthy data sources and storing them with the highest possible quality, cleaning and checking for errors or inconsistencies. The aim is therefore to fix redundant, incomplete or incorrect data, and to form the right data structure to ease the mining process. Resulting data is therefore the first stage in which raw data begins to take the form of usable information. This data structure is finally considered by data mining algorithms to produce useful insights [42], that is, it is usable to non-data scientists. All in all, what is represented by data is essential to be known beforehand so their processing is done accordingly to achieve the right results. The following are different data representations that do not necessarily have a fix connection with data types provided in the previous section.

3.1 Databases

Databases are by far the most widely used data representation to be processed. Input data generally come as normal databases on different data forms (databases, transactional data, data warehouses or even data lake). Here, data is somehow static where records are represented as strings including attributes that feature such data record. Depending on the task to be carried out, that is, the knowledge to be extracted, some databases (tabular representation) slightly vary including useful (extra) information to be considered in the mining process. As a matter of example, let us consider the market basket analysis in which it is useful to determine which patterns, useful patterns, produce a low/high profit to the company regardless they are frequently/infrequently purchased [28]. For this problem, it is therefore required to include non-binary purchase quantities for the items in the transactions and it is also required to consider that all items are not always equally important (in terms of profit). Such datasets, known as quantitative databases, are represented as a finite set of distinct items $I = \{i_1, i_2, ..., i_n\}$ and every single item is associated with a positive integer that represents its utility or profit. Such datasets may also include information about the purchased quantity for every single item. All this information should properly be processed so the right insights can be obtained.

Sometimes, and it generally happens in real-life data, some kind of uncertainty is present due to different reasons [1] including limitations in the understanding of reality or just by limitations in the data-gathering devices. The major feature of uncertain data is that users do not have any real conviction about the presence or absence of an item i_j in a transaction t_i. This uncertainty is mainly expressed in terms of a probability $P(i_j, t_i)$, denoting the likelihood of i_j being present in t_i. The probability may be expressed as a percentage or in per unit basis so a value close to 0 denotes that i_j has an insignificantly low chance to be present in data, whereas a value of 1 states that i_j is present with no doubt. Processing such input data may be considered somehow similar to traditional data (precise data), denoting items in precise data with a probability of 1.

Up to this point, we have presented a database in a tabular data representation where each row unequivocally identifies a single record including a set of elements or items that characterizes a data object. In some problems, though, data information is ambiguous so a data object may be described by an undetermined number of different descriptions (data records) [21]. This ambiguity is required to be properly processed so the information that describes a specific object is accurately taken from the set of descriptions associated with such data object [32]. Formally speaking, a database Ω is defined as a set of n bags $\Omega = \{B^1, B^2, ..., B^n\}$ and each bag B^j comprises an undetermined number of transactions that describe the data object. It is not so difficult to understand that every bag describes different data objects. The way in which such input data is processed depending on the goal and the users' requirements since a feature may be taken as a good descriptor if it appears at least once in the bag or if it appears within a range in the bag.

In many application domains, data represent any kind of sequentiality among items or events [36]. A single event is defined as a collection of items that appear together and, therefore, they do not have a temporal ordering. However, there exists a temporal order in the events. In a formal way [1] and considering a databases Ω comprising a set of items $I = \{i_1, i_2, ..., i_n\}$, an event e_j is defined as a non-empty unordered collection of items, i.e. $\{e_j = \{i_k, ..., i_m\} \subseteq I, 1 \leq k, m \leq n\}$. Each transaction t_j is denoted as a sequence of events in the form $t_j = \langle e_1 \rightarrow ... \rightarrow e_n \rangle$ and each event e_i is described as an itemset $\{i_i, ..., i_j\}$ defined in the set of items $I \in \Omega$.

3.2 Data Streams

Data streams [16] are known as continuous flows of data that are generated from disparate sources in real time. Streaming [9] is therefore a term that denotes continuous, never-ending sets of data (streams) with no beginning or end. A data stream is similar to a constantly provided data input, and it is a simple analogy to how water flows through a river, coming from various sources, in different volumes, and flow into a single combined stream. Formally speaking, a streaming database Ω is a sequence of transactions of indefinite length occurring at a time t_j. In other words, the database Ω can be defined as $\Omega = \{t_1, t_2, ..., t_n\}$. When working with data streams it is useful to deal with slicing windows, that is, sequences of transactions occurring from time t_i to t_j.

Data stream applications [16] have been widely studied and it is not so hard to find in the specialized literature data generated by sensor networks, meteorological analysis, stock market analysis and computer network traffic monitoring. The main feature of all these applications is input data are far too large to fit in main memory and usually require to keep them into a secondary storage device. It makes it extremely challenging to extract useful knowledge from data streams since most data mining techniques assume a finite amount of data to be analysed. Besides, far from following a stationary data distribution, it is unpredictable when referring to data streams. According to Gama et al. [9], any successful development of algorithms in data streams has to take into account the following restrictions:

- Data arrive continuously.
- There is no control over the order in which the data should be processed.
- The size of a stream is (potentially) unbounded.
- Data are discarded after they have been processed. In practice, one can store part of the data for a given period of time, using a forgetting mechanism to discard them later.
- The unknown data probability distribution may change over time.

Amain problem that is required to be addressed when working with data streams is the concept drift phenomenon [17]. It refers to changes in the conditional distribution of the output (target variable to be studied) given the input features, while the distribution of the input may stay unchanged. Let us consider a typical example of concept drift that is related to a change in users' interests when following an online news stream. Suppose that the user is searching for a new apartment so dwelling houses are relevant for him/her, whereas holiday homes are not relevant. In a specific moment, the user has bought a house and starts looking for a holiday destination. From that moment on, dwelling houses become not relevant, and holiday homes become relevant. This scenario is what it is known as concept drift. As a result, a learning algorithm needs to be adapted to unexpected changes to continue working.

The ability to process data streams is a key procedure in many fields and it can be seen as a natural extension for the incremental learning systems [9], which build models by considering example by example. Adaptive learning algorithms can be seen as advanced incremental learning algorithms that can adapt to the evolution of the data-generating process over time.

3.3 Graphs

Existing data mining approaches are mainly based on flat and tabular representations where data is defined through rows (data records) and columns (data features or items). Nowadays, however, there exists considerable interest in graph structures [27] arising in technological, sociological and scientific settings. Paying attention to social networks, it is possible to form a network that represents who trusts whom, who have any kind of connection (familiar, friendship, etc.), who talks to whom, etc. The study of such networks is of high interest in many fields to determine the importance of each node (the number of edges incident to each node) as well as the distances between pairs of nodes (the shortest-path length).

In a formal way, a graph can be defined as a 4-tuple $G = (V, E, \mu, v)$, where V states for the finite set of nodes, $E \subseteq V \times V$ denotes the set of edges, $\mu : V \to L_V$ is a node labelling function and $v : E \to L_E$ is an edge labelling function. Developing algorithms that discover any subgraph that frequently occur in data (considering a graph database) is particularly challenging and computationally intensive. The frequency of a given subgraph g is calculated by considering all the graphs included in a graph dataset $\Omega = \{G_1, G_2, ..., G_n\}$. The frequency or support of g is defined as $support(g) = |\Omega_g|/|\Omega|$, where $\Omega_g = \{G_i : g \subseteq G_i, G_i \in \Omega\}$. Additionally, it is said that the subgraph g is frequent if its support value is no less than a minimum pre-defined support threshold value.

Several algorithms have been properly designed to extract interesting subgraphs. These solutions, however, can only cope with static graphs structures (see Fig. 4) or graphs that do not change over time [14]. Social connections, for instance, are not a static issue and heavily change over time (see Fig. 5). Being able to capture the dynamics of the graph or how the graph evolves over time is very important to

Fig. 4 Sample static graph structure

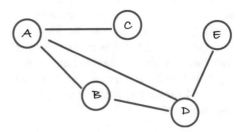

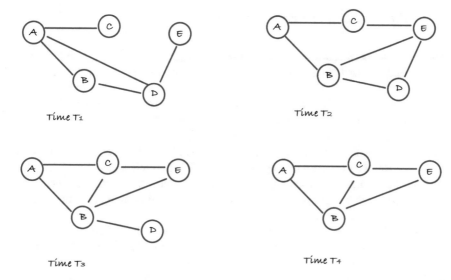

Fig. 5 Sample dynamic graph structure

determine how friendship relationships are formed. In this sense, some authors [14] have considered a time-ordered sequence of graph snapshots where edges and nodes can be inserted, removed and attribute values may change at each timestamp. However, not only the proper graph structure (topology) is important but also the weights of the vertexes. Jin et al. [23] considered a dynamic network and aimed to discover connected subgraphs whose vertices show the same trend during a time interval of two consecutive timestamps. The idea was to reveal important changes occurring in a dynamic system.

4 Data Mining Techniques

We have already described some data types and data processing approaches or ways in which raw data is transformed into meaningful structures so data mining can be performed. At this point, it is important to denote which data mining techniques can be applied to the input (already transformed) data so different types of patterns can be obtained [19]. The following are some of the most widely used techniques. However, we are aware that many new and trending techniques are being proposed for different purposes nowadays.

4.1 Mining Frequent Patterns and Associations

The key element in data analysis is the pattern, which represents any type of homogeneity and regularity in data, and serves as a good descriptor of intrinsic and important properties of data [1]. A pattern is generally defined as a set of elements (items) that are somehow related in a database. In a formal way, a pattern P in a database Ω is defined as a subset of items $I = \{i_1, ..., i_n\} \in \Omega$. In another way, $P \subseteq I$ that describes valuable features of data [55], and its size or length is calculated as the number of single items that it comprises. The task responsible for mining interesting and useful patterns is known as pattern mining and it comprises multiple and varied techniques with different purposes, depending on the type of patterns to be extracted. The most well-known pattern mining technique is the one for mining frequent patterns, also known as frequent itemset mining [33].

A typical example of frequent itemset mining is market basket analysis, and it is perhaps the first application domain in which it was correctly applied. A high number of purchases are usually bought on impulse when shopping, so it is of high interest for managers to analyse in-depth the shopping behaviour to obtain valuable information about which specific items tend to be strongly related [5]. This analysis might allow shopkeepers to increase sales by re-locating the products on the shelves, or even it might allow managers to plan diverse advertising strategies. In general, frequent itemset mining aims to make sense of data, arranging elements of data to obtain those sets that most frequently appear [34]. Such kind of algorithms requires high efficient processes paying special attention to their computational cost. To clarify the complexity of analysing itemsets in data and considering n different items, a total of $2^n - 1$ different patterns can be found. Hence, any straightforward approach becomes extremely complex with the increasing number of items.

Frequent itemset mining is highly related to the task of mining associations among itemsets, which is known as association rule mining. Association rules were proposed by Agrawal et al. [2] as a way of describing correlations among items within a pattern (frequent itemset). Let us consider a frequent itemset or pattern P defined in a database Ω as a subset of items $I = \{i_1, ..., i_n\} \in \Omega$, that is, $P \subseteq I$. Let us also consider two different subsets for such pattern P, that is, $X \subset P \subseteq I$ and $Y = P \setminus X$. It is important to highlight that X and Y do not have any common item, that is, $X \cap Y = \emptyset$. An association rule is formally defined as an implication of the form $X \rightarrow Y$, denoting that if the antecedent X is satisfied, then it is highly probable that the consequent Y is also satisfied [62]. The problem of mining association rules includes a huge large search space, even higher than the one of mining frequent itemsets. Given a dataset comprising n items, a total of $3^n - 2^{n+1} + 1$ different association rules can be found. This daunting process of mining association rules can be viewed as a two-step process: (1) Find all frequent itemsets, that is, those itemsets that occur at least as frequently as a predetermined minimum value; (2) Generate strong association rules from the frequent itemsets, that is, rules that satisfy a predefined minimum reliability or conditional probability $P(Y|X)$.

Association rules considering the frequency and reliability are not mandatorily interesting. The use of these two measures in isolation may produce a large and useless set of generated rules. The reduction of such number of generated rules is essential to obtain an interesting set of rules that can be easily applied to the problem at hand. An example can be the adding of the correlation between the sets X and Y to help to decide when an association is useful or not. Much different quality measures [35] have been proposed so far and it has been stated that a good interestingness measure should not be affected by data records that do not contain the itemsets of interest. This problem is denoted as the null-invariant property, and a measure is null-invariant if its value is free from the influence of null-transactions.

4.2 Predictive Analysis

The predictive analysis [58] aims to make predictions about future outcomes based on historical data by generating future insights with a significant degree of precision. The idea is that any organization/institution can forecast trends in the future. One of the most common predictive analytic models is classification. These models were designed to categorise information based on historical data, which are denoted as training data (data objects for which the class or concept are known). The classification task might be described as a two-step process: learning and classification. The learning step is responsible to build the classification model, whereas the classification step is used to predict class labels for given data. One of the major problems that any classification model has to face up is the data overfitting. This is the process by which a model learns the training data so good that the resulting model incorporates some particular anomalies of the training data that are not present int he general dataset. In this regard, it is key to consider a test set, which is an independent set containing tuples that were not considered to build the model. Thanks to this test set the accuracy of the model is obtained as the percentage of test set tuples that are correctly classified by the model. Last but not least, it is important to pay attention to the first step of classification, which was related to learning (build a classification model). The following are some methods considered in the literature to build classification models.

A popular solution to build classification models is through decision trees. They are flowchart-like structures in which internal nodes represent tests on attributes, branches represent the outcome of the tests, and leaf nodes denote the provided class label. This class label can be seen as the decision taken after testing all attributes. Any path from the root of the tree to a leaf represents a classification rule. The use of decision trees are really popular in classification since their construction does not require any domain knowledge or parameter setting and it is fast and simple. Additionally, their representation is highly intuitive and easily understood by humans. One of the first algorithms based on decision trees was proposed in the early 80s under the name of ID3 and some improvements have been recently proposed [7]. Its successor, known as C4.5 [48], has become a benchmark in the field and it is usually considered to compare any new proposed algorithm. The CART algorithm [29], which is based on binary decision trees, was also proposed on those years in which decision trees were built in a top-down recursive divide-and-conquer manner. Once a decision tree is built, many of its branches may include anomalies of the training data (overfitting) and a tree pruning process is therefore required. Pruned trees tend to be smaller, less complex, faster and better at correctly classifying independent test data than unpruned trees. Two ways of pruning trees are considered by researchers: pre-pruning (pruning by stopping its construction early) and post-pruning (removing subtrees from the already constructed tree. The branches to be removed are replaced by a leaf, which is labelled with the most frequent class from the replaced subtree).

Bayesian classifiers [10] are also popular solutions to build classification models. These models are based on classification, and they can predict the probability that a given tuple belongs to a particular class. These methods are based on Bayes' theorem, and the resulting classifiers tend to be highly accurate and fast (even for large databases). Bayesian classification is, therefore, a probabilistic (statistical) approach that can learn and make inferences based on a different view of what it is known as learning from data. Before data are analysed, prior opinions about data can be expressed in a probability distribution, and a posteriori distribution is then expressed after analysing data. Bayesian learning can produce the probability distributions of the quantities of interest so optimal decisions can be reached together with observed data.

Another type of classifiers is rule-based models [19], where the classifiers are represented as sets of IF–THEN rules. These rules are similar to those described in association analysis, including two main parts: antecedent (left side) and consequent (right side). In this case, rules may include one or more attribute to be tested in the antecedent part of the rule, whereas the consequent is the class prediction [13]. If the condition, that is, all the attributes belonging to the antecedent part are true for a given data record, then the rule is triggered. In those situations where more than one rule is triggered, a strategy is required to determine which rule is the right to be considered. Many different strategies are considered by different researchers. For example, it is possible to assign the highest priority to the more restrictive rule, or even that including a higher number of attributes in the antecedent. Associative classification [43] is a kind rule-based classifiers. The aim is to produce association rules in a first step and, then, build a classifier according to the previously mined rules. Recent studies have shown that associative classification has specific advantages over other classification methods: they are often capable of building efficient and accurate classification models since all possible relationships among the attribute values are obtained during the extraction of association rules; they are easily updated and tuned without affecting the complete ruleset; like any rule-based classifier, the final model is easily understandable and interpretable.

4.3 Supervised Descriptive Analysis

In pattern analysis, each pattern represents a data subset and its frequency is related to the size of such a subset. A pattern represents a small/big part of data, denoting a specific internal structure, and a distribution that is different from the whole data. Sometimes, patterns are used to describe important properties of two or more data subsets previously identified or labelled, transforming, therefore, the pattern mining concept into a more specific one, *supervised descriptive pattern mining* [40]. The main aim now is to understand underlying phenomena (according to a single or multiple target variables) and not to classify new examples. Supervised descriptive discovery originally gathered three main tasks: contrast set mining [11], emerging pattern mining [18] and subgroup discovery [4]. Nowadays, however, many additional tasks can be grouped under the supervised descriptive pattern mining concept [56]. The following are some examples of important tasks in the field aiming at producing interpretable, non-redundant, potentially actionable and expressive knowledge or information.

Contrast set mining was described by Dong et al. [11] as a way of describing differences and similarities among datasets that the user needs to contrast. Each dataset may correspond to a target variable of a more general dataset, or it may represent subsets satisfying various conditions. In any case, such subsets are independent and do not share any data record. Thus, contrast set mining aims to provide statistical pieces of evidence that denote a data record as a member of a class (target variable value). Formally speaking, a contrast set is defined as a pattern P which distribution highly differ among data subsets. In general terms, contrast sets quantify the difference in frequency for each subset S_i. P is denoted as a contrast set if and only if $\exists i j : max(|support(P, S_i) - support(P, S_j)|) \geq \alpha \in [0, 1]$, denoting $support(P, S_i)$ as the support (frequency) of a pattern P on this subset.

Emerging pattern mining is closely related to contrast set mining [11] presenting slight differences. Instead of comparing multiple data subsets, emerging patterns make a comparison between two data types. It can be viewed as contrast patterns between two kinds of data whose support changes significantly between the two data types. This kind of patterns was formally defined in [56] as follows: given a pattern P defined on two datasets Ω_1 and Ω_2, this pattern P is denoted as an emerging pattern if its support (frequency) on Ω_1 is significantly higher than its support (frequency) on Ω_2 or vice versa. This difference in support is quantified by the growth rate, that is, the ratio of the two supports ($support(P, \Omega_1)/support(P, \Omega_2)$ or $support(P, \Omega_2)/support(P, \Omega_1)$), and values greater than 1 denotes an emerging pattern. Here, the growth rate is 1 if $support(P, \Omega_1) = support(P, \Omega_2)$ and 0 if both P does not satisfy any record from either Ω_1 or Ω_2. In those cases where $support(P, \Omega_1) = 0$ and $support(P, \Omega_2) \neq 0$, or $support(P, \Omega_1) \neq 0$ and $support(P, \Omega_2) = 0$, then the growth rate is defined as ∞ and such patterns are denoted as jumping emerging patterns.

Subgroup discovery [4] aims to describe important features (distributional unusualness) for a specific property of interest (target variable). Such descriptions are provided in the form of rules ($P \rightarrow Target$) denoting an unusual statistical distribution of a pattern P (set of features) concerning the target variable $Target$. The problem was first introduced by *Klösgen* [25] and *Wrobel* [59] as follows: Given a population of individuals (customers, objects, etc.) and a property of those individuals that we are interested in, the task of subgroup discovery is to find population subgroups that are statistically most interesting for the user, e.g. subgroups that are as large as possible and have the most unusual statistical characteristics concerning a target attribute of interest. The interest of the rules that describe subgroups are quantified according to a wide variety of metrics [20] and can be grouped according to the type of target variable. A discrete target variable includes measures that are divided into four main types: measures of complexity (interpretability of the discovered subgroups and simplicity of the extracted knowledge); generality (quality of the subgroups according to their coverage); precision (reliability of each subgroup); and interest (significance and unusualness of the subgroups). A numerical target variable includes measures that are categorized as mean-based measures, variance-based measures, median-based measures and distribution-based measures.

Exceptional model mining [26] is considered as a multi-target generalization of subgroup discovery [20]. This task aims to search for data subsets on a pair of predefined target variables (t_x and t_y) in which there is an unusual interaction among the target variables. This interaction was originally quantified in terms of the *Pearson's* standard correlation coefficient ρ between t_x and t_y for both the data subset and the whole dataset. Exceptional model mining also considers the complement of the data subset instead of the whole dataset [12]. The complement of a data subset comprises all the transactions that are not included in such subset. Recently, some authors [31] have proposed an extension of exceptional model mining that does not require to preset the target variables before running the algorithm. Multiple unusual interactions may be present in a larger set of target variables simultaneously. This extension of exceptional model mining aims to extract subgroups where several pairwise correlations are exceptional.

4.4   Cluster Analysis

Clustering is a process in which data objects group into different groups or clusters [49]. Any object within the same cluster presents a high similarity with the rest, whereas it is very dissimilar to any object belonging to other clusters. It is important to highlight that not every clustering method produces the same clusters on the same dataset and it does not only depend on the metrics considered to form the clusters but on the algorithm itself. Cluster analysis is widely used to know the data distribution, to extract the main characteristics of each cluster, as well as just to determine clusters to be studied in depth in the future.

It is possible to find many different clustering algorithms in the specialized literature [47] and a fixed categorization of such algorithms is hardly possible because the categories may overlap in a high degree. Nevertheless, the major clustering algorithms can be classified into four categories: partitioning methods, hierarchical methods, density-based methods and grid-based methods. The first category, related to partitioning methods, determine that k partitions of the data are given (the number of partitions needs to be lower or equal to the number of objects to be described) and each partition must contain at least one object. In this category, most existing methods are based on distance, cannot find complex cluster shapes, adopt popular heuristic approaches and they suffer from large databases. An example of this category is the k-means algorithm, which works by progressively improving the quality of the clusters.

Hierarchical methods categorize the data objects through a hierarchical decomposition according to two processes: bottom-up and top-down. The first one considers each object as a single group, and it successively merges groups according to their similarity. The top-down approach, on the contrary, all the objects belong to a unique cluster which is successively split into smaller clusters. The quality of the clusters in this category is mainly distance-based or density-based. It is finally important to denote that the main problem of hierarchical methods is that the merge/split-step cannot be undone once it is done.

Density-based methods form clusters according to the distance between objects. As a result, clusters are dense regions of objects in space that are separated by low-density regions. These methods work by increasing the size of the cluster until the density exceeds a predefined threshold value.

4.5 Outliers Analysis

Anomaly detection or outlier detection [19] is a daunting process by which data objects with behaviours that are very different from what it was expected are discovered. An outlier is therefore defined as a data object that deviates significantly from the rest. At this point, anyone may consider outliers as noise, but it is important to clarify that these two concepts are not synonyms. Noise is an error in the input data, which is uninteresting from any data analysis and they should be removed before any method for anomaly detection (outlier detection) is performed. It is possible, however, to consider outlier detection as a task related to novelty detection when it is applied to data evolving. Here, some outliers in the evolving data may appear as novel features that will become as frequent in the future. It is obvious that, once a new feature is confirmed by novelty detection, it is not treated as outlier any more. Outlier detection is important in many application domains such as medicine, fraud detection or public security. Depending on the domain to be applied, outliers can be categorized into three main groups [6]: global outliers, contextual outliers and collective outliers. A global outlier is the simplest type of outliers and it is defined as a data object that deviates significantly from the rest of the dataset. An important point for this type of outliers is the use of an appropriate quality measure that determine such deviation. As for contextual outliers, they are defined as data objects that deviate significantly concerning a specific context. It is therefore required to specify the context beforehand, which is usually a specific attribute (or set of attributes). The data object needs to be an outlier for the data subset that satisfies such attribute. Hence, unlike global outliers, which consider the whole dataset, contextual outliers only consider the data subset for the predefined context. Finally, a collective outlier is a different concept that determines a set of data objects as outliers if they, as a whole, deviate significantly from the entire data. Here, it is possible that each individual data object does not behave as an outlier.

There are many approaches in the literature to address the outlier detection problem. In such problems in which an expert labels the examples as normal or outlier, the task can be performed by a classification model that can recognize outliers. In other problems, however, data objects are not labelled as normal or outlier, and the task is, therefore, to search for features that are mainly shared by the majority of data objects so outliers can be identified. Finally, it is also possible that some other problems include some labelled examples but such a number is not so high. Here, a small set of normal/outlier data objects are labelled. Here, it is possible to use some of the labelled objects to look for unlabelled data objects that are close in some way.

5   Importance of Finding Periodic Patterns in Databases

Frequent and periodic pattern mining aims to detect whether a pattern occurs frequently and regularly, or mostly in a specific time interval in data. In other words, its goal is to extract the occurrence behaviour of patterns. This task was first proposed by Tanbeer et al. [52] as a way of determining the interestingness of frequent patterns in terms of the shape of occurrence. When analysing periodic patterns [15], it is said that a pattern is interesting if it is frequent and regular according to two different threshold values (frequency and periodicity).

Let us consider the sample transactional database shown in Table 4. Here, the following patterns or itemsets equally appear in data: $\{Item2, Item3\}$, $\{Item5, Item7\}$ and $\{Item1, Item6\}$. They can be considered as frequent patterns since they appear in half of the transactions. However, these patterns may not be periodic-

Table 4 Sample transactional database

tid	Sets of items
#1	Item1, Item2, Item3, Item6
#2	Item2, Item3, Item5
#3	Item2, Item3, Item4, Item5
#4	Item4, Item5, Item7
#5	Item1, Item2, Item5, Item6, Item7
#6	Item1, Item5, Item6, Item7

frequent because of non-similar occurrence periods. For example, $\{Item2, Item3\}$ and $\{Item5, Item7\}$ appear more frequently at a certain part of the database. $\{Item2, Item3\}$ occurs during the three first transactions, whereas $\{Item5, Item7\}$ during the last three transactions. On the contrary, $\{Item1, Item6\}$ does not follow a regular interval since it appears in the first, fifth and sixth transactions. Here, it is important to take into account that $\{Item2, Item3\}$ and $\{Item5, Item7\}$ may not be of interest if we are looking for regular items (appearing along the whole dataset). On the other hand, they are useful if the aim is to look for local periods in which a pattern is frequent (they appear in all the first/last data transactions).

According to Tanbeer et al. [52] the mining of periodic patterns in databases is of high importance in many application domains. Considering the market basket analysis, extracting all frequently sold products can be useless if some of such products are bought in specific periods. Thus, it is important to extract only those products that were regularly sold compared to the rest. As for a web administrator, he/she may be interested in the click sequences of his/her web, so the frequency is not useful but the periodicity of the clicks. Additionally, in bioinformatics, it is important to extract set of genes that not only appear frequently but also co-occur at regular interval in DNA sequences.

Many different approaches conform the state-of-the-art in frequent-periodic pattern mining. Tanbeer et al. [52] proposed a pattern-growth approach that generates periodic-frequent patterns by applying depth-first search. Kiran et al. [24] and Surana et al. [50] addressed the periodicity problem by considering rare or infrequent itemsets. Amphawan et al. [3] proposed an efficient algorithm for mining top-k periodic frequent patterns.

References

1. C.C. Aggarwal, J. Han, *Frequent Pattern Mining* (Springer International Publishing, Berlin, 2014)
2. R. Agrawal, T. Imielinski and A.N. Swami, Mining association rules between sets of items in large databases. Proceedings of the 1993 ACM SIGMOD International Conference on Management of Data, SIGMOD Conference'93 (ACM, New York, 1993), pp. 207–216
3. K. Amphawan, P. Lenca and A. Surarerks, Mining top-K periodic-frequent pattern from transactional databases without support threshold. Advances in Information Technology—Third International Conference, IAIT 2009 (Springer, New York, 2009), pp. 18–29
4. M. Atzmueller, Subgroup discovery—advanced review. *WIREs: Data Mining Knowl. Dis.*, **5**, 35–49 (2015)
5. Michael J. Berry, Gordon Linoff, *Data Mining Techniques: For Marketing, Sales, and Customer Support* (John Wiley & Sons Inc., New York, 2011)
6. V. Chandola, A. Banerjee, V. Kumar, Anomaly detection: A survey. ACM Comput. Surv. **41**(3), 1–58 (2009)
7. J. Chen, D.-L. Luo, and F.-X. Mu. An improved ID3 decision tree algorithm. 2009 4th International Conference on Computer Science and Education (IEEE, New York, 2009), pp. 127–130
8. M. Chen, S. Mao, Y. Liu, Big data: A survey. Mobile Networks and Applications **19**(2), 171–209 (2014)
9. J.A. Silva, E.R. Faria, R.C. Barros, E.R. Hruschka, A.C. Carvalho, J. Gama, Data stream clustering: A survey. ACM Comput. Surv. **46**(1), 13:1-13:31 (2013)

10. P. Domingos, M. Pazzani, On the optimality of the simple Bayesian classifier under zero-one loss. Mach. Learn. **29**(2–3), 103–130 (1997)
11. G. Dong, J. Bailey (eds.), *Contrast Data Mining: Concepts, Algorithms, and Applications* (CRC Press, Boca Raton, 2013)
12. W. Duivesteijn, A. Feelders, A.J. Knobbe, Exceptional model mining—supervised descriptive local pattern mining with complex target concepts. Data Min. Knowl. Disc. **30**(1), 47–98 (2016)
13. Pedro G. Espejo, Sebastián Ventura, Francisco Herrera, A survey on the application of genetic programming to classification. IEEE Trans. Syst. Man Cybern. Part C **40**(2), 121–144 (2010)
14. P. Fournier-Viger, C. Cheng, Z. Cheng, J.C. Lin, N. Selmaoui-Folcher, Mining significant trend sequences in dynamic attributed graphs. Knowl. Based Syst. **182**, 104797 (2019)
15. P. Fournier-Viger, P. Yang, R.U. Kiran, S. Ventura, J.M. Luna, Mining local periodic patterns in a discrete sequence. Information Science **544**, 519–548 (2021)
16. M.M. Gaber, A. Zaslavsky, S. Krishnaswamy, Mining data streams: A review. SIGMOD Record **34**(2), 18–26 (2005)
17. J. Gama, I. Zliobaite, A. Bifet, M. Pechenizkiy, A. Bouchachia, A survey on concept drift adaptation. ACM Comput. Surv. **46**(4), 44:1-44:37 (2014)
18. Á.M. García-Vico, C.J. Carmona, D. Martín, M. García-Borroto, M.J. del Jesus, An overview of emerging pattern mining in supervised descriptive rule discovery: taxonomy, empirical study, trends, and prospects. Wiley Interdisciplinary Reviews: Data Mining and Knowledge Discovery **8**(1), e1231 (2018)
19. J. Han, M. Kamber, *Data Mining: Concepts and Techniques* (Morgan Kaufmann, Burlington, 2000)
20. F. Herrera, C.J. Carmona, P. González, M.J. del Jesus, An overview on subgroup discovery: Foundations and applications. Knowl. Inf. Syst. **29**(3), 495–525 (2011)
21. Francisco Herrera, Sebastián Ventura, Rafael Bello, Chris Cornelis, Amelia Zafra, *Dánel Sánchez Tarragó, and Sarah Vluymans* (Springer, Multiple Instance Learning - Foundations and Algorithms, 2016)
22. Geoffrey Holmes, Andrew Donkin, and Ian H. Witten. Weka: A machine learning workbench. Proceedings of ANZIIS'94-Australian New Zealand Intelligent Information Systems Conference (IEEE, New York, 1994), pp. 357–361
23. R. Jin, S. McCallen and E. Almaas, Trend motif: A graph mining approach for analysis of dynamic complex networks. Seventh IEEE International Conference on Data Mining (ICDM 2007) (IEEE, New York, 2007), pp. 541–546
24. R.U. Kiran and P. Krishna Reddy, Towards efficient mining of periodic-frequent patterns in transactional databases. Database and Expert Systems Applications, 21th International Conference, DEXA 2010, Proceedings, Part II (Springer, Berlin, 2010), pp. 194–208
25. W. Klösgen, Explora: A multipattern and multistrategy discovery assistant, in *Advances in Knowledge Discovery and Data Mining*. ed. by U.M. Fayyad, G. Piatetsky-Shapiro, P. Smyth, R. Uthurusamy (American Association for Artificial Intelligence, Reston, 1996), pp. 249–271
26. D. Leman, A. Feelders and A. J. Knobbe, Exceptional model mining. Proceedings of the European Conference in Machine Learning and Knowledge Discovery in Databases, vol. 5212 of ECML/PKDD 2008 (Springer, Berlin, 2008), pp. 1–16
27. J. Leskovec, J. Kleinberg and C. Faloutsos. Graphs over time: Densification laws, shrinking diameters and possible explanations. Proceedings of the Eleventh ACM SIGKDD International Conference on Knowledge Discovery in Data Mining (ACM, New York, 2005), pp. 177–187
28. J.C.-W. Lin, W. Gan, P. Fournier-Viger, T.-P. Hong, V.S. Tseng, Efficient algorithms for mining high-utility itemsets in uncertain databases. Knowledge Based Systems **96**, 171–187 (2016)
29. W.-Y. Loh, Classification and regression trees. Wiley Interdisciplinary Reviews: Data Mining and Knowledge Discovery **1**(1), 14–23 (2011)
30. J. Lu, *An Introduction to XML Query Processing and Keyword Search*, vol. 9783642345555 (Springer, Berlin, 2013)
31. J.M. Luna, M. Pechenizkiy, W. Duivesteijn, S. Ventura, Exceptional in so many ways discovering descriptors that display exceptional behaviour on contrasting scenarios. IEEE Access **8**, 82–94 (2020)

32. J.M. Luna, A. Cano, V. Sakalauskas, S. Ventura, Discovering useful patterns from multiple instance data. Inf. Sci. **357**, 23–38 (2016)
33. J.M. Luna, P. Fournier-Viger and S. Ventura, Frequent itemset mining: A 25 years review. *Wiley Interdiscip. Rev. Data Min. Knowl. Discov.*, **9**, e1329 (2019)
34. J.M. Luna, P. Fournier-Viger and S. Ventura, Extracting user-centric knowledge on two different spaces: Concepts and records. IEEE Access **8**, 134782–134799 (2020)
35. J.M. Luna, M. Ondra, H.M. Fardoun, S. Ventura, Optimization of quality measures in association rule mining: An empirical study. International Journal of Computational Intelligence Systems **12**(1), 59–78 (2018)
36. N.R. Mabroukeh, C.I. Ezeife, A taxonomy of sequential pattern mining algorithms. ACM Comput. Surv. **43**(1), 1–41 (2010)
37. V. Marx, The big challenges of big data. Nature **498**(7453), 255–260 (2013)
38. N. Miloslavskaya, A. Tolstoy, Big data, fast data and data lake concepts. Procedia Computer Science **88**, 300–305 (2016)
39. J. Morrow, *Be Data Literate: The Data Literacy Skills Everyone Needs to Succeed* (KoganPage, London, 2021)
40. P.K. Novak, N. Lavrač, G.I. Webb, Supervised descriptive rule discovery: A unifying survey of contrast set, emerging pattern and subgroup mining. J. Mach. Learn. Res. **10**, 377–403 (2009)
41. N. Nurseitov, M. Paulson, R. Reynolds, C. Izurieta, Comparison of JSON and XML data interchange formats: A case study. Caine **9**, 157–162 (2009)
42. F. Padillo, J.M. Luna, A. Cano and S. Ventura, A data structure to speed-up machine learning algorithms on massive datasets. Hybrid Artificial Intelligent Systems—11th International Conference, HAIS 2016 (Springer, Cham, 2016), pp. 365–376
43. F. Padillo, J.M. Luna, S. Ventura, LAC: library for associative classification. Knowledge Based Systems **193**, 105432 (2020)
44. B.K. Patle, L.B. Ganesh, A. Pandey, D.R.K. Parhi, A. Jagadeesh, A review: On path planning strategies for navigation of mobile robot. Defence Technology **15**(4), 582–606 (2019)
45. C.L.P. Chen, C.-Y. Zhang, Data-intensive applications, challenges, techniques and technologies: A survey on big data. Inf. Sci. **275**, 314–347 (2014)
46. S.K. Rahimi, F.S. Haug, *Distributed Database Management Systems: A Practical Approach* (Wiley, New York, 2010)
47. Hermes Robles-Berumen, Amelia Zafra, Habib M. Fardoun, Sebastián Ventura, LEAC: an efficient library for clustering with evolutionary algorithms. Knowledge Based Systems **179**, 117–119 (2019)
48. S. Ruggieri, Efficient c4.5. *IEEE Trans. Knowl. Data Eng.*, **14**(2):438–444, 2002
49. R. Xu, D. Wunsch, Survey of clustering algorithms. IEEE Trans. Neural Networks **16**(3), 645–678 (2005)
50. A. Surana, R.U. Kiran and P. Krishna Reddy, An efficient approach to mine periodic-frequent patterns in transactional databases. New Frontiers in Applied Data Mining—PAKDD 2011 International Workshops, Revised Selected Papers (Springer, Berlin, 2011), pp. 254–266
51. P.N. Tan, M. Steinbach, V. Kumar, *Introduction to Data Mining* (Addison Wesley, Harlow, 2005)
52. S.K. Tanbeer, C.F. Ahmed, B.-S. Jeong and Y.-K. Lee, Discovering periodic-frequent patterns in transactional databases. Advances in Knowledge Discovery and Data Mining, 13th Pacific-Asia Conference, PAKDD 2009 (ringer, Berlin, 2009), pp. 242–253
53. A. Thusoo, J.S. Sarma, N. Jain, Z. Shao, P. Chakka, N. Zhang, S. Antony, H. Liu and R. Murthy. Hive—a petabyte scale data warehouse using hadoop. 2010 IEEE 26th International Conference on Data Engineering (ICDE 2010) (IEEE, New York, 2010), pp. 996–1005
54. A. van't Spijker. *The New Oil: Using Innovative Business Models to Turn Data Into Profit* (Technics Publications, Basking Ridge, 2014)
55. S. Ventura and J.M. Luna. *Pattern Mining with Evolutionary Algorithms* (Springer International Publishing, Berlin, 2016)
56. S. Ventura and J.M. Luna, *Supervised Descriptive Pattern Mining* (Springer, Berlin, 2018)

57. William P. Wagner, Trends in expert system development: A longitudinal content analysis of over thirty years of expert system case studies. Expert Syst. Appl. **76**, 85–96 (2017)

58. I.H. Witten, E. Frank, M.A. Hall and C.J. Pal. *Data Mining: Practical Machine Learning Tools and Techniques* (2016)

59. S. Wrobel, An algorithm for multi-relational discovery of subgroups. Proceedings of the 1st European Symposium on Principles of Data Mining and Knowledge Discovery, PKDD '97 (Springer, Berlin, 1997), pp. 78–87

60. X. Wu, X. Zhu, G.-Q. Wu, W. Ding, Data mining with big data. IEEE Trans. Knowl. Data Eng. **26**(1), 97–107 (2014)

61. C. Zhang, J. Naughton, D. DeWitt, Q. Luo and G. Lohman, On supporting containment queries in relational database management systems. Proceedings of the 2001 ACM SIGMOD International Conference on Management of Data (ACM, New York, 2001), pp. 425–436

62. C. Zhang, S. Zhang, *Association Rule Mining: Models and Algorithms* (Springer, Berlin, 2002)

Discovering Frequent Patterns in Very Large Transactional Databases

Jose M. Luna

Abstract Finding frequent patterns in very large transactional databases is a challenging problem of great concern in many real-world applications. In this chapter, we first introduce the model of frequent patterns. Second, we describe the search space for finding the desired patterns. Third, we present four popular algorithms to find the patterns. Finally, we present the extensions of frequent patterns.

1 Introduction

Technological advances in the field of Information and Communication Technologies have enabled organizations to collect and store big data effectively. Useful knowledge that can empower users with the competitive information to achieve socioeconomic development lies in this data. The field of data analytics (or data mining) has emerged to discover the hidden knowledge in big data.

Frequent pattern mining (FPM) is an important knowledge discovery technique in data mining. It involves finding all frequently occurring patterns in a transactional database. A classic application is market basket analysis. It involves finding the itemsets that were frequently purchased by the customers in the data. An example of a frequent pattern is $\{Cheese, Beer\}$ $[support = 10\%]$, which provides the information that 10% of the customers have purchased the items 'Cheese' and 'Beer' together. Such an information may be found to be extremely useful to the users for various purposes, such as product recommendation and inventory management. Other applications of FPM may be found at [2].

The basic model of frequent pattern is as follows: Let $I = \{i_1, ..., i_n\}$ be the set of items. Let $X \subseteq I$ be a pattern (or an itemset). A pattern containing k number of items is called as k-pattern. A transaction $t_{tid} = (tid, Y)$, where $tid \in \mathbb{R}^+$ represents the transaction identifier and $Y \subseteq I$ represents a pattern. A transactional database, $DB = \{t_1.t_2, \cdots, t_m\}$, $m \geq 1$, is a set of transactions. The *support* of pattern X in

J. M. Luna (✉)
Department of Computer Science and Numerical Analysis, Andalusian Research Institute in Data Science and Computational Intelligence (DaSCI), University of Cordoba, 14071 Córdoba, Spain
e-mail: jmluna@uco.es

© The Author(s), under exclusive license to Springer Nature Singapore Pte Ltd. 2021
R. Uday Kiran et al. (eds.), *Periodic Pattern Mining*,
https://doi.org/10.1007/978-981-16-3964-7_2

DB, denoted as $sup(X)$, represents the number of transactions containing X in DB. The pattern X is said to be a frequent pattern if $sup(X) \geq minSup$, where $minSup$ represents the user-specified *minimum support* threshold value. The **problem definition** of frequent pattern mining is to find all patterns that have *support* no less than $minSup$ in DB. (Please note that the *support* of a pattern can also expressed in percentage of database size. However, we employ the former definition of *support* throughout this book for ease of explanation.)

Example 1 Let $I = \{a, b, c, d, e\}$ be a set of items. A hypothetical transactional database generated by the items in I is shown in Table 1. The set of items a and b, i.e., $\{a, b\}$ (or ab in short) is a pattern. This pattern contains 2 items. Therefore, it is a 2-pattern. In Table 1, the pattern ab appears in the transactions whose *tids* are 2, 4, and 5. Thus, the *support* of ab in Table 1 is 3 or 60% (= (3/5) × 100). If the user-specified $minSup = 2$, then ab is said to be a frequent pattern because $sup(ab) \geq minSup$. The complete set of frequent patterns generated from Table 1 is shown in Table 2.

It is important to remark that setting a minimum support value is a non-trivial task and generally requires a profound background in the application field. Inexpert and many expert users need to try different thresholds by guessing and re-executing the algorithms once and again until the results are good for them. As it has been demonstrated, a small change in the threshold value may lead to very few or an extremely large set of solutions.

Table 1 Transactional database

tid	Items
1	ac
2	abc
3	bde
4	abe
5	$abcd$

Table 2 Frequent patterns found in Table 1

Pattern	Support	Pattern	Support
a	4	ac	3
b	4	ae	2
c	3	bc	2
d	2	bd	2
e	2	abc	2
ab	3		

Fig. 1 Search space
representation methods. **a**
Itemset lattice. **b** Set
enumeration tree

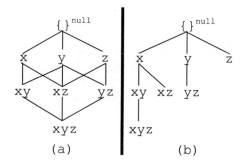

(a) | (b)

2 Search Space

The space of items in a database gives rise to an itemset lattice (or a set enumeration tree[1]). This itemset lattice represents the search space for finding frequent patterns in a database. The size of this search space is $2^n - 1$, where n represents the total number of distinct items in the database. Reducing this huge search space is a non-trivial and challenging task in frequent pattern mining. When confronted with this problem in the real-world applications, researchers employed *anti-monotonic property*[2] (see Property 1) to effectively reduce the search space. In other words, this property makes the frequent pattern mining practicable in the real-world applications.

Example 2 Let x, y, and z be three items in a hypothetical transactional database. The itemset lattice produced by the combinations of these three items is shown in Fig. 1a. The alternative representation of this lattice as a set enumeration tree is shown in Fig. 1b. The size of this lattice is 7 ($= 2^3 - 1$). This lattice represents the search space for finding frequent patterns. Frequent pattern mining algorithms search this enormous lattice using the anti-monotonic property. For instance, a frequent pattern mining algorithm will check the interestingness of the pattern xy if and only if all of its non-empty subsets, i.e., x and y, are also frequent in the database.

Property 1 *(Anti-monotonic property.) If $X \subset Y$, then $sup(X) \geq sup(Y)$. Thus, if $sup(X) \ngeq minSup$, then $\forall Y \supset Y$, $sup(Y) \ngeq minSup$.*

Example 3 The *support* of pattern cd in Table 1, i.e., $sup(cd) = 1$. Since $sup(cd) \ngeq minSup$, cd is an infrequent pattern. Moreover, all supersets of cd will also be infrequent patterns because their *supports* cannot also be more than $minSup$.

---

[1] The set enumeration tree is a high-performance data representation technique, which resembles the depth-first search on the itemset lattice

[2] Other names of this property are: apriori property and downward closure property.

3 Popular Algorithms

Several algorithms have been described in the literature to find frequent patterns. A recent survey on the past 25 years of frequent pattern mining may be found at [19]. In this chapter, we present four of the widely studied frequent pattern mining algorithms, namely Apriori [3], FP-Growth [13], Eclat [46], and LCM [37].

3.1 Apriori

Apriori is one of the fundamental algorithms to find frequent patterns in a database. It is a breadth-first search algorithm that finds all frequent patterns by employing *"level-wise candidate-generate-and-test paradigm."* This paradigm briefly involves the following three steps: (i) find frequent k-patterns from the candidate k-patterns, (ii) generate candidate k-patterns by joining frequent $(k-1)$-patterns among themselves, and (iii) repeat the above two steps until no more candidate k-patterns can be generated. The pseudocode of this algorithm is presented in Algorithm 1.

Algorithm 1 Pseudo-code of the Apriori algorithm.

Require: $I, DB, minSup$ {set of items, dataset and minimum support value}
Ensure: F
1: $F = \emptyset$
2: $L_1 = \{i \in I \mid support(i, DB) \geq minSup\}$
3: $F = F \cup L_1$
4: **for** $(k = 2; L_k \neq \emptyset; k++)$ **do**
5:     $C =$ set of candidate patterns produced by L_{k-1}
6:     $L_k = \{p \in C \mid support(p, DB) \geq minSup\}$
7:     $F = F \cup L_k$
8: **end for**
9: **return** $F = \cup_k F_k$

Let us consider the sample transactional database DB shown in Table 1. Let us also consider a minimum support value of three. The following is the set of frequent patterns that can be extracted from DB: a, b, c, ab, and ac. This set highly varies when the minimum support value is modified. Thus, considering a minimum support value of four, then the resulting set of frequent patterns is reduced: a and b. Similarly, if the minimum support value is decreased, then the resulting set of frequent patterns is increased. The following is the set of frequent patterns that can be extracted from DB with a minimum support value of two: a, b, c, d, e, ab, ac, ac, bc, bd, be, and abc. Finally, it is important to remark that determining the exact minimum support value is not trivial and generally requires a profound background in the application field. Inexpert and many expert users need to try different thresholds by guessing and re-executing the algorithms once and again until the results are good for them. As it has been demonstrated, a small change in the threshold value may lead to very few or an extremely large set of solutions.

3.2   Frequent Pattern-Growth

The popular adoption and successful industrial application of the Apriori algorithm has been hindered by the following two limitations: (i) Apriori algorithm generates too many candidate patterns and (ii) Apriori algorithm requires multiple scans on the database. Han et al. [13] introduced Frequent Pattern-growth (FP-growth) algorithm to address the limitations of Apriori algorithm. It is a depth-first search algorithm that finds the desired patterns by employing the following two steps: (i) compress the given database into a *tree* structure known as Frequent Pattern-tree (FP-tree) and (ii) find all frequent patterns by recursively mining the FP-tree.

A really efficient algorithm for mining frequent patterns was proposed by Han et al. [13]. This algorithm, named FP-Growth, achieves a high performance by representing the data as a tree structure [31] in which nodes denote items together with their frequency, and paths among nodes represent the patterns (set of connected nodes). This compressed representation of the input dataset drastically reduces the number of scans required to compute the patterns and their frequencies.

To understand how the FP-Growth algorithm works, let us explain first how to construct the tree structure (see Fig. 2) by taking the sample transactional dataset DB (see Table 1). The first step is to calculate the support value for each item: $support(a, DB) = 4$, $support(b, DB) = 4$, $support(c, DB) = 3$, $support$

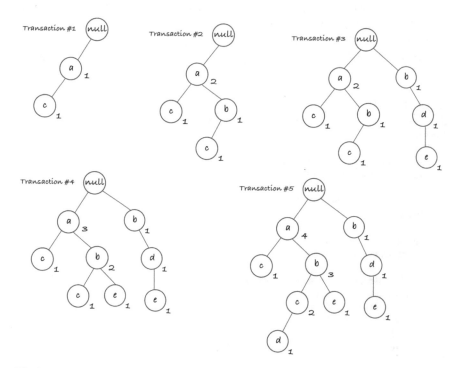

Fig. 2 Building a tree structure for the sample dataset shown in Table 1

$(d, DB) = 2, support(e, DB) = 2$. Items are, therefore, sorted in descending order
of support in each transaction: $a \prec b \prec c \prec d \prec e$. After that, the process starts with
the empty node *null*, and each transaction of DB is analyzed. Taking the first transaction, it comprises items ac and they are included as a branch from the root (the
empty node). Each time a node is added to the tree, the value 1 is assigned to it as the
frequency. If a transaction shares items of any existing branch in the tree, then the
inserted transaction will be in the same path from the root to the common prefix. The
values of the common nodes increase by one. Let us see the second transaction. The
common path is a, so its value is increased by one (see transaction #2 in Fig. 2). The
rest of the items is added as a new branch from the common path. The frequencies
for the new nodes are initialized to one. The third transaction does not share any path
with the tree, so it is added as a new branch from the root, that is, the *null* node. The
process iterates till all the transactions are analyzed, and the resulting tree is obtained
as a compressed data format. It is important to highlight that a fixed order among the
items is required or the resulting tree structure will be meaningless.

The pseudocode of the FP-Growth algorithm is illustrated in Algorithm 2. To
construct the tree, the algorithm takes frequent items (singletons) from data. Considering a minimum frequency value of 2, the following items are considered: a appears
four times, b appears four times, c appears three times, d appears two times, and e
appears two times. Then, the tree is constructed following that order as previously
described. Once the tree is obtained, FP-Growth first takes the lowest node, that is,
item e, which occurs in 2 branches: abe with frequency 1 and bde with frequency 1.
From these branches, only one frequent pattern can be obtained: be, with a frequency
of 2. The process is repeated for each item. Taking now the item d, two branches
are analyzed: $abcd$ and bd. Combining them, the frequent pattern bd is obtained,

---

Algorithm 2 Pseudo-code of the FP-Growth algorithm.

Require: $T, \alpha, minSup$ {Tree-structure, initial node, set of items and minimum support}
Ensure: F
1: $F = \emptyset$
2: **if** T contains a single path P **then**
3:    **for** each combination β of nodes in P **do**
4:       generate the pattern $X = \alpha \cup \beta$
5:       support of X is the minimum support of nodes in β
6:    **end for**
7: **else**
8:    **for** each a in the header of T **do**
9:       generate the pattern $X = \alpha \cup a$
10:      support of X is the support of a
11:      construct a conditional pattern base of X
12:      construct a conditional FP-Tree T_X from X
13:      **if** T_X is not empty **then**
14:        recursive call of FP-Growth($T_X, X, minSup$)
15:      **end if**
16:    **end for**
17: **end if**
18: **return** F

with a frequency of 2. As for the item c, two branches are analyzed: ac and abc. The resulting frequent pattern is ac. The process iterates over all the items. As it is demonstrated, once the tree structure is built, no further passes over the dataset are necessary. Any frequent pattern can be obtained directly from the tree by exploring the tree from the bottom-up (considering the items as suffixes).

Research studies on FP-Growth determined that this algorithm is efficient and scalable. Its performance is calculated as an order of magnitude faster than Apriori. Its main drawback is building and traversing of the tree structure, which is not trivial in huge datasets.

3.3 ECLAT

Eclat (Equivalent CLAss Transformation) was proposed in 1997 [46] as the first algorithm for mining frequent patterns that work on a vertical data representation. This data representation represents the items as lists including the transactions (tids) in which each item appears. Back to the sample transactional database DB shown in Table 1, a vertical data representation of DB is illustrated in Fig. 3. The list formed by item a includes transactions number 1, 2, 4, and 5. As for the item b, it includes transactions number 2, 3, 4, and 5. Eclat considers the vertical data representation as a set of tidsets or pointers to the transaction tids including each item. This algorithm computes the support of each item by simply calculating the length of the tidsets. Thus, the items a and b have a support value of 4 since they appear in 4 transactions. The support of c is 3, and the support of both d and e is 2.

Eclat works by combining tidsets, so for each frequent pattern of length k, a candidate pattern of length $k + 1$ is produced by adding the singleton that lexicographically follows according to those items included in the original frequent pattern (see Algorithm 3). A new tidset is obtained from the intersection of both tidsets (singleton and original frequent pattern) and its size is denoted as the frequency of the resulting pattern. The algorithm follows a breadth-first search strategy taking patterns and doing intersections with the next item in lexicographical order. For

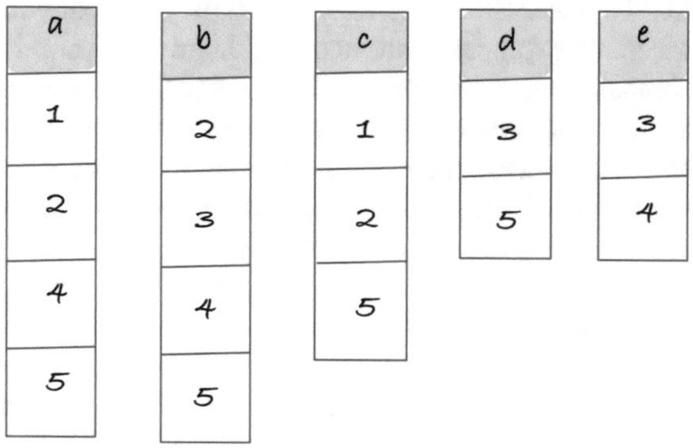

Fig. 3 Tidsets for the sample dataset shown in Table 1

Algorithm 3 Pseudo-code of the Eclat algorithm.

Require: $I, minSup$ {set of items, and minimum support value}
Ensure: F
1: $TidList = $ compute the tidsets of all the items in I
2: $L = \{l \in TidList \mid length(l) \geq minSup\}$
3: $F = F \cup L$
4: **for** $\forall l \in L$ **do**
5:    $c = l$
6:    **for** $\forall m \in L \mid m > j$ **do**
7:       $c' = c \cap m$
8:       **if** $length(c') \geq minSup$ **then**
9:          $c = c'$
10:          $F = F \cup c$
11:       **end if**
12:    **end for**
13: **end for**
14: **return** F

---

example, taking the item a and considering a minimum frequency value of 2, it is combined with a to form the pattern ab. The intersection of their tidsets is the new set $\{2, 4, 5\}$, and therefore, it is a frequent pattern since its frequency is 3 (length of the tidset). In an iterative process, the pattern ab is combined with the next item in lexicographical order, that is, c. The resulting tidset is $\{2, 5\}$, so its frequency is 2. The resulting pattern abc is now combined with d, obtaining the tidset $\{5\}$. Hence, this pattern is infrequent since its frequency is lower than 2. As a result, the process does not continue with the next item and it goes back to combine a with c.

Finally, let us summarize the shortcomings of Eclat. This algorithm does not require to scan the dataset to find the frequency of new patterns as Apriori does. However, the bottleneck comes when the number of transactions increases, requiring huge memory and computational time for intersecting the tidsets.

3.4 LCM

LCM (Linear Closed itemset Miner) [37] is a really fast algorithm proposed in 2003 for mining frequent patterns. This highly optimized algorithm won the FIMI2003 competition in which many efficient algorithms were proposed. LCM does not follow a single data representation but two of them, so it horizontally stores transactions and it also keeps the id of the transactions for each item (vertical data representation). What makes LCM be so fast and efficient is not the data representation but some alluring ideas that it includes. Authors of this algorithm clearly stated [37] that the pruning process usually included in FPM algorithms is not complete, and it is common to operate unnecessary frequent patterns. Taking it into consideration, they overcome the problem with a hypercube decomposition, also known as perfect extension pruning. An occurrence deliver schema is an additional promising

---
Algorithm 4 Pseudo-code of the occurrence deliver schema.

Require: P, DB_P {Pattern to be analyzed, conditional dataset}
Ensure: $Bucket$
1: **for** $\forall i \in P$ **do**
2:    Set $Bucket[i] = \emptyset$
3: **end for**
4: **for** $\forall t \in DB_P$ **do**
5:    **for** $\forall e \in t$ **do**
6:       Insert t into $Bucket[e]$
7:    **end for**
8: **end for**

---

idea included in LCM to easily calculate the frequency of each pattern. The occurrence deliver schema (see Algorithm 4) takes as input a pattern P and a conditional database given P. It creates the buckets for each item in P and adds into buckets those transactions including the item. First, the algorithm orders the items based on their frequencies and removing infrequent items from data. To understand how this procedure works, let us take the transactional database DB shown in Table 1 in which items are sorted in ascending order of support in each transaction: $d \prec e \prec c \prec a \prec b$. Following the occurrence deliver schema illustrated in Algorithm 4, the following buckets are obtained (see Figure 4): $Bucket[a] = \{1, 2, 4, 5\}$, $Bucket[b] = \{2, 3, 4, 5\}$, $Bucket[c] = \{1, 2, 5\}$, $Bucket[d] = \{3, 5\}$, and $Bucket[e] = \{3, 4\}$. Thus, it is easy to know that the pattern a is satisfied by 4 transactions, pattern b by 4 transactions, pattern c by 3 transactions, and so on. Taking the first item in the list of sorted items, that is, item d, a projection is performed and new buckets are obtained for that projected database (see Algorithm 4) as it is illustrated in Figure 4. Thus, given the prefix d, it is easy to check that the itemset da appears once ($Bucket[a] = \{5\}$), whereas the itemset db appears 2 times ($Bucket[b] = \{3, 5\}$). The process is repeated recursively (a depth-first search strategy is considered) for the remaining items so the conditional database are smaller and smaller.

As for the hypercube decomposition technique, it aims to improve the searching process by stopping the recursive process of forming patterns. Once a perfect extension is detected, then it is directly reported. A perfect extension of a pattern I is an item i that satisfies $i \notin I$, and the frequency of I and $I \cup i$ is exactly the same. Perfect extensions have the following properties: If the item i is a perfect extension of a pattern I, then i is also a perfect extension of any pattern Q such as $I \subseteq Q$ as long as $i \notin Q$; If E is the set of all perfect extensions of the pattern I, then all sets $I \cup Q$ with $Q \in 2^E$ (the power set of the set E) have the same support as I. Back to the sample transactional dataset shown in Table 1, the item a is a perfect extension of the pattern bc since the frequency of bc is exactly the same as abc, that is, 2. Consequently, the item a is also a perfect extension of the pattern b and c.

A new LCM version was proposed one year later [39], and it was granted for a second year in a row with the best implementation award in FIMI2004 competition. The new version introduced an improvement in runtime by reducing data. The algorithm deletes an item if it is not present in at least α data records or if it is present in

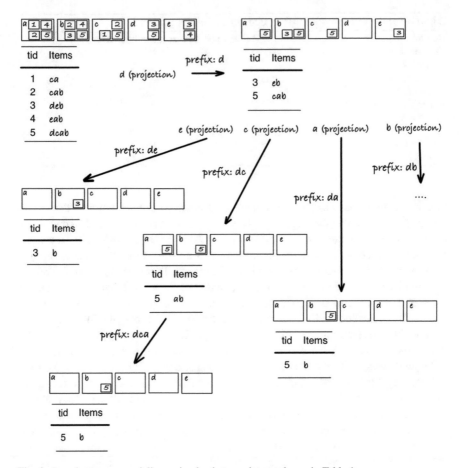

Fig. 4 Sample occurrence deliver using buckets on dataset shown in Table 1

all data records. Additionally, identical data records are merged into a single one. A third LCM version was also proposed one year later [38], and it is considered as the fastest version up to date. This version includes really efficient structures in the FPM task: bitmaps, array lists, and prefix trees. A bitmap is efficient for dense datasets since it enables fast intersections/unions to be performed with not so much memory consumption. The k most frequent items are kept, known as k-*items machine* in the literature. Those items not included in the bitmap representation are stored as array lists and considered to build a prefix tree. It is essential to highlight that this third version also comprises an occurrence deliver technique, the database reduction, and the hypercube decomposition technique proposed in previous versions.

4   Types of Patterns

The growing interest in pattern mining has encouraged the definition of a new type of pattern according to the analysis required by experts in different application fields.

4.1   Maximal Frequent Patterns and Closed Frequent Patterns

Since the objective of FPM is to find all frequently occurring patterns in the database, this model often produces too many patterns most of which may be uninteresting to the users. Moreover, the computational cost of finding these huge number of frequent patterns may not be non-trivial.

Example 4 The frequent pattern model not only finds abc as a frequent pattern in Table 1, but also finds all of its non-empty subsets, i.e., a, b, c, ab, ac, and bc as frequent patterns. Thus, producing too many patterns, most of which are uninteresting or redundant to the user.

When encountered with this problem in real-world applications, researchers have tried to find a reduced set of frequent patterns, namely *maximal frequent patterns* and *closed frequent patterns*. We now briefly discuss both of these patterns.

Definition 1 (Maximal frequent pattern X.) A frequent pattern X is said to be a *maximal frequent pattern* if $sup(X) \geq minSup$ and $\forall Y \supset X,\ sup(Y) \not\geq minSup$.

Example 5 The frequent pattern abc in Table 1 is a maximal frequent pattern because all of its supersets are infrequent patterns. Moreover, all non-empty subsets of abc cannot be maximal frequent patterns. Thus, maximal frequent pattern mining significantly reduces the number of patterns being discovered in a database.

Maximal frequent pattern mining helps us to find long patterns in a database effectively. However, they lead to a loss of information as they do not record the *support* information of its subsets. This motivated researchers to find closed frequent patterns in a database. A closed frequent pattern is a frequent pattern that is not strictly included in another pattern having the same frequency. Thus, closed frequent patterns are lossless by nature as they preserve the *support* information of all frequent patterns in a database.

Definition 2 (Closed frequent pattern X.) Let X and Y be two frequent patterns such that $X \subset Y$. The frequent pattern X is said to be a closed frequent pattern if $sup(X) \neq sup(Y), sup(X) \geq minSup$.

Example 6 Consider the frequent patterns a, c, and ac in Table 2. Since $sup(a) \neq sup(ac)$, a is a closed frequent pattern. In contrast, c is not a closed frequent pattern because $sup(c) = sup(ac)$.

The relationship between the set of frequent patterns (F), the set of closed frequent patterns (C), and the set of maximal frequent patterns (M) is $F \supseteq C \supseteq M$.

4.2   Condensed Patterns

As previously stated, the mining of frequent patterns [1] is the keystone in data analysis and the extraction of useful patterns from data. Many efficient approaches have been proposed but it is still expensive to find the complete set of solutions (frequent patterns). The daunting process may be eased by computing a small subset of frequent patterns that can be used to approximate the frequency values of arbitrary frequent patterns. These frequent patterns used to approximate further solutions are known as condensed patterns [25]. The mining of these patterns drastically reduces the runtime but, sometimes, the error in the approximation is not enough from the domain/problem point of view. On some occasions, this process (considering a maximal error bound) is enough even when no full precision is achieved.

The support approximations are calculated through a function $\mathcal{F}$ defined on a transactional database DB. The following can be an approximation function for any pattern P: $\mathcal{F}(P) = 0$ if there exists no superset $P' \supseteq P$ such as P' is defined as a condensed pattern; whereas $\mathcal{F}(P) = [support(P', DB) - 3, support(P', DB)]$ being $support(P')$ the minimum support for any $P' \subset P$ such as P' is defined as a condensed pattern. Considering such a function, the support of any pattern can be estimated. Back to the sample transactional database DB (see Table 1), $\mathcal{F}(abcde) = 0$ since there is no $P' \in DB | abcde \subseteq P'$. On the contrary, $\mathcal{F}(ac) = [3 - 3, 3] = [0, 3]$ since the minimum support of any subset of ac is $support(c, DB) = 3$.

This way of dealing with patterns is sometimes preferable to the mining of all the patterns. The following are the main reasons: (1) When dealing with really large datasets, small deviations often have minor effects on the analysis; (2) Computing condensed patterns leads to more efficient approaches since it is only required to operate with and access to a small portion of frequent patterns [29].

4.3   Top-k Patterns

In the task of frequent pattern mining, it is common to consider a frequency threshold value to split the search space into useful and useless solutions (patterns). This frequency threshold mainly depends on the users' expectations and the data themselves. Thus, it is the user who has to specify the boundary to consider a pattern a valid solution or not. This boundary will produce a large set of patterns (if the frequency threshold is too low) or a reduced set of patterns (if the frequency threshold is high enough). However, the problem to properly specify such a threshold value is not easy and it also depends on the data distribution. Some datasets may not produce any pattern for a specific frequency value, whereas other datasets may produce tons of patterns for the same frequency value. As a result, it is not trivial to know beforehand the right boundary value for each dataset. Additionally, two main dangers of working with frequency-based algorithms are: 1) setting up an incorrect threshold value may cause an algorithm to fail in finding the interesting patterns; 2) the algorithm may report spurious patterns that do not really exist or are not interesting at all.

To precisely control the output size and discover the patterns with the highest frequency are of vital importance in many application fields [27]. In this regard, it is necessary to extract not only frequent patterns but those having a high significance and low redundancy. Salam et al. [28] proposed an algorithm for mining frequent patterns without any minimum frequency threshold. To achieve this, they proposed two novel algorithms for mining top-most and top-k frequent patterns. The top-k parameter has been differently used by researchers: Wang et al. [41] employed a top-k parameter to extract frequent closed patterns; Tzvetkov et al. [36] considered sequential databases; Cheung et al. [43] retrieved the k most frequent patterns; and Chuang et al. [7] used a top-k parameter for mining top few significant maximal frequent patterns. Fournier-Viger et al. [35] proposed to redefine the task of mining high utility patterns as mining top-k high utility patterns. Luna et al. [23] also proposed a free-parameter algorithm for mining patterns in the form of association rules.

4.4 User-Centric Patterns

Up to date, it is possible to find many research studies in the specialized literature related to different and efficient ways of speeding up the mining of frequent patterns [19]. Currently, more and more research studies are paying attention to extract patterns according to the users' needs or expectations. The user, as the final consumer of the data insights, should play a relevant role in the mining process and it is not just enough to find any pattern that overcomes some minimum frequency values, but those patterns that provide useful information to the user. Here, it is essential to highlight comprehensibility and flexibility as two challenging research issues in the pattern mining field. Comprehensibility describes the degree by which a user can understand the provided information. This is a subjective measure since a pattern can be little comprehensible for a specific user and, at the same time, too much comprehensible for others. Nevertheless, it is possible to define this metric as an objective measure with a fixed formula: the fewer the number of items included in the extracted pattern, the more comprehensible the pattern is. Flexibility, on the contrary, refers to the ability to adapt the solutions to the users' requirements by introducing subjective knowledge into the mining process. These two features, namely comprehensibility and flexibility, have been widely studied and some authors [24] have proposed the use of grammars in pattern mining to introduce subjective knowledge in the mining process and to produce more flexible and expressive results.

Recently, there are some signs of progress in supplying existing pattern mining approaches [19] with methods to extract more actionable insights [40]. Thus, the user is playing a crucial role in the mining process restricting the search space with various constraints based on his/her subjective knowledge of the problem [18]. Other proposals are focused on using new and more flexible forms of information [12]. New methods are also being considered to handle context-sensitive concepts to avoid any discriminative behavior [21], as well as to extract exceptional behavior [22]. As a result, more and more researchers are paying special attention to the extraction of the appropriate knowledge type [20], accomplishing the users' aims or requirements.

4.5   Weighted Frequent Patterns

In traditional frequent pattern mining, items within the solutions are uniformly treated and present the same degree of importance. However, in a real-world scenario, items usually have different importance and it is, therefore, required to weigh them. A simple example is related to the market basket analysis, where expensive items may contribute a large portion of overall revenue even though it does not appear in many data records. Weighted frequent pattern mining [45] has been suggested as a promising task to find important frequent patterns by considering the weights of patterns.

Based on the previous ideas, multiple weighted frequent pattern mining algorithms have been proposed up to date. First proposals, such as MINWAL [6], WARM [33], and WAR [42], were proposed taking the Apriori [3] algorithm as a baseline. All these algorithms follow a breadth-first search methodology, considering a level-wise paradigm in which all the candidate patterns of length $k+1$ are obtained by using all the extracted patterns of length k. A major drawback of these algorithms is they require multiple database scans, giving rise to poor performance. Yun et al. [44] proposed a more efficient algorithm, named WFIM (Weighted Frequent Itemset Mining), that was the first weighted frequent pattern algorithm based on an FP-tree. This algorithm considered a minimum weight value and a weight range, and the FP-Tree is arranged in weight ascending order. A similar algorithm, known as WCloset, was also proposed but instead of mining weighted frequent patterns, it aimed to extract closed weighted frequent patterns. Ahmed et al. [4] proposed an approach to keep track of the varying weights of each item in a prefix tree. They proposed the DWFPM (dynamic weighted frequent pattern mining) algorithm, which is able to handle dynamic weights during the mining process.

Recently, Uday et al. [14] introduced two pattern-growth algorithms: Sequential Weighted Frequent Pattern-growth and Parallel Weighted Frequent Pattern-growth. These two algorithms, which were designed to discover weighted frequent patterns efficiently, employ three novel pruning techniques to reduce the computational cost effectively. The first technique, called cutoff-weight, prunes uninteresting items in the database. The second pruning technique, called conditional pattern base elimination, eliminates the construction of conditional pattern bases if a suffix item is an uninteresting item. The third pruning technique, called pattern-growth termination, proposes a new terminating condition for the pattern-growth technique.

4.6   High Utility Patterns

Frequent pattern mining is a widely known and useful task, but it has three essential limitations. First, the item purchase quantities are not considered in the data records. As a result, any item is considered equally important than buying a single unit. In market basket analysis, this is a useless analysis since customers tend to buy more than a single unit. Second, all items are considered as equally important, which does not show a real situation if a real-world problem is considered. It is more important to sell a laptop than a keyboard since the former yields a much higher profit. Third, the frequency of the patterns could be useless for the user, which is more interested in the amount of profit. High utility pattern mining has recently emerged to address all these limitations [34].

The high utility problem is, therefore, a generalization of frequent pattern mining, considering that every single item has a different utility or relative importance. Additionally, each item in the high utility problem can appear more than once in a data record or transaction. The aim of high utility pattern mining is to find all patterns having a utility higher than a predefined threshold value. Research studies have proposed really interesting and efficient approaches [9]. However, all of them have an important limitation since in the high utility pattern mining problem, the anti-monotonicity property does not hold for the utility measure. The first algorithm for mining high utility patterns is called Two-Phase [10]. This algorithm is an extension of the Apriori algorithm [3] considering an upper-bound on the utility, called Transaction Weighted Utilization (TWU), which is anti-monotonic. After that, really efficient algorithms were proposed such as UP-Growth [34]. However, these algorithms overestimate their utility, so it is computationally hard to calculate their real utility. To overcome such limitations, different researchers have proposed to ease the utility calculation: HUI-Miner [17], FHM+ [8], and EFIM [47], to list a few. Recently, high utility pattern mining algorithms are considering the time at which transactions were made [11]. Additionally, some research studies [16] focus on finding high utility patterns that periodically appear in data.

4.7 Periodic-Frequent Patterns

Periodic-frequent patterns were introduced by Tanbeer et al. [32] as a way of determining the interestingness of frequent patterns in terms of the shape of occurrence. In other words, whether frequent patterns occur periodically, irregularly, or mostly in the specific time interval in the dataset. For example, the shopkeeper in a retail market may be interested only in those products that were regularly sold compared to the rest.

A frequent pattern is defined as a periodic-frequent pattern if it appears and maintains a similar period/interval in a database. To put it in another way, a frequent pattern is said to be periodic-frequent if it occurs at regular intervals specified by the user in data. More technically, a pattern is called a periodic-frequent pattern if it satisfies both of the following two criteria: 1) its periodicity is no greater than a user-given maximum periodicity threshold value; 2) its support is no less than a user-given minimum support threshold value. As a result, the periodic-frequent pattern mining problem is to discover the complete set of periodic-frequent patterns in a database satisfying the two aforementioned criteria.

Tanbeer et al. [32] proposed the first algorithm for mining periodic-frequent patterns. This algorithm, known as PFP-growth, is a pattern-growth approach that generates periodic-frequent patterns by applying depth-first search in the pattern lattice. From a single item i that satisfies the requirements (frequency and periodicity), PFP-growth obtains larger patterns by adding one item at a time. Additionally, thanks to the periodicity measure, which ensures the anti-monotonic property, the algorithm is quite efficient in discovering the complete set of periodic-frequent patterns. After this first approach, many research studies have paid attention to this interesting task giving rise to really efficient approaches. Kiran et al. [15] and Surana et al. [30] enhanced the PFP-growth algorithm to address the problem of rare or infrequent pattern mining. Amphawan et al. [5] proposed an efficient algorithm for mining top-k periodic-frequent patterns, reducing the resulting set and making it more understandable for the end-user. Rashid et al. [26] employed standard deviation of periods as an alternative criterion to assess the periodic behavior of frequent patterns. They considered extensions of the well-known PFP-growth algorithm [32] to obtain a resulting set of solutions known as regular frequent patterns.

References

1. C.C. Aggarwal, J. Han, *Frequent Pattern Mining* (Springer International Publishing, Berlin, 2014)
2. C.C. Aggarwal, J. Han, *Frequent Pattern Mining* (Springer Publishing Company, Berlin, 2014)
3. R. Agrawal, T. Imielinski, and A.N. Swami, Mining association rules between sets of items in large databases. Proceedings of the 1993 ACM SIGMOD International Conference on Management of Data, SIGMOD Conference'93 (ACM, New York, 1993), pp. 207–216
4. C.F. Ahmed, S.K. Tanbeer, B.-S. Jeong and Y.-K. Lee, Handling dynamic weights in weighted frequent pattern mining. *IEICE Trans. Inf. Syst.*, **91-D**(11), 2578–2588 (2008)
5. K. Amphawan, P. Lenca and A. Surarerks, Mining top-k periodic-frequent pattern from transactional databases without support threshold. Advances in Information Technology—Third International Conference, IAIT 2009 (Springer, Berlin, 2009), pp. 18–29
6. C.H. Cai, A.W.C. Fu, C.H. Cheng and W.W. Kwong, Mining association rules with weighted items. Proceedings of the 1998 International Database Engineering and Applications Symposium, IDEAS 1998 (IEEE, New York, 1998), pp. 68–77
7. Kun-Ta. Chuang, Jiun-Long. Huang, Ming-Syan. Chen, Mining top-k frequent patterns in the presence of the memory constraint. The VLDB Journal **17**(5), 1321–1344 (2008)
8. Philippe Fournier-Viger, Jerry Chun-Wei Lin, Quang-Huy Duong, and Thu-Lan Dam. FHM + : Faster high-utility itemset mining using length upper-bound reduction. In *Trends in Applied Knowledge-Based Systems and Data Science - 29th International Conference on Industrial Engineering and Other Applications of Applied Intelligent Systems, IEA/AIE 2016, Morioka, Japan, August 2-4, 2016, Proceedings*, pages 115–127, 2016
9. Philippe Fournier-Viger, Jerry Chun-Wei Lin, Roger Nkambou, Bay Vo, and Vincent S. Tseng. *High-Utility Pattern Mining: Theory, Algorithms and Applications*. Springer Publishing Company, Incorporated, 1st edition, 2019
10. Philippe Fournier-Viger, Cheng-Wei Wu, Souleymane Zida, and Vincent S. Tseng. FHM: faster high-utility itemset mining using estimated utility co-occurrence pruning. In *Foundations of Intelligent Systems - 21st International Symposium, ISMIS 2014, Roskilde, Denmark, June 25-27, 2014. Proceedings*, pages 83–92, 2014
11. Philippe Fournier-Viger and Souleymane Zida. FOSHU: faster on-shelf high utility itemset mining - with or without negative unit profit. In *Proceedings of the 30th Annual ACM Symposium on Applied Computing, Salamanca, Spain, April 13-17, 2015*, pages 857–864, 2015

12. J. Han, Y. Fu, Mining multiple-level association rules in large databases. IEEE Transactions on Knowledge and Data Engineering **11**(5), 798–804 (1999)
13. Jiawei Han, Jian Pei, Yiwen Yin, Mining frequent patterns without candidate generation. SIGMOD Rec. **29**(2), 1–12 (2000)
14. R. Uday Kiran, Amulya Kotni, P. Krishna Reddy, Masashi Toyoda, Subhash Bhalla, and Masaru Kitsuregawa. Efficient discovery of weighted frequent itemsets in very large transactional databases: A re-visit. In Naoki Abe, Huan Liu, Calton Pu, Xiaohua Hu, Nesreen K. Ahmed, Mu Qiao, Yang Song, Donald Kossmann, Bing Liu, Kisung Lee, Jiliang Tang, Jingrui He, and Jeffrey S. Saltz, editors, *IEEE International Conference on Big Data, Big Data 2018, Seattle, WA, USA, December 10-13, 2018*, pages 723–732. IEEE, 2018
15. R. Uday Kiran and P. Krishna Reddy. Towards efficient mining of periodic-frequent patterns in transactional databases. In *Database and Expert Systems Applications, 21th International Conference, DEXA 2010, Bilbao, Spain, August 30 - September 3, 2010, Proceedings, Part II*, pages 194–208, 2010
16. Yu-Feng. Lin, Wu. Cheng-Wei, Chien-Feng. Huang, Vincent S. Tseng, Discovering utility-based episode rules in complex event sequences. Expert Syst. Appl. **42**(12), 5303–5314 (2015)
17. Mengchi Liu and Jun-Feng Qu. Mining high utility itemsets without candidate generation. In *21st ACM International Conference on Information and Knowledge Management, CIKM'12, Maui, HI, USA, October 29 - November 02, 2012*, pages 55–64, 2012
18. J.M. Luna, J.R. Romero, S. Ventura, Design and behavior study of a grammar-guided genetic programming algorithm for mining association rules. Knowledge and Information Systems **32**(1), 53–76 (2012)
19. José María Luna, Philippe Fournier-Viger, and Sebastián Ventura. Frequent itemset mining: A 25 years review. *Wiley Interdiscip. Rev. Data Min. Knowl. Discov.*, 9(6), 2019
20. J.M. Luna, P. Fournier-Viger and S. Ventura, Extracting user-centric knowledge on two different spaces: Concepts and records. *IEEE Access* **8**, 134782–134799 (2020)
21. José María Luna, Mykola Pechenizkiy, María José del Jesus, and Sebastián Ventura. Mining context-aware association rules using grammar-based genetic programming. *IEEE Trans. Cybern.*, 48(11):3030–3044, 2018
22. José María Luna, Mykola Pechenizkiy, Wouter Duivesteijn, and Sebastián Ventura. Exceptional in so many ways - discovering descriptors that display exceptional behavior on contrasting scenarios. *IEEE Access*, 8:200982–200994, 2020
23. José María Luna, José Raúl Romero, Cristóbal Romero, and Sebastián Ventura. Reducing gaps in quantitative association rules: A genetic programming free-parameter algorithm. *Integr. Comput. Aided Eng.*, 21(4):321–337, 2014
24. José María Luna, José Raúl Romero, and Sebastián Ventura. Design and behavior study of a grammar-guided genetic programming algorithm for mining association rules. *Knowl. Inf. Syst.*, 32(1):53–76, 2012
25. J. Pei, G. Dong, W. Zou, J. Han, Mining Condensed Frequent-Pattern Bases. Knowledge and Information Systems **6**(5), 570–594 (2004)
26. Md. Mamunur Rashid, Md. Rezaul Karim, Byeong-Soo Jeong, and Ho-Jin Choi. Efficient mining regularly frequent patterns in transactional databases. In *Proceedings of the 17th International Conference on Database Systems for Advanced Applications - Volume Part I, DASFAA'12*, page 258-271, Berlin, Heidelberg, 2012. Springer-Verlag
27. Cristóbal Romero, Amelia Zafra, José María Luna, and Sebastián Ventura. Association rule mining using genetic programming to provide feedback to instructors from multiple-choice quiz data. *Expert Syst. J. Knowl. Eng.*, 30(2):162–172, 2013
28. Abdus Salam and M. Sikandar Hayat Khayal. Mining top-k frequent patterns without minimum support threshold. *Knowledge and Information Systems*, 30(1):57-86, 2012
29. A. Soulet, B. Crémilleux, Adequate condensed representations of patterns. Data Mining and Knowledge Discovery **17**(1), 94–110 (2008)
30. Akshat Surana, R. Uday Kiran, and P. Krishna Reddy. An efficient approach to mine periodic-frequent patterns in transactional databases. In *New Frontiers in Applied Data Mining - PAKDD 2011 International Workshops, Shenzhen, China, May 24-27, 2011, Revised Selected Papers*, pages 254–266, 2011

31. P. N. Tan, M. Steinbach, and V. Kumar. *Introduction to Data Mining*. Addison Wesley, 2005
32. Syed Khairuzzaman Tanbeer, Chowdhury Farhan Ahmed, Byeong-Soo Jeong, and Young-Koo Lee. Discovering periodic-frequent patterns in transactional databases. In *Advances in Knowledge Discovery and Data Mining, 13th Pacific-Asia Conference, PAKDD 2009, Bangkok, Thailand, April 27-30, 2009, Proceedings*, pages 242–253, 2009
33. Feng Tao, Fionn Murtagh, and Mohsen Farid. Weighted association rule mining using weighted support and significance framework. In *Proceedings of the Ninth ACM SIGKDD International Conference on Knowledge Discovery and Data Mining*, KDD '03, page 661-666, New York, NY, USA, 2003. Association for Computing Machinery
34. Vincent S. Tseng, Bai-En. Shie, Wu. Cheng-Wei, S Yu. Philip, Efficient algorithms for mining high utility itemsets from transactional databases. IEEE Trans. Knowl. Data Eng. **25**(8), 1772–1786 (2013)
35. Vincent S. Tseng, Wu. Cheng-Wei, Philippe Fournier-Viger, S Yu. Philip, Efficient algorithms for mining top-k high utility itemsets. IEEE Trans. Knowl. Data Eng. **28**(1), 54–67 (2016)
36. Petre Tzvetkov, Xifeng Yan, Jiawei Han, TSP: mining top-k closed sequential patterns. Knowl. Inf. Syst. **7**(4), 438–457 (2005)
37. Takeaki Uno, Tatsuya Asai, Yuzo Uchida, and Hiroki Arimura. Lcm: An efficient algorithm for enumerating frequent closed item sets. In *Fimi*, volume 90. Citeseer, 2003
38. Takeaki Uno, Masashi Kiyomi, and Hiroki Arimura. Lcm ver. 3: collaboration of array, bitmap and prefix tree for frequent itemset mining. In *Proceedings of the 1st international workshop on open source data mining: frequent pattern mining implementations*, pages 77–86, 2005
39. Takeaki Uno, Masashi Kiyomi, Hiroki Arimura, et al. Lcm ver. 2: Efficient mining algorithms for frequent/closed/maximal itemsets. In *Fimi*, volume 126, 2004
40. S. Ventura and J. M. Luna. *Pattern Mining with Evolutionary Algorithms*. Springer International Publishing, 2016
41. Jianyong Wang, Jiawei Han, Lu. Ying, Petre Tzvetkov, TFP: an efficient algorithm for mining top-k frequent closed itemsets. IEEE Trans. Knowl. Data Eng. **17**(5), 652–664 (2005)
42. Wei Wang, Jiong Yang, S Yu. Philip, WAR: weighted association rules for item intensities. Knowl. Inf. Syst. **6**(2), 203–229 (2004)
43. Yin-Ling Cheung and Ada Wai-Chee Fu, Mining frequent itemsets without support threshold: with and without item constraints. IEEE Transactions on Knowledge and Data Engineering **16**(9), 1052–1069 (2004)
44. Unil Yun and John J. Leggett. WFIM: weighted frequent itemset mining with a weight range and a minimum weight. In *Proceedings of the 2005 SIAM International Conference on Data Mining, SDM 2005, Newport Beach, CA, USA, April 21-23, 2005*, pages 636–640, 2005
45. Unil Yun and Keun Ho Ryu, Approximate weighted frequent pattern mining with/without noisy environments. Knowl. Based Syst. **24**(1), 73–82 (2011)
46. Mohammed J Zaki, Srinivasan Parthasarathy, Mitsunori Ogihara, and Wei Li. Parallel algorithms for discovery of association rules. *Data mining and knowledge discovery*, 1(4):343–373, 1997
47. Souleymane Zida, Philippe Fournier-Viger, Jerry Chun-Wei Lin, Cheng-Wei Wu, and Vincent S. Tseng. EFIM: A highly efficient algorithm for high-utility itemset mining. In *Advances in Artificial Intelligence and Soft Computing - 14th Mexican International Conference on Artificial Intelligence, MICAI 2015, Cuernavaca, Morelos, Mexico, October 25-31, 2015, Proceedings, Part I*, pages 530–546, 2015

Discovering Full Periodic Patterns in Temporal Databases

Pamalla Veena and R. Uday Kiran

Abstract Periodic patterns are an important class of regularities that exist in a temporal database. These patterns were broadly classified into two types: full periodic patterns and partial periodic patterns. In this chapter, we will first discuss these two basic types of periodic patterns. Second, we describe a model to discover a class of full periodic patterns, called periodic-frequent patterns, in a temporal database. Third, we present an algorithm to find the desired patterns. Fourth, we describe different extensions of a periodic-frequent pattern. Finally, we end this chapter by describing the procedure to execute the periodic-frequent pattern mining algorithm in the PAMI Python kit.

1 Introduction

A temporal database is a collection of transactions and their timestamps. Three fundamental properties of a temporal database are: (i) transactions were ordered by their timestamps, (ii) uneven time gap may exist in-between the transactions, and (iii) multiple transactions can share a common timestamp. These three properties differentiate a temporal database from a widely studied transactional database, which is basically an unordered collection of transactions. Many real-world applications, such as intelligent transportation systems, eCommerce, and social networking applications, naturally produce a temporal database. The classic application is the air crafts incident data produced by the Federal Aviation Authority, USA [2]. Table 1 lists some of the incidents reported to the Federal Aviation Authority. We can observe that the first four incidents (or transactions) happened at irregular time intervals, while

P. Veena (✉)
Sri Balaji PG College, Ananthapur, Andhra Pradesh, India
e-mail: rage.vinny@gmail.com

R. Uday Kiran
The University of Aizu, Aizu-Wakamatsu, Fukushima, Japan
e-mail: udayrage@u-aizu.ac.jp

© The Author(s), under exclusive license to Springer Nature Singapore Pte Ltd. 2021   41
R. Uday Kiran et al. (eds.), *Periodic Pattern Mining*,
https://doi.org/10.1007/978-981-16-3964-7_3

Table 1 Few aircraft incidents in the Federation Aviation Authority database. The format of Event Date is "month/date/year." The terms "US," "IS," and "AD" represent "The United States," "Injury Severity," and "Aircraft Damage," respectively

Event ID	Event Date	Location	Country	IS	AD	...
20001218X45448	06/19/1977	EUREKA, CA	US	Fatal(2)	Destroyed	...
20041105X01764	08/02/1979	Canton, OH	US	Fatal(1)	Destroyed	...
20170710X52551	09/17/1979	Boston, MA	US	Non-fatal	Substantial	...
20001218X45446	08/01/1981	COTTON, MN	US	Fatal(4)	Destroyed	...
20020909X01558	01/01/1982	TUSKEGEE, AL	US	Non-fatal	Substantial	...
20020909X01559	01/01/1982	HOBBS, NM	US	Non-fatal	Substantial	...
20020909X01560	01/01/1982	JACKSONVILLE, FL	US	Non-fatal	Substantial	...

the next three incidents happened on the same day. Thus, representing a temporal database.

A temporal database generalizes the transactional database by considering the items' temporal occurrence information within a transaction. More important, we must not transform a temporal database into a transactional database by merging the transactions sharing a common timestamp. It is because such a process will lead to the following errors:

- **Type-I error.** Merging the transactions sharing the same timestamp can result in losing the actual *support* of a pattern. Consequently, we may miss discovering an exciting pattern as a periodic-frequent pattern.

Example 1 The last three transactions in Table 1 have occurred on the same date, which is "01/01/1982." If we merge these three transactions into a single transaction, then we will lose the actual *support* information on the number of "non-fatal incidents" and "substantial aircraft damages" happened on that day. Consequently, we may miss the interesting patterns containing "non-fatal incidents" and "substantial aircraft damages."

- **Type-II error.** Merging the transactions sharing the same timestamp may create false correlations (or associations) between the items, thus generating an uninteresting pattern as a periodic-frequent pattern.

Example 2 Continuing with the previous example, merging the last transactions may induce incorrect correlation between the locations. Consequently, an uninteresting pattern {Tsukege, Al: Hobbs, NM: JacksonVillage,FL} may be generated as an interesting pattern.

The data generated by many real-world applications naturally exist as a temporal database. Useful information that can empower the users with competitive information lies within this database. However, finding this information in temporal databases is non-trivial and challenging because the renowned frequent pattern

Table 2 A temporal database

tid	ts	Items	tid	ts	Items	tid	ts	Items	tid	ts	Items
101	1	ab	104	4	cef	107	7	abce	110	10	abef
102	3	abd	105	5	ab	108	8	cd	111	11	cdg
103	3	cdgh	106	7	h	109	9	cd	112	12	aef

mining algorithms completely disregard the items' temporal occurrence information in the database. When confronted with this problem in the real-world applications, researchers have extended the frequent pattern model to discover periodic patterns in a database [1, 5, 8]. Han et al. [5] divided periodic patterns into two types of patterns, full periodic and partial periodic patterns. The former considers that every point in the period contributes to the cycle behavior of the time series, such as all the hours (days) in a day (year). According to the latter, some but not all points in the period contribute to the cycle behavior of the time series. Thus, partial periodic patterns are a looser kind of full periodic patterns and exist ubiquitously in real world. In this chapter, we will study about a class of full periodic patterns, called periodic-frequent patterns [11], that may exist in a temporal database.

2 A Model of Full Periodic Pattern

A periodic-frequent pattern is a frequent pattern that is occurring at regular intervals in a temporal database. A periodic-frequent pattern indicates there exists something predictable within the database. Thus, there is value in finding these patterns in real-world applications. The model of periodic-frequent patterns is as follows.

Let I be the set of items. Let $X \subseteq I$ be a **pattern** (or an itemset). A pattern containing k number of items is called a k-**pattern**. A **transaction**, $t_{tid} = (tid, ts, Y)$ is a tuple, where $tid \in \mathbb{R}$ represents transactional identifier, $ts \in \mathbb{R}$ represents the timestamp at which the pattern Y has occurred. A **temporal database** $TDB = \{t_1, \cdots, t_m\}$, $m = |TDB|$, where $|TDB|$ represents the number of transactions in TDB.

Example 3 Table 2 shows the temporal database with the set of items $I = \{a, b, c, d, e, f, g, h\}$. The set of items "$a$" and "$b$," i.e., $\{a, b\}$ (or ab, in short) is a pattern. This pattern contains only two items. Therefore, this is a 2-pattern. In the first transaction, $t_1 = (101, 1, ab)$, 101 represents the transaction identifier, 1 denotes the occurrence timestamp, and "ab" represents the pattern occurring in this transaction.

Definition 1 (The *support* **of pattern** X.) If $X \subseteq Y$, it is said that X occurs in Y or Y contains X. Let ts_i^X, $i \geq 1$, denote the timestamp of a transaction containing X. Let $TS^X = \{ts_j^X, \cdots, ts_k^X\}$, $1 \leq j \leq k \leq m$, be an **ordered set of timestamps** in which the pattern X has occurred in TDB. The number of transactions containing X in TDB is defined as the **support** of X and denoted as $sup(X)$. That is, $sup(X) = |TS^X|$.

Example 4 In Table 2, the pattern *ab* appears in the transactions whose timestamps are 1, 2, 5, 7, and 10. Therefore, $TS^{ab} = \{1, 2, 5, 7, 10\}$. The *support* of "*ab*," i.e., $sup(ab) = |TS^{ab}| = |1, 2, 5, 7, 10| = 5$.

Definition 2 (A frequent pattern X.) The pattern X is said to be a frequent pattern if $sup(X) \geq minSup$, where $minSup$ represents the user-specified *minimum support* threshold value.

Example 5 If the user-specified $minSup = 3$, then "*ab*" is a frequent pattern because $sup(ab) \geq minSup$.

Definition 3 (The periods of pattern X.) Let $ts_{initial}$ and ts_{final} denote the initial and final timestamps in TDB, respectively. The *periods* of a pattern X in TDB are calculated in the following three ways:

1. The time taken for the initial appearance of X in TDB, i.e., $p_1^X = (ts_j^X - ts_{initial})$.
2. The inter-arrival times of pattern X in TDB, i.e., $p_z^X = (ts_p^X - ts_q^X)$, $1 < z < (sup(X) + 1)$ and $ts_p^X, ts_q^X \in TS^X$, where ts_p^X and ts_q^X are consecutive timestamps in TS^X.
3. The time elapsed after the final occurrence of X in TDB, i.e., $p_{sup(X)+1}^X = (ts_{final} - ts_k^X)$.

The first and the last periods, i.e., p_1^X and $p_{sup(X)+1}^X$, are crucial in determining whether a pattern is occurring periodically in the entire database or not. These two *periods* also ensure that the generated periodic-frequent patterns satisfy the *downward closure property*. This property makes the periodic-frequent pattern mining practicable in the real-world applications. (The *downward closure property* of periodic-frequent patterns was described in the next section.)

Example 6 Let the initial and final timestamps of all transactions in Table 2 are 0 and 12, respectively. That is, $ts_{initial} = 0$ and $ts_{final} = 12$. The complete set of *periods* for "*ab*" in Table 2 are as follows: $p_1^{ab} = 1 \ (= 1 - ts_{initial})$, $p_2^{ab} = 1 \ (= 2 - 1)$, $p_3^{ab} = 3 \ (= 5 - 2)$, $p_4^{ab} = 2 \ (= 7 - 5)$, $p_5^{ab} = 3 \ (10 - 7)$, and $p_6^{ab} = 2 \ (= ts_{final} - 10)$. In our example, we have set $ts_{initial} = 0$. However, if the application demands $ts_{initial}$ can be set to the minimal timestamp of all the transactions in the database.

Definition 4 (The periodicity of pattern X.) Let $P^X = \{p_1^X, p_2^X, \cdots, p_r^X\}$ be the set of all *periods* for pattern X. The **periodicity** of X, denoted as $per(X) = max(p_1^X, p_2^X, \cdots, p_r^X)$.

Example 7 Continuing with the previous example, the complete set of *periods* of "*ab*" in Table 2, i.e., $P^{ab} = (0, 1, 3, 2, 3, 2)$. The *periodicity* of "*ab*," i.e., $per(ab) = max(0, 1, 3, 2, 3, 2) = 3$. It means the pattern "*ab*" has appeared at least once in every 3 units of timestamps.

Definition 5 (A periodic-frequent pattern X.) A frequent pattern X is said to be a periodic-frequent pattern if $per(X) \leq maxPer$, where $maxPer$ refers to the user-specified *maximum periodicity* constraint.

Table 3 Periodic-frequent patterns generated from Table 2. The terms "pat," "sup," and "per" represent "pattern," "support," and "periodicity," respectively

pat	sup	per	pat	sup	per	pat	sup	per
a	6	3	d	5	5	ab	5	3
b	5	3	e	4	4	ef	3	6
c	6	3	f	3	6	cd	4	5

Example 8 If the user-defined $maxPer = 6$, then the frequent pattern "ab" is said to be a periodic-frequent pattern because $per(ab) \leq maxPer$. The complete set of periodic-frequent patterns generated from Table 2 are shown in Table 3.

Definition 6 (Problem definition.) Given a temporal database TDB and the user-specified *minimum support* ($minSup$) and *maximum periodicity* ($maxPer$) constraints, find all patterns in TDB that satisfy the user-specified $minSup$ and $maxPer$ constraints.

The *support* of a pattern can be expressed in percentage of $|TDB|$. Similarly, the *period* and *periodicity* of a pattern can be expressed in percentage of $(ts_{final} - ts_{initial})$. However, we employ the former definitions of *support*, *period*, and *periodicity* throughout this chapter for brevity.

A key element that makes periodic-frequent pattern mining practicable in the real world is $maxPer$ constraint. It is used to prune the search space and limit the number of patterns being generated. Since $maxPer$ controls the maximum time interval within which a pattern must reoccur in the entire temporal database, the generated periodic-frequent patterns represent full periodic patterns in a database. In the next section, we will discuss an algorithm to find the desired patterns.

3 Periodic-Frequent Pattern Growth

The space of items in a database gives rise to an itemset lattice. This itemset lattice represents the search space while finding the periodic-frequent patterns in a temporal database. This search space's size is $2^n - 1$, where n presents the total number of items in a database. This colossal search space raises computational challenges while finding periodic-frequent patterns in large temporal databases.

The Periodic-Frequent Pattern-growth (PFP-growth) algorithm [11] performs the depth-first search on the itemset lattice to find the complete set of periodic-frequent patterns in a database. More importantly, PFP-growth smartly addresses the huge search space issue by exploiting the *downward closure property* of periodic-frequent patterns. The downward closure property of periodic-frequent patterns is shown in property 1. Example 9 illustrates the downward closure property of periodic-frequent patterns. The correctness of periodic-frequent patterns satisfying the *downward clo-sure property* is based on Property 2 and shown in Lemma 1.

Property 1 *(The downward closure property of periodic-frequent pattern.)* All *non-empty subsets of a periodic-frequent pattern must also periodic-frequent pat-terns. That is, if Y is a periodic-frequent pattern, then* $\forall X \subset Y$ *and* $X \neq \emptyset$, *X is also a periodic-frequent pattern.*

Example 9 The pattern "*ab*" is a periodic-frequent pattern in Table 2. Thus, its non-empty subsets, i.e., "*a*" and "*b*," must also be periodic-frequent patterns.

Property 2 *If* $X \subset Y$, *then* $TS^X \supseteq TS^Y$. *Thus,* $sup(X) \geq sup(Y)$ *and* $per(X) \leq per(Y)$.

Lemma 1 *Let X and Y be two patterns such that* $X \subset Y$. *If* $sup(X) < minSup$ *or* $per(X) > maxPer$, *then neither X nor Y can be periodic-frequent patterns.*

Proof If $sup(X) < minSup$ or $per(X) > maxPer$, then X cannot be a periodic-frequent pattern as per Definition 5. Based on Property 2, it turns out that Y cannot also be a periodic-frequent pattern as $sup(Y) < minSup$ or $per(Y) > maxPer$. Hence proved.

The PFP-growth algorithm briefly involves the following steps: (i) scan the database and identify periodic-frequent items (or 1-patterns), (ii) perform another scan on the database and construct Periodic-Frequent tree (PF-tree) constituting of only periodic-frequent items, and (iii) find all periodic-frequent patterns by recur-sively mining the PF-tree. Before we discuss the above three steps, we describe the structure of the PF-tree.

3.1 Structure of PF-Tree

A PF-tree has two components: a PF-list and a prefix tree. The PF-list consists of each distinct *item* (I), *support* (S), *periodicity* (P), and a pointer pointing to the first node in the prefix tree carrying the item.

The prefix tree in PF-tree resembles the prefix tree in FP-tree [6]. However, to record the temporal occurrence information of the patterns, the nodes in PF-tree explicitly maintain the occurrence information for each transaction by keeping an occurrence timestamp list, called a ts-**list**. To achieve memory efficiency, only the last node of every transaction maintains the ts-list. Hence, two types of nodes are maintained in a PF-tree: **ordinary node** and **tail node**. The former is a type of node similar to that used in an FP-tree, whereas the latter represents the last item of any sorted transaction. Therefore, the structure of a *tail* node is $i[ts_p, ts_q, ..., ts_r]$, $1 \leq p \leq q \leq r$, where i is the node's item name and $ts_i \in \mathbb{R}$ is the timestamp of a transaction containing the items from *root* up to the node i. The conceptual structure of PF-tree is shown in Fig. 1. Like an FP-tree, each node in PF-tree maintains parent,

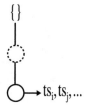

Fig. 1 Conceptual structure of prefix tree in PF-tree. Dotted ellipse represents ordinary node, while other ellipse represents tail node of sorted transactions with timestamps $ts_i, ts_j \in R$

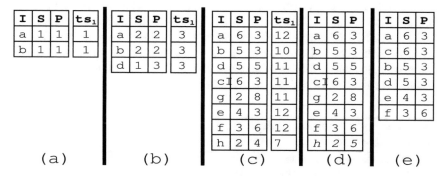

I	S	P	ts₁
a	1	1	1
b	1	1	1

(a)

I	S	P	ts₁
a	2	2	3
b	2	2	3
d	1	3	3

(b)

I	S	P	ts₁
a	6	3	12
b	5	3	10
d	5	5	11
c	6	3	11
g	2	8	11
e	4	3	12
f	3	6	12
h	2	4	7

(c)

I	S	P
a	6	3
b	5	3
d	5	5
c	6	3
g	2	8
e	4	3
f	3	6
h	2	5

(d)

I	S	P
a	6	3
c	6	3
b	5	3
d	5	3
e	4	3
f	3	6

(e)

Fig. 2 Construction of PF-list. **a** After scanning first transaction. **b** After scanning second transaction. **c** After scanning entire database. **d** Final list of periodic-frequent items

children, and node traversal pointers. Please note that no node in PF-tree maintains the support count as in an FP-tree. To facilitate a high degree of compactness, items in the prefix tree are arranged in support-descending order.

3.2 Step 1: Identifying Periodic-Frequent Patterns

Since periodic-frequent patterns satisfy the downward closure property, periodic-frequent items will play an important role in effective mining of these patterns. The set of periodic-frequent items in a temporal database for the user-defined $minSup$ and $maxPer$ can be discovered by populating the PF-list with a scan on the database. Algorithm 1 describes the discovery of periodic-frequent items by populating the PF-list. We now illustrate this algorithm using the database shown in Table 2. Let the user-specified $minSup = 3$ and $maxPer = 6$.

Let "ts_l" be a temporary list that records the timestamp of last occurrence of an item $i_j \in I$. The scan on the first transaction, "101 : 1 : ab," inserts the items "a" and "b" in the PF-list and sets their S, P, and ts_l values to 1, 1, and 1, respectively (lines 1 to 5 in Algorithm 1). Figure 2(a) shows the PF-list generated after scanning the first transaction. The scan on the second transaction, "102 : 3 : abd," inserts the item "d"

in PF-list by setting its S, P, and ts_l values to $1, 3$ ($= 3 - 0$), and 3, respectively. In addition, the S, P, and ts_l values of the already existing items "a" and "b" are updated to $2, 2$, and 3, respectively (line 6 in Algorithm 1). Figure 2(b) shows the PF-list generated after scanning the second transaction. Similar process is followed for remaining transactions in the database and PF-list is updated accordingly. Figure 2(c) shows the PF-list generated after scanning the entire database. Next, the actual *periodicity* of all the items in the PF-list is once again calculated using the ts_{final} value (lines 9 to 11 in Algorithm 1). Figure 2(d) shows the PF-list with the updated *periodicity* values. It can be observed that the *periodicity* of h got updated from 4 to 5 to reflect its actual *periodicity* in the database. The uninteresting items that have *support* less than *minSup* or *periodicity* more than *maxPer* are pruned from the PF-list. The remaining items are considered as periodic-frequent items and are sorted in descending order of their *support* values (line 12 in Algorithm 1). Figure 2(e) shows the final PF-list containing sorted list of periodic-frequent items. Let L denote this sorted list of periodic-frequent items.

Algorithm 1 findingPeriodicFrequentItems(TDB: Temporal database, $minSup$: minimum support, $maxPer$: maximum periodicity)

1: Let ts_l be a temporary array that records the timestamp of the last appearance of each item in TDB. Let $t = \{tid, ts_{cur}, X\}$ denote the current transaction with tid, ts_{cur} and X representing the transaction identifier, timestamp of the current transaction and pattern, respectively.
2: **for** each transaction $t \in TDB$ **do**
3:    **if** an item i occurs for the first time **then**
4:       Insert i into the PF-list with $sup^i = 1$, $per^i = (ts_{initial} - ts_{cur})$ and $ts_l^i = ts_{cur}$.
5:    **else**
6:       Set $sup^i = sup^i + 1$, $per^i = max(per^i, (ts_{cur} - ts_l^i))$ and $ts_l^i = ts_{cur}$.
7:    **end if**
8: **end for**
9: **for** each item i in PF-list **do**
10:    Set $per^i = max(per^i, (ts_{final} - ts_l^i))$.
11: **end for**
12: Remove items from the PF-list that do not satisfy $minSup$ and $maxPer$. Sort the remaining items in PF-list in descending order of their *support*. Let L denote this sorted list of periodic-frequent items.

3.3 Step 2: Construction of PF-Tree

After finding periodic-frequent items items, we conduct another scan on the database and construct the prefix tree of PF-tree as in Algorithms 2 and 3. These algorithms are the same as those for constructing an FP-tree [6]. However, the major difference is that no node in PF-tree maintains the *support* count as in an FP-tree.

A PF-tree is constructed as follows. First, create the root node of the tree and labeled it as "*null*." Scan the database second time. The items in each transaction are

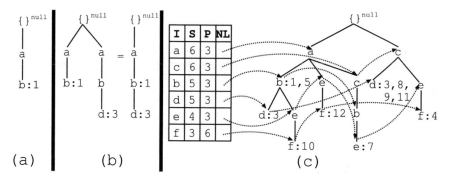

Fig. 3 Construction of PF-tree. **a** After scanning first transaction. **b** After scanning second transaction. **c** After scanning entire database

processed in L order (i.e., sorted according to descending support count), and a branch is created for each transaction such that only the tail nodes record the timestamps of transactions. For example, the scan of the first transaction, "101 : 1 : ab," which contains two items (a and b in L order), leads to the construction of the first branch of the tree with two nodes, $\langle a \rangle$ and $\langle b : 1 \rangle$, where a is linked as a child of the root and $b : 1$ is linked to a. The PF-tree generated after scanning the first transaction is shown in Fig. 3a. The scan on the second transaction, "101 : 3 : abd," containing the items a, b, and d in L order, would result in a branch where a is linked to the root, b is linked to a, and $d : 3$ is linked to b. However, this branch would share a common prefix, a and b, with the existing path for first transaction. Therefore, we create a new node $\langle d : 3 \rangle$, and link d to b as shown in Fig. 3b. A similar process is repeated for the remaining transactions and the tree is updated accordingly. Node links were maintained between the items in the PF-tree for tree traversal. Figure 3c shows the final PF-tree generated after scanning the entire database.

Algorithm 2 constructPFtree(TDB: Temporal database, PF-list: periodic-frequent list)

1: Create the root of PF-tree, T, and label it "*null*".
2: **for** each transaction $t \in TDB$ **do**
3:   Set the timestamp of the corresponding transaction as ts_{cur}.
4:   Select and sort the periodic-frequent items in t according to the order of L. Let the sorted candidate item list in t be $[p|P]$, where p is the first item and P is the remaining list.
5:   Call $insert\_tree([p|P], ts_{cur}, T)$.
6: **end for**

Algorithm 3 insert_tree($[p|P]$, ts_{cur}, T)

1: **while** P is non-empty **do**
2:    **if** T has a child N such that $p.itemName \neq N.itemName$ **then**
3:       Create a new node N. Let its parent link be linked to T. Let its node-link be linked to nodes with the same itemName via the node-link structure. Remove p from P.
4:    **end if**
5: **end while**
6: Add ts_{cur} to the leaf node.

3.4   Step 3: Recursive Mining of PF-Tree

Although PF-tree and FP-tree arrange items in support-descending order, we cannot directly apply FP-growth mining on PF-tree as this tree handles the ts-lists at the *tail* nodes. In this context, PF-growth employs another pattern growth-based bottom-up mining technique to mine periodic-frequent patterns from PF-tree.

The PF-tree is mined as follows. Start from each periodic-frequent item (as an initial suffix pattern), construct its conditional pattern base (a subdatabase, which consists of the set of prefix paths in the PF-tree co-occurring with the suffix pattern), then construct its (conditional) PF-tree, and perform mining recursively on such a tree. The pattern growth is achieved by the concatenation of the suffix pattern with the periodic-frequent patterns generated from a conditional PF-tree.

The procedure to discover periodic-frequent patterns from PF-tree is shown in Algorithm 4. The working of these algorithm is as follows. We proceed to construct the conditional pattern base (or prefix tree) for each periodic-frequent item in the PF-list, starting from the bottom-most item, say i. To construct the prefix tree for i, the prefix sub-paths of nodes i are accumulated in a tree structure, PT_i. Since i is the bottom-most item in the PF-list, each node labeled i in the PF-tree must be a tail node. While constructing PT_i, we map the ts-list of every node of i to all items in the respective path explicitly in the temporary array (one for each item). This temporary array facilitates the calculation of *support* and *periodicity* for each item in PT_i (line 2 in Algorithm 4). If an item j in PT_i has $sup(j) \geq minSup$ and $per(j) \leq maxPer$, we construct its conditional tree and mine it recursively to discover periodic-frequent patterns (lines 3 to 8 in Algorithm 4). Moreover, to enable the construction of the prefix tree for the next item in the PF-list, the ts-lists are pushed up to the respective parent nodes in the original PF-tree and in PT_i as well. All nodes of i in the original PF-tree and i's entry in the PF-list are deleted thereafter (line 9 in Algorithm 4).

Consider item "f," which is the last item in the PF-list. This item appears in the following three branches of original PF-tree: $\langle a, b, e, f : 10 \rangle$, $\langle a, e, f : 12 \rangle$ and $\langle c, e, f : 4 \rangle$ (see Fig. 3c). Considering f as the suffix item, we construct its conditional pattern base (or prefix tree) as follows: $\langle a, b, e : 10 \rangle$, $\langle a, e : 12 \rangle$ and $\langle c, e : f \rangle$. Figure4a shows the prefix tree of f, i.e., PT_f. Among all the four items in PF_f, only item e satisfies the condition of *support* no less than $minSup$ and *periodicity* no more than $maxPer$. Therefore, the conditional tree CT_f from PT_f

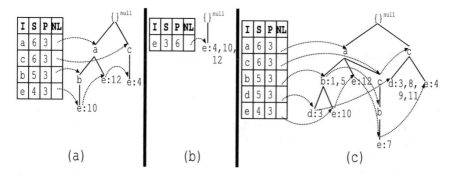

Fig. 4 Recursive mining of PF-tree. (a) Conditional pattern base of f, (b) conditional PF-tree of f, and (c) main PF-tree generated after pruning f

Table 4 Mining the PF-tree by creating conditional (sub-)pattern bases

item	Conditional Pattern base	Conditional PF-tree	PFPs
f	$\{abe : 9, 11\}, \{ae : 4\}, \{dce : 8\}$	$\langle e : 4, 8, 9, 11\rangle$	$\{fe : 2\}$
e	$\{ab : 9, 11\}, \{a : 4\}, \{ad : 9\}\{dc : 8\}$	$\langle a:9,9,11\rangle$	$\{ea : 2\}$
c	$\{abd : 12, 12\}, \{ad : 1\}, \{d : 6, 8, 10\}$	$\langle d : 1, 6, 8, 10, 12, 12\rangle$	$\{cd : 4\}$
d	$\{ab : 12, 12\}, \{a : 1, 9\}$	$-$	$-$
b	$\{a : 1, 3, 5, 9, 11, 12, 12\}$	$\langle a : 1, 3, 5, 9, 11, 12, 12\rangle$	$\{ab : 5\}$

is constructed with only one item "e," as shown in 4b. The ts-list of "e" in CT_f' generates TS^{ef}. The *support* and *periodicity* of "ef" are measured from TS^{ef}. Since $sup(ef) \geq minSup$ and $per(ef) \leq maxPer$, "ef" will be generated as a periodic-frequent pattern. Next, "f" is pruned from the original PF-tree and its ts-lists are pushed to its parent nodes, as shown in 4c. All the above processes are once again repeated until the PF-list $\neq \emptyset$. Mining of PF-tree in Fig. 3(c) is shown in Table 4.

Algorithm 4 PF-growth($Tree, \alpha$)

1: **for** each a_i in the header of Tree **do**
2:     Generate pattern $\beta = a_i \cup \alpha$. Collect all of the $a_i's$ ts-lists into a temporary array, TS^β, and calculate $sup(\beta)$ and $per(\beta)$.
3:     **if** $sup\beta \geq minSup$ and $per(\beta) \leq maxPer$ **then**
4:         Construct β's conditional pattern base then β's conditional PF-tree $Tree_\beta$.
5:         **if** $Tree_\beta \neq \emptyset$ **then**
6:             call PF-growth($Tree_\beta, \beta$);
7:         **end if**
8:     **end if**
9:     Remove a_i from the $Tree$ and push the a_i's ts-list to its parent nodes.
10: **end for**

---

4   Additional Topics

4.1   Maximal Periodic-Frequent Patterns

Since the rationale of periodic-frequent pattern model is to discover all patterns that are occurring regularly in a temporal database, this model often produces too many patterns, most of which may be uninteresting to the user. This problem is known as *combinatorial explosion problem*. The notion of "maximal periodic-frequent pattern" was exploited to address this problem. A maximal periodic-frequent pattern is a periodic-frequent pattern whose supersets are not periodic-frequent patterns. The set of maximal periodic-frequent patterns is generally a minimal subset of periodic-frequent patterns. More importantly, the maximal periodic-frequent patterns are representative since they can be used to recover all periodic-frequent patterns. The exact *support* and *periodicity* of these patterns, if needed, can later be retrieved by performing a single scan on the database.

Definition 7   (Maximal periodic-frequent pattern X.**)** A periodic-frequent X is said to be a maximal periodic-frequent pattern if all of its supersets, say $Y \supset X$, are not periodic-frequent patterns. That is, if X is a maximal periodic-frequent pattern, then $\forall Y \supset X,\ sup(Y) \leq minSup$ or $per(Y) \geq maxPer$.

Example 10   The complete set of periodic-frequent patterns generated from Table 2 are shown in Table 3. Among all of the nine generated periodic-frequent patterns, only three periodic-frequent patterns, i.e., ab, ef, and cd, are considered as maximal periodic-frequent patterns it is because their supersets are not periodic-frequent patterns. In contrast, the periodic-frequent items from a to f are not considered as maximal periodic-frequent patterns. It is because their supersets are periodic-frequent patterns.

A variation of PFP-growth algorithm can be employed to find maximal periodic-frequent patterns. More details on this algorithm can be found at [7].

4.2   Closed Periodic-Frequent Patterns

The limitation of maximal periodic-frequent pattern model is that it fails to preserve the *support* and *periodicity* information of the subsets of a maximal periodic-frequent pattern. The concept of *closed periodic-frequent patterns* represents a concise lossless subset that uniquely describes the complete set of periodic-frequent patterns along with their *support* and *periodicity*.

Definition 8   (Closed periodic-frequent pattern X.**)** Let X and Y be two periodic-frequent patterns such that $X \subset Y$. The periodic-frequent pattern X is said to be a closed periodic-frequent pattern if $sup(X) \neq sup(Y)$, $sup(X) \geq minSup$.

Example 11 Consider the patterns *a*, *b*, and *ab* in Table 3. Among these three patterns, *a* and *ab* are closed periodic-frequent patterns. It is because the *support* and *periodicity* of *a* are not same as those of its superset *ab*, In contrast, *b* is not a closed periodic-frequent pattern because its *support* and *periodicity* are same as those of its superset *ab*.

The algorithm to find the complete set of closed periodic-frequent patterns may be found at [7].

4.3 Stable Periodic-Frequent Patterns

A key element that makes periodic-frequent pattern mining viable in the real-world databases is the *maxPer* constraint. Since *maxPer* controls the maximum time interval within which a pattern must appear in the database, it considers an occurrence of a pattern to be aperiodic if there exists even one *period* that is more than the user-specified threshold value. In other words, *maxPer* prunes all those interesting patterns that have exhibited partial periodic behavior in the database. To confront this problem, Philippe et al. [4] introduced a new measure, called *liability*, to find all those patterns that have exhibited partial periodic behavior in the database. A variant of PFP-growth algorithm was also presented to find all desired patterns in a database.

4.4 Periodic High-Utility Patterns

High-utility patterns are an important class of regularities that exist in a quantitative transactional database. Each high-utility pattern represents a set of items that have recorded high value (or utility) in a quantitative transactional database. Philippe et al. [3] explored the notion of periodicity to discover a class of user interest-based patterns, called periodically occurring high-utility patterns, in a quantitative temporal database. An efficient algorithm named PHM (periodic high-utility itemset miner) was also proposed to efficiently enumerate all periodic high-utility patterns. The popular adoption and successful industrial application of this model have been hindered by its computational cost as the generated patterns do not satisfy the downward closure property. Efforts are being put forth by the researchers [10] to develop efficient algorithms to find periodic high-utility patterns.

4.5 Regular Frequent Patterns

Rashid et al. [9] have employed the *variance* of *periods* as an interestingness criterion to discover a class of periodically occurring frequent patterns, called *regular frequent patterns.* A pattern growth algorithm employing a variant of PF-tree was employed to discover the complete set of regular patterns in a temporal database. The main limitation of this study is its huge search space and computational cost as the generated regular frequent patterns do not satisfy the downward closure property.

5 Finding Periodic-Frequent Patterns Using PAMI Software

In this section, we first describe the procedure to create a temporal database. Next, we describe the procedure to execute PFP-growth algorithm in PAMI software. Finally, we describe the procedure to utilize PFP-growth algorithm as a library in a Python program.

5.1 Format to Create a Temporal Database

In order to discover periodic-frequent patterns from the data using PAMI software, one needs to construct a temporal database such that each transaction exists in the following format: $\langle timestamp \rangle$ $\langle items\ appearing\ in\ a\ transaction \rangle$. Please note that the *timestamp* and the items appearing in a transaction must be separated by a tab space. Moreover, the *timestamp* must be provided in the relative context, and thus its initial value must always be equal to 1. Multiple transactions can share the same timestamp and irregular time gaps may exist between two consecutive timestamps. An example of a transaction is as follows: "1<tab>a<tab>b<tab>c<tab>d<tab>e," where 1 represents the timestamp of a transaction and a, b, c, d, and e represent the items occurring in the corresponding transaction.

5.2 Executing PFP-Growth in PAMI

The step-by-step procedure to execute PFP-growth algorithm in PAMI software is as follows:

1. Clone the Github repository on your system by executing the following command on the terminal: git clone https://github.com/udayRage/PAMI.git.
2. The above command will create a directory, named PAMI, and will download all of the available algorithms.
3. Enter into PF-growth directory by executing the following command: *cd PAMI/periodicFrequentPattern/basic*
4. Execute the PFP-growth algorithm on the terminal using the following command: $python PFPGrowth.py\langle input File \rangle$ $\langle output File \rangle$ $\langle minSup \rangle$ $\langle maxPer \rangle$

E.g., *python PFPGrowth.py sampleTDB.txt patterns.txt 0.3 0.4*

Please note that if you enter *minSup* or *maxPer* in float type, we calculate it as *len(Database) * minSup* or *len(Database) * maxPer*. If you enter *minSup* or *maxPer* in *int* we take it as it is.

5. The output file contains patterns and their respective *support* and *periodicity* values. The format of the output file is ⟨*Pattern : support : periodicity*⟩. The items within a pattern are separated by a tab space.

5.3 Utilizing the PF-Growth Algorithm in Your Python Program

Please download the recent version of PAMI library from the Python Package Index (PyPI) repository. This task can be easily achieved by executing the following command: *pip install pami*. Once the PAMI library was successfully installed on your machine, utilize the PFP-growth algorithm in your Python code as shown in Algorithm 5. For more information, please refer to the documentation code provided within the PFP-growth algorithm.

Algorithm 5 Utilizing code

```
1: from PAMI.periodicFrequentPattern.basic import PFPGrowth as pfgrowth
2: pf = pfgrowth.PFPGrowth("inputFileName", "minSup", "maxPer")
3: pf.startMine()
4: periodicFrequentPatterns = pf.getPatterns()
5: print("Total number of PeriodicFrequent Patterns:", len(periodicFrequentPatterns))
6: pf.savePatterns("outpuFileName")
7: memUSS = pf.getMemoryUSS()
8: print("Total Memory in USS:", memUSS)
9: memRSS = pf.getMemoryRSS()
10: print("Total Memory in RSS", memRSS)
11: rTime = pf.getRuntime()
12: print("Total ExecutionTime in seconds:", rTime)
```

References

1. Walid G. Aref, Mohamed G. Elfeky, Ahmed K. Elmagarmid, Incremental, online, and merge mining of partial periodic patterns in time-series databases. IEEE TKDE **16**(3), 332–342 (2004)
2. FAA. Federal Aviation Authority. http://www.asias.faa.gov. Accessed 17 Jun 2020
3. P. Fournier-Viger, J.C. Lin, Q.-H. Duong and T.-L. Dam, PHM: mining periodic high-utility itemsets, in *Advances in Data Mining. Applications and Theoretical Aspects - 16th Industrial Conference, ICDM 2016*, volume 9728 of *Lecture Notes in Computer Science*, ed. by P. Perner (Springer, Berlin, 2016), pp. 64–79
4. P. Fournier-Viger, P. Yang, J.C. Lin and R. Uday Kiran, Discovering stable periodic-frequent patterns in transactional data. Advances and Trends in Artificial Intelligence. From Theory to Practice—32nd International Conference on Industrial, Engineering and Other Applications of Applied Intelligent Systems, IEA/AIE 2019, volume 11606 of Lecture Notes in Computer Science (Springer, Berlin, 2019), pp. 230–244
5. Jiawei Han, Wan Gong, Yiwen Yin, Mining segment-wise periodic patterns in time-related databases. KDD **98**, 214–218 (1998)
6. Jiawei Han, Jian Pei, Yiwen Yin, Runying Mao, Mining frequent patterns without candidate generation: A frequent-pattern tree approach. Data Min. Knowl. Discov. **8**(1), 53–87 (2004). (January)
7. R. Uday Kiran, Y. Watanobe, B. Chaudhury, K. Zettsu, M. Toyoda and M. Kitsuregawa, Discovering maximal periodic-frequent patterns in very large temporal databases. 7th IEEE International Conference on Data Science and Advanced Analytics, DSAA 2020 (IEEE, New York, 2020), pp. 11–20
8. B. Özden, Sridhar Ramaswamy and A. Silberschatz, Cyclic association rules. Proceedings 14th International Conference on Data Engineering (IEEE, New York, 1998), pp. 412–421
9. M.M. Rashid, M.R. Karim, B.-S. Jeong and H.-J. Choi, Efficient mining regularly frequent patterns in transactional databases, in *Database Systems for Advanced Applications*, ed. by S. Lee, Z. Peng, X. Zhou, Y.-S. Moon, R. Unland and J. Yoo (Springer, Berlin, 2012), pp. 258–271
10. T. Yashwanth Reddy, R. Uday Kiran, M. Toyoda, P. Krishna Reddy, M. Kitsuregawa, Discovering partial periodic high utility itemsets in temporal databases, in *Database and Expert Systems Applications—30th International Conference, DEXA 2019, Linz, Austria, August 26-29, 2019, Proceedings, Part II*. Lecture Notes in Computer Science, vol. 11707 (Springer, 2019), pp. 351–361
11. S.K. Tanbeer, C.F. Ahmed, B.S. Jeong and Y.K. Lee, Discovering periodic-frequent patterns in transactional databases. Pacific-Asia Conference on Knowledge Discovery and Data Mining (Springer, Berlin, 2009), pp. 242–253

Discovering Fuzzy Periodic Patterns in Quantitative Temporal Databases

Pamalla Veena, R. Uday Kiran, Penugonda Ravikumar, and Sonali Aggrawal

Abstract Finding periodic patterns in very large databases is a challenging problem of great importance in many real-world applications. Most previous studies focused on discovering these patterns in binary temporal databases by disregarding the quantity information of the items within the database. In this chapter, we will first explore the notion of "fuzzy sets" to discover a class of periodic patterns, called fuzzy periodic-frequent patterns, in a quantitative temporal database. Second, we present an algorithm to find the desired patterns. Finally, we end this chapter by describing the procedure to execute the fuzzy periodic-frequent pattern mining algorithm in the PAMI python kit.

1 Introduction

Periodic-frequent patterns are an important class of periodic patterns that exist in a temporal database. Since it was first introduced in [5], the problem of finding these patterns has received a great deal of attention [2–4, 6]. The classic application is market-basket analytics. It analyzes how regularly the itemsets are being purchased by the customers. An example of a periodic-frequent pattern is as follows:

$$\{Bread, Jam\} \quad [support = 8\%, periodicity = 1\ hour]. \tag{1}$$

P. Veena
Sri Balaji PG College, Ananthapur, Andhra Pradesh, India
e-mail: rage.vinny@gmail.com

R. Uday Kiran (✉) · P. Ravikumar
The University of Aizu, Aizu-Wakamatsu, Fukushima, Japan
e-mail: udayrage@u-aizu.ac.jp

P. Ravikumar
e-mail: raviua138@gmail.com

S. Aggrawal
Indian Institute of Information Technology Allahabad, Prayagraj, India
e-mail: sonali@iiita.ac.in

© The Author(s), under exclusive license to Springer Nature Singapore Pte Ltd. 2021
R. Uday Kiran et al. (eds.), *Periodic Pattern Mining*,
https://doi.org/10.1007/978-981-16-3964-7_4

57

The above pattern says that 8% of the customers have purchased the items "Bread" and "Jam," and the maximum duration between any two consecutive purchases containing both of these items is no more than an hour. This predictive behavior of the customers' purchases may facilitate the user in product recommendation and inventory management.

The popular adoption and successful industrial application of periodic-frequent pattern model has been hindered by the following obstacle: "*Most studies have aimed at finding periodic-frequent patterns in a binary temporal database. Consequently, they are inadequate to find those interesting patterns that are occurring regularly (or periodically) in a quantitative temporal database.*" When confronted with this problem in the real-world applications, researchers explored the notion of "fuzzy sets" to discover a class of full periodic patterns, called fuzzy periodic-frequent patterns [1], in a quantitative transactional database. We now describe the model of fuzzy periodic-frequent pattern.

2 Model of Fuzzy Periodic-Frequent Pattern

Let $I = \{i_1, i_2, \cdots, i_m\}$, $m \geq 1$, be a finite set of m distinct items. A quantitative temporal database, QTD, is an ordered collection of transactions and their associated timestamps. Each transaction in this database contains items and their associated quantities. That is, $QTD = \{(1, T_1), (2, T_2), \cdots, (ts, T_{ts})\}$, where $ts \in \mathbb{R}^+$ represents the timestamp and each transaction $T_q \in QTD$, $1 \leq q \leq ts$, is a subset of I, containing several items with its purchase quantities v_{i_q}. A set of items $Y \subseteq I$ is called an itemset (or a pattern). An itemset containing k, $k \geq 1$, number of items is called a k-itemset. An itemset Y is said to be contained in a transaction T_q if $Y \subseteq T_q$.

Example 1 Let $I = \{a, b, c, d, e, f\}$ be the set of items. An hypothetical quantitative temporal database generated from the recording of the items in I is shown in Table 1. This database contains 8 transactions. Each transaction in this database is associated with a transactional identifier (tid) and a timestamp (ts). In the first transaction, $(101, \ 1, \ \{a : 5, b : 1, c : 10, d : 2, e : 9\})$, 101 represents the transactional identifier, 1 represents the timestamp, and $\{a : 5, b : 1, c : 10, d : 2, e : 9\}$ represents the transaction containing items and their associated quantities.

Definition 1 Let $\mu = \{\mu_1, \mu_2, \cdots, \mu_h\}$ be the set of h distinct membership functions. The linguistic variable R_i is an attribute of a quantitative temporal database whose value is the set of fuzzy terms (or items) represented in natural language as $\{R_{i1}, R_{i2}, \cdots, R_{ih}\}$ and can be defined in the membership functions μ. That is, $R_i = \{R_{i1}, R_{i2}, \cdots, R_{ih}\}$

Example 2 Given the fuzzy membership functions as shown in Fig. 1, the linguistic variable that can be generated for an item a, i.e., $R_a = \{a.L, a.M, a.H\}$. Similar linguistic variables (or fuzzy items) can be generated for the remaining items in I.

Table 1 Quantitative temporal database

tid	ts	Items	tid	ts	Items
101	1	$a:5, b:1, c:$ $10, d:2, e:9$	105	5	$a:7, c:9, d:$ 3
102	2	$a:8, b:2, c:$ $3, f, 1$	106	6	$b:2, c:6, d:$ 3
103	3	$b:d, c:$ $9, f:10$	107	7	$a:5, b:2, c:$ 5
104	4	$a:5, b:3, c:$ $10, e:3$	108	8	$a:3, c:$ $10, d:2, e:2$

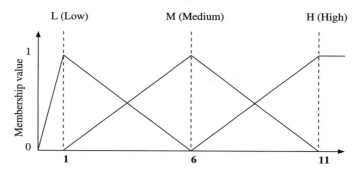

Fig. 1 The linear membership functions of linguistic 3-terms

Definition 2 The fuzzy set of an item i in T_q, denoted as f_{iq}, is the set of fuzzy items with their membership degrees transformed from the quantitative value v_{iq} by the membership functions μ as follows:

$$f_{iq} = \mu_i(v_{iq}) = \frac{f v_{iq1}}{R_{i1}} + \frac{f v_{iq2}}{R_{i2}} + \cdots + \frac{f v_{iqh}}{R_{ih}}, \qquad (2)$$

where h is the number of fuzzy terms of i transferred by μ, R_{il} is the l-th fuzzy terms of i, $f v_{iql}$ is the membership degree of v_{iq} of i in the l-th fuzzy terms R_{il} and $f v_{iql} \in [0, 1]$.

Example 3 The value of a in the first transaction of Table 1, i.e., $v_{a1} = 5$. Based on the membership function shown in Fig. 1, the fuzzy set of a in T_1, i.e., $f_{a1} = \frac{0.2 \left(= \frac{6-5}{6-1}\right)}{a.L} + \frac{0.8 \left(= \frac{5-1}{6-1}\right)}{a.M} + \frac{0}{a.H} = \frac{0.2}{a.L} + \frac{0.8}{a.H}$. For the purpose of simplicity, we represent $f_{a1} = \{a.L : 0.2, a.M : 0.8\}$. Similarly, for the items b, c, d and e in T_1, $f_{b1} = \{b.L : 1\}$, $f_{c1} = \{c.M : 0.2, c.H : 0.8\}$, $f_{d1} = \{d.L : 0.8, d.M : 0.2\}$ and $f_{e1} = \{e.M : 0.4, e.H : 0.6\}$.

Table 2 Fuzzy temporal database generated from Table 1. The "*tid*" information was removed from the database for brevity

ts	*itemset*
1	$a.L : 0.2, a.M : 0.8, b.L : 1, c.M : 0.2, c.H : 0.8, d.L : 0.8, d.M : 0.2, e.M : 0.4, e.H : 0.6$
2	$a.M : 0.6, a.H : 0.4, b.L : 0.8, b.M : 0.2, c.L : 0.6, c.M : 0.4, f.L : 1, f.M : 0$
3	$b.L : 0.6, b.M : 0.4, c.M : 0.4, c.H : 0.6, f.M : 0.2, f.H : 0.8$
4	$a.L : 0.2, a.M : 0.8, b.L : 0.6, b.M : 0.4, c.M : 0.2, c.H : 0.8, e.L : 0.6, e.M : 0.4$
5	$a.M : 0.8, a.H : 0.2, c.M : 0.4, c.H : 0.6, e.L : 0.6, e.M : 0.4$
6	$b.L : 0.8, b.M : 0.2, c.L : 0, c.M : 1.0, d.L : 0.6, d.M : 0.4$
7	$a.L : 0.2, a.M : 0.8, b.L : 0.8, b.M : 0.2, c.L : 0.2, c.M : 0.8$
8	$a.L : 0.6, a.M : 0.4, c.M : 0.2, c.H : 0.8, d.L : 0.8, d.M : 0.2, e.L : 0.8, e.M : 0.2$

Definition 3 Given a set of fuzzy membership functions μ, a transaction $T_q \in QTD$ can be transformed into a fuzzy transaction FT_q such that $FT_q = \bigcup\limits_{i \in T_q} f_{iq}$. Thus, the fuzzy quantitative temporal database, $FQTD = \bigcup\limits_{q=1}^{ts} FT_q$.

Example 4 The fuzzy transaction generated from T_1, i.e., $FT_1 = f_{a1} \cup f_{b1} \cup f_{c1} \cup f_{d1} \cup f_{e1} = \{a.L : 0.2, a.M : 0.8, b.L : 1.0, c.M : 0.2, c.H : 0.8, d.L : 0.8, d.M : 0.2, e.M : 0.4, e.H : 0.6\}$. The fuzzy quantitative temporal database generated by performing similar process on every transaction of Table 1 is shown in Table 2.

Definition 4 Let $R = \bigcup\limits_{i \in I} R_i$ denote the set of all fuzzy items generated from I. A fuzzy pattern $X \subseteq R$. If X contains k number of distinct fuzzy items, then X is called fuzzy k-pattern. The *support* of a fuzzy item (or 1-pattern) R_{il}, denoted $sup(R_{il})$, is the summation of scalar cardinality of the fuzzy values of fuzzy term R_{il} in $FQTD$. That is,

$$sup(R_{il}) = \sum_{R_{il} \subseteq FT_q \wedge FT_q \in FQTD} fv_{ilq}. \tag{3}$$

Example 5 The complete set of fuzzy items in Table 2, i.e., $R = \{a.L, a.M, a.H, b.L, - \cdots, f.H\}$. The set of fuzzy items $a.M$ and $b.L$, i.e., $\{a.M, b.L\}$ is a fuzzy pattern. This pattern contains 2 fuzzy items. Therefore, it is a fuzzy 2-pattern. In Table 2, the fuzzy item $a.M$ appears in the transactions whose timestamps are 1, 2, 4, 5, 7 and 8. Thus, the *support* of fuzzy item $a.M$, i.e., $sup(a.M) = fv_{a.M_1} + fv_{a.M_2} + fv_{a.M_4} + fv_{a.M_5} + fv_{a.M_7} + fv_{a.M_8} = 0.8 + 0.6 + 0.8 + 0.8 + 0.8 + 0.4 = 4.2$. Similarly, $sup(b.L) = 4.6$.

Definition 5 The support of fuzzy k-pattern, $k \geq 2$, denoted as $sup(X)$, is the summation of scalar cardinality of the fuzzy values for X, which can be defined as

$$sup(X) = \{X \in R_{il}| \sum_{R_{il} \subseteq T_q \wedge T_q \in FQTD} min(fv_{aql}, fv_{bql}) \tag{4}$$

where $a, b \in X$ and $a \neq b$.

Example 6 In Table 2, the pattern $\{a.M, b.L\}$ appears in the transactions whose timestamps are 1, 2, 4, and 7. Thus, the *support* of $\{a.M, b.L\}$ in Table 2, i.e., $sup(a.M, b.L) = min(0.8, 1) + min(0.6, 0.8) + min(0.8, 0.6) + min(0.8, 0.8) = 0.8 + 0.6 + 0.6 + 0.8 = 2.8$.

Definition 6 The fuzzy pattern X is said to be a fuzzy frequent pattern if $sup(X) \geq minSup$, where $minSup$ represents the user-specified minimum support value.

Example 7 If the user-specified $minSup = 2$, then the fuzzy pattern $\{a.M, b.L\}$ is said to be a fuzzy frequent pattern because $sup(\{a.M, b.L\}) \geq minSup$.

Definition 7 (**A period of** X) If $X \subseteq FT_q$, it is said that X occurs in the transaction T_q. Let ts_q^X denote the timestamp of the fuzzy transaction FT_q containing X. Let ts_i^X and ts_j^X, $ts_{min} \leq i \leq j \leq ts_{max}$, denote two consecutive timestamps at which X has occurred in $FQTD$. A *period* of X in $FQTD$, denoted as $p_k = ts_j^X - ts_i^X$.

Example 8 The pattern $\{a.M, b.L\}$ appears in the transactions whose timestamps are 1, 2, 4, and 7. Thus, the periods of this pattern are $1 (= 1 - ts_{initial})$, $1(= 2 - 1)$, $2(= 4 - 2)$, $3(= 7 - 4)$, and $1(= ts_{final} - 7)$, where $ts_{initial} = 0$ represents the initial stamp and ts_{final} represents the final timestamps of all transactions in $FQTD$.

Definition 8 (*Periodicity* of X.) Let $P^X = \{p_1^X, p_2^X, \cdots, p_k^X\}$, $k = sup(X) + 1$, denote the set of all *periods* of X in the database. The *periodicity* of X, denoted as $per(X)$, represents the maximum value among all of its *periods*. That is, $per(X) = max(p_1^X, p_2^X, \cdots, p_k^X)$.

Example 9 Continuing with the previous example, the set of all periods of $\{a.M, b.L\}$, i.e., $P^{\{a.M, b.L\}} = \{1, 1, 2, 3, 1\}$. Thus, the *periodicity* of $\{a.M, b.L\}$, i.e., $per(\{a.M, b.L\}) = max(1, 1, 2, 3, 1) = 3$.

Definition 9 (**Fuzzy periodic-frequent pattern** X.) A pattern X is called a fuzzy periodic-frequent pattern if its *periodicity* is no greater than the user-specified *maximum periodicity* ($maxPer$) and *support* is no less than the user-specified *minimum support* ($minSup$). In other words, X is a fuzzy periodic-frequent pattern if $per(X) \leq maxPer$ and $sup(X) \geq minSup$.

Example 10 If the user-specified $minSup = 2$ and $maxPer = 3$, then the itemset $\{a.M, b.L\}$ is said to be a fuzzy periodic-frequent pattern because $per(\{a.M, b.L\}) \leq maxPer$ and $sup(\{a.M, b.L\}) \geq minSup$. The complete set of fuzzy periodic-frequent patterns generated from Table 1 are shown in Table 3.

Table 3 FPFPs generated for working example

S.No.	*Pattern*	*Support*	*Periodicity*
1	$\{a.M\}$	4.2	2
2	$\{a.M, b.L\}$	2.8	3
3	$\{a.M, c.M\}$	2.2	2
4	$\{b.L\}$	4.6	2
5	$\{b.L, c.M\}$	2.8	2
6	$\{c.M\}$	3.6	1

Since the $maxPer$ controls the maximum time interval within which a pattern must occur in a database, the generated fuzzy periodic-frequent patterns also represent fuzzy full periodic patterns that may exist in a quantitative temporal database.

Definition 10 (Problem definition.) Given the quantitative temporal database (QTD) and the user-specified fuzzy membership function (μ), *minimum support* ($minSup$) and *maximum periodicity* ($maxPer$), the problem of fuzzy periodic-frequent pattern mining involves discovering all patterns in QTD that have $sup(X) \geq minSup$ and $per(X) \leq maxPer$.

3 Proposed Algorithm

The proposed FPFP-Miner employs a fuzzy-list structure to record the fuzzy information of the items in a $FQTD$. The fuzzy-list structure can be used to efficiently and effectively speed up the computations for directly discovering FPFPs. The phases of the proposed FPFP-Miner algorithm are described below.

3.1 Fuzzy Periodic Frequent 1-Patterns (FPFP-1)

To find FPFP-1's, an improved maximum scalar cardinality strategy is adopted, thus making the number of transformed terms used in later processing equal to the number of the original items. This strategy can be used to find the most represented term of each item in the original database.

Definition 11 Improved maximum scalar cardinality: For a linguistic variable i, the fuzzy terms R_{il} with the maximum scalar cardinality (support) among the **transformed periodic fuzzy terms** are used to present the linguistic variable (item).

After that, the fuzzified quantitative database is then used to find the FPFP-1's. The represented fuzzy terms are considered as the FPFP-1's if the items *support* is greater than or equal to $minSup$ and *periodicity* is no more than the $maxPer$. The

FPFP-1's of each transformed transaction are sorted in their support-ascending order. If two or more items have same *support*, then those items are sorted in *periodicity* descending order. This strategy can be used to easily find the fuzzy values between the transformed fuzzy terms based on the designed fuzzy-list structures explained in the next subsection

Definition 12 The support-ascending order. For the remaining fuzzy terms with their fuzzy values in a transaction T_q, the fuzzy terms are sorted in their support-ascending order to perform the intersection operation for discovering their support values among the fuzzy k-items ($k \geq 2$)

Based on the proposed improved maximum scalar cardinality and the support-ascending order strategies, the original databases can be transformed as the fuzzified databases.

3.2 Periodic Fuzzy List Structure

After the original quantitative database is transformed, the FPFP-1's are used to build their own fuzzy-list structures for keeping the fuzzy information. The definitions used in the fuzzy-list structure are respectively given below

Definition 13 A fuzzy term R_{il} in transaction T_q, and $R_{il} \subseteq T_q$. The set of fuzzy terms after R_{il} in T_q is denoted as $\frac{T_q}{R_{il}}$.

Definition 14 The internal fuzzy value of a fuzzy term R_{il} in transaction T_q is denoted as $if(R_{il}, T_q)$.

Definition 15 The resting fuzzy value of a fuzzy term R_{il} in transaction T_q is denoted as $rf(R_{il}, T_q)$ by performing the union operation to get the maximum fuzzy value of all the fuzzy terms as the upper-bound value in T_q/R_{il} in T_q, which is defined as

$$rf(R_{il}, T_q) = max(if(z, T_q)|z \in T_q/R_{il}). \qquad (5)$$

In the constructed fuzzy-list structure, each element consists of three attributes as

- **Timestamp** (*ts*), which indicates temporal occurence of T_q containing R_{il}.
- **Internal fuzzy value** (*if*), which indicates the fuzzy value of R_{il} in T_q.
- **Resting fuzzy value** (*rf*), which indicates the maximum fuzzy value of the resting fuzzy terms after R_{il} in T_q.

The initial fuzzy-list structures of the fuzzy terms in L_1 are first constructed. The support-ascending order of the fuzzy terms in L_1 is ($c.H < a.M < b.L$). The construction algorithm of fuzzy-list structure is shown in Algorithm 1.

Algorithm 1 Construct

Input : $P_X.FL$, $P_Y.FL$ the fuzzy-lists of P_X and P_Y.
Output : $P_{XY}.FL$ the fuzzy-list of P_{XY}
1: initialize $P_{XY}.FL$ to **Null**
2: **for** each $E_Y \in P_X$ **do**
3:     **if** $\exists E_Y \in P_Y.FL$ and $E_X.tid = E_Y.tid$ **then**
4:         $E_{XY}.tid \leftarrow E_X.tid$
5:         $E_{XY}.if \leftarrow min(E_X.if, E_Y.if)$
6:         $E_{XY}.rf \leftarrow E_Y.rf$
7:         $E_{XY} \leftarrow \langle E_{XY}.tid, E_{XY}.if, E_{XY}.rf \rangle$
8:         append E_{XY} to $P_{XY}.FL$
9: **return** P_{XY}

Fig. 2 Mining process

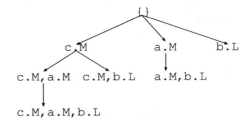

3.3 Search Space of FPFP-Miner

Based on the designed fuzzy-list structure, the search space of the proposed FPFP-Miner algorithm can be represented as an enumeration tree according to the developed support-ascending order strategy. In this example, the search space of the enumeration tree is shown in Fig. 2. Since the complete search space of the enumeration tree is very huge for discovering all fuzzy periodic frequent patterns, it is necessary to reduce the search space but still can completely find the fuzzy periodic frequent patterns

Strategy 1. For a pattern X, if its $SUM(X.if)$ is no less than the $minSup$ and $periodicity(X)$ less than $maxPer$, it is considered as a Fuzzy periodic frequent pattern. Also, if $min(SUM(X.if), SUM(X.rf))$ of X is no less than the minimum support count, the supersets of X are required to be generated and determined

Strategy 2. If the summation of the resting fuzzy values of the itemset X is no larger than the minimum support count, any extensions of X will not be a periodic fuzzy frequent itemsets and can be directly ignored to avoid the construction phase of the fuzzy-list structures of the extensions of X.

The approach for this is clearly stated in Algorithm 2

Algorithm 2 $FPFP\text{-}Miner$

Input: FL_{prefix}, prefix, FPFP-1, $maxPer$ and $minSup$.
Output : Fuzzy periodic patterns
1: **while** FPFP-1 $\neq$ Null **do**
2:    $FPFP_{suffix}$ = FPFP-1.pop()
3:    $FL_{(prefix\|suffix)}=construct(FL_{prefix}, FPFP_{suffix})$
4:    **if** $(sum(FL_{(prefix\|suffix)}.if) \geq minSup)$ AND $period((FL_{prefix\|suffix}) \leq maxPer)$
    then
5:       Generate $(prefix \| suffix)$ as a FPFP.
6:    **if** $sum(FL_{(prefix\|suffix)}.rf) \geq minSup$ **then**
7:       $FPFP\text{-}Miner(FL_{(prefix\|suffix)},$
       $(prefix \| suffix), FPFP\text{-}1, maxPer, minSup)$

---

4 Finding Fuzzy Periodic-Frequent Patterns Using PAMI Software

In this section, we first describe the procedure to create a quantitative temporal database. Next, we describe the procedure to execute FPFP-Miner algorithm in PAMI software. Finally, we describe the procedure to utilize FPFP-Miner algorithm as a library in a python program.

4.1 Format to Create a Quantitative Temporal Database

In order to discover fuzzy periodic-frequent patterns from the data using PAMI software, one needs to construct a temporal database such that each transaction must exist in the following format: $\langle timestamp \rangle$: $\langle items\ are\ separated\ by\ a\ space \rangle$: $\langle quantities\ of\ the\ respective\ items\ separated\ by\ space \rangle$. An example of a transaction is as follows: "1 : $a\ b\ c\ d$: 10 13 15 9," where 1 is the timestamp, a, b, c and d are the items occurring in a transaction with the respective values of 10, 13, 15, and 9.

4.2 Executing FPFP-Miner in PAMI

The step-by-step procedure to execute FPFP-Miner algorithm in PAMI software is as follows:

1. Clone the Github repository on your system by executing the following command on the terminal: git clone https://github.com/udayRage/PAMI.git.
2. The above command will create a directory, named PAMI, and will download all of the available algorithms.

3. Enter into FPFP-Miner directory by executing the following command:
   cd PAMI/periodicFrequentPattern/basic/
4. Execute the FPFPMiner algorithm on the terminal using the following command:
   python F P F P Miner.py ⟨input File⟩ ⟨output File⟩ ⟨minSup⟩ ⟨max Per⟩.
   Please note that the user can specify *minSup* and *max Per* constraints either in counts or proportion of the database size. For example, if the user executes the following command:
   python FPFPMiner.py sampleTDB.csv patterns.txt 0.3 0.4,
   we consider the *minSup* and *max Per* were specified with respect to the proportion of database size. If the user executes the following command:
   python FPFPMiner.py sampleTDB.csv patterns.txt 30 40,
   we consider the *minSup* and *max Per* in counts.
5. The output file contains patterns and their respective *support* and *periodicity* values. The format of the output file is ⟨*Pattern : support : periodicity*⟩. The items within a pattern are separated by tab space.

4.3   Utilizing the FPFPMiner Algorithm in Your Python Program

Please download the recent version of PAMI library from the Python Package Index (PyPI) repository. This task can be easily achieved by executing the following command: *pip install pami*. Once the PAMI library was successfully installed on your machine, utilize the FPFPMiner Algorithm in your python code as shown in Algorithm 3. For more information, please refer to the documentation code provided within the FPFPMiner Algorithm.

Algorithm 3 Utilizing code

```
1: from PAMI.periodicFrequentPattern.basic import FPFPMiner as alg
2: fpfp =alg.FPFPMiner("inputFileName",0.3,0.4)
3: fpfp.startMine()
4: periodicPatterns = fpfp.getPatterns()
5: print("Total number of Periodic Frequent Patterns:",len(periodicPatterns))
6: fpfp.savePatterns("outputFileName")
7: memUSS = fpfp.getMemoryUSS()
8: print("Total Memory in USS:", memUSS)
9: memRSS =fpfp.getMemoryRSS()
10: print("Total Memory in RSS", memRSS)
11: run = fpfp.getRuntime()
12: print("Total ExecutionTime in seconds:", run)
```

References

1. R. Uday Kiran, C. Saideep, P. Ravikumar, K. Zettsu, M. Toyoda, M. Kitsuregawa and P. Krishna Reddy, Discovering fuzzy periodic-frequent patterns in quantitative temporal databases. 29th IEEE International Conference on Fuzzy Systems, FUZZ-IEEE 2020 (IEEE, New York, 2020), pp. 1–8
2. R. Uday Kiran, H. Shang, M. Toyoda and M. Kitsuregawa, Discovering partial periodic itemsets in temporal databases. Proceedings of the 29th International Conference on Scientific and Statistical Database Management, SSDBM '17 (ACM, New York, 2017), pp. 30:1–30:6
3. R. Uday Kiran, J.N. Venkatesh, P. Fournier-Viger, M. Toyoda, P. Krishna Reddy and M. Kitsuregawa, Discovering periodic patterns in non-uniform temporal databases. Advances in Knowledge Discovery and Data Mining—21st Pacific-Asia Conference, Part II (Springer, Berlin, 2017), pp. 604–617
4. R. Uday Kiran, Y. Watanobe, B. Chaudhury, K. Zettsu, M. Toyoda and M. Kitsuregawa, Discovering maximal periodic-frequent patterns in very large temporal databases. 7th IEEE International Conference on Data Science and Advanced Analytics, DSAA 2020 (IEEE, New York, 2020), pp. 11–20
5. S.K. Tanbeer, C.F. Ahmed, B.S. Jeong and Y.K. Lee, Discovering periodic-frequent patterns in transactional databases. Pacific-Asia Conference on Knowledge Discovery and Data Mining (Springer, Berlin, 2009), pp. 242–253
6. S.K. Tanbeer, M.M. Hassan, A. Almogren, M. Zuair, B.-S. Jeong, Scalable regular pattern mining in evolving body sensor data. Future Generation Comp. Syst. **75**, 172–186 (2017)

Discovering Partial Periodic Patterns in Temporal Databases

Pamalla Veena, Palla Likhitha, B. Sai chithra, and R. Uday Kiran

Abstract In the previous chapters, we have studied the models and algorithms to discover different types of full periodic patterns in a temporal database. A key limitation of these studies is that they fail to discover those interesting patterns that have exhibited partial periodic behavior in a database. In this chapter, we will first study the novel model of partial periodic pattern that may exist in a temporal database. Second, we discuss an algorithm to find the desired patterns. Finally, we end this chapter by describing the procedure to execute the partial periodic pattern mining algorithm in the PAMI python kit.

1 Introduction

Periodic-frequent pattern mining is an important knowledge discovery technique in data mining. It involves discovering patterns that satisfy the user-specified *minimum support* (*minSup*) and *maximum periodicity* (*maxPer*) constraints. The *minSup* controls the minimum number of transactions that a pattern must cover in a database. The *maxPer* controls the maximum time interval within which a pattern must reappear in the entire database. A classic application is market basket analytics. It involves identifying the set of items that were frequently and regularly purchased by the customers. An example of a periodic-frequent pattern is as follows:

P. Veena
Sri Balaji PG College, Ananthapur, 515002, Andhra Pradesh, India
e-mail: rage.vinny@gmail.com

P. Likhitha · B. Sai chithra
IIIT-Idupulapaya, Andhra pradesh, 516330, India
e-mail: likhithapalla7@gmail.com

B. Sai chithra
e-mail: saichithra.b@gmail.com

R. Uday Kiran (✉)
The University of Aizu, Aizu-Wakamatsu, Fukushima, 965-8580, Japan
e-mail: udayrage@u-aizu.ac.jp

© The Author(s), under exclusive license to Springer Nature Singapore Pte Ltd. 2021
R. Uday Kiran et al. (eds.), *Periodic Pattern Mining*,
https://doi.org/10.1007/978-981-16-3964-7_5

$$\{Bat, Ball\} \quad [support = 5\%, periodicity = 1 \ hour] \tag{1}$$

The above pattern says that 5% of the customers have purchased the items 'Bat' and 'Ball,' and the maximum duration between any two consecutive purchases containing both of these items is no more than an hour. This predictive behavior of the customers' purchases may facilitate the user in product recommendation and inventory management. Other real-world applications of periodic-frequent pattern mining includes accident data analytics [8] and body sensor data analytics [7]. Mining periodic-frequent patterns has inspired other data mining tasks such as high-utility periodic pattern mining [2], recurring pattern mining [4].

A key element that makes periodic-frequent pattern mining practicable in the real-world applications is $maxPer$ constraint. It is used to prune the search space and limit the number of patterns being generated from the database. Since $maxPer$ controls the maximal time interval within which a pattern must reappear in the database, the basic model of periodic-frequent pattern prunes all those interesting frequent patterns that may have exhibited partial periodic behavior in the database. When confronted with this problem in the real-world applications, researchers have employed alternative periodicity determining measures, such as *periodic-ratio*, *variance*, and *liability*, to discover regularities in a temporal database. Unfortunately, these measures do not satisfy the *downward closure property*. Consequently, finding periodically occurring patterns using these measures is computationally expensive process (or impracticable on large databases).

Against this background, Uday et al. [5] proposed the practicable model of partial periodic pattern by introducing a new measure, called *period-support*, that satisfies the *downward closure property*. Furthermore, the authors have also presented a pattern-growth algorithm, called partial periodic pattern-tree (3P-tree), to find all of the desired patterns. In the subsequent sections, we will discuss the model of partial periodic pattern, its mining algorithm, and the implementation of the algorithm in the PAMI library.

2 The Model of Partial Periodic Pattern

Let $I = \{i_1, i_2, \cdots, i_n\}$ be the set of 'n' items appearing in a database. A set of items $X \subseteq I$ is called a **pattern**. A pattern containing k items is called a k-pattern. The length of this pattern is k. A transaction tr consists of transaction identifier, timestamp, and pattern. That is, $tr = (tid, ts, Y)$, where tid represents the transactional identifier, $ts \in \mathbb{R}$ represents the transaction time (or timestamp) and Y is a pattern. A temporal database TDB is an ordered set of transactions, i.e., $TDB = \{tr_1, tr_2, \cdots, tr_m\}$, where $m = |TDB|$ represents the database size (the number of transactions). Let ts_{min} and ts_{max} denote the minimum and maximum timestamps in TDB, respectively. For a transaction $tr = (tid, ts, Y)$, such that $X \subseteq Y$, it is said that X occurs in tr and such a timestamp is denoted as ts^X. Let $TS^X = (ts_a^X, ts_b^X, \cdots, ts_c^X), a \leq b \leq c$, be the **ordered list of timestamps** of trans-

Table 1 A temporal database

tid	ts	Items	tid	ts	Items	tid	ts	Items	tid	ts	Items
101	1	abg	105	5	abg	109	9	$abef$	113	12	$abcd$
102	1	acd	106	6	cd	110	9	ade	114	12	$abcd$
103	3	ab	107	7	bg	111	10	cdg			
104	4	aef	108	8	$cdef$	112	11	$abef$			

actions in which X appears in TDB. The number of transactions containing X in TDB (i.e., the size of TS^X) is defined as the *support* of X and denoted as $sup(X)$. That is, $sup(X) = |TS^X|$.

Example 1 Table 1 shows a temporal database with $I = \{abcdefg\}$. The set of items 'a' and 'b,' i.e., 'ab' is a pattern. This pattern contains two items. Therefore, it is a two-pattern. In the first transaction, $tr_1 = (101, 1, abg)$, '101' represents the *tid* of the transaction, '1' represents the timestamp of this transaction and 'abg' represents the items occurring in this transaction. Other transactions in this database follow the same representation. This database contains 14 transactions. Therefore, $m = 14$. The minimum and maximum timestamps in this database are 1 and 12, respectively. Therefore, $ts_{min} = 1$ and $ts_{max} = 12$. The pattern 'ab' appears in the transactions whose timestamps are 1, 3, 5, 9, 11, 12, and 12. Therefore, $TS^{ab} = \{1, 3, 5, 9, 11, 12, 12\}$. The *support* of '$ab$,' i.e., $sup(ab) = |TS^{ab}| = 7$.

Definition 1 (Periodic appearance of pattern X.) Let ts_j^X, $ts_k^X \in TS^X$, $1 \leq j < k \leq m$, denote any two consecutive timestamps in TS^X. The time difference between ts_k^X and ts_j^X is referred as **an inter-arrival time** of X, and denoted as iat^X. That is, $iat^X = ts_k^X - ts_j^X$. Let $IAT^X = \{iat_1^X, iat_2^X, \cdots, iat_k^X\}$, $k = sup(X) - 1$, be the list of all inter-arrival times of X in TDB. An inter-arrival time of X is said to be **periodic** (or interesting) if it is no more than the user-specified *period*. That is, a $iat_i^X \in IAT^X$ is said to be **periodic** if $iat_i^X \leq per$.

Example 2 The pattern 'ab' has initially appeared at the timestamps of 1 and 3. The difference between these two timestamps gives an inter-arrival time of 'ab.' That is, $iat_1^{ab} = 2\ (= 3 - 1)$. Similarly, other inter-arrival times of 'ab' are $iat_2^{ab} = 2\ (= 5 - 3)$, $iat_3^{ab} = 4\ (= 9 - 5)$, $iat_4^{ab} = 2\ (= 11 - 9)$, $iat_5^{ab} = 1\ (= 12 - 11)$ and $iat_6^{ab} = 0\ (= 12 - 12)$. Therefore, the resultant $IAT^{ab} = \{2, 2, 4, 2, 1, 0\}$. If the user-specified $per = 2$, then $iat_1^{ab}, iat_2^{ab}, iat_4^{ab}, iat_5^{ab}$ and iat_6^{ab} are considered as the periodic occurrences of 'ab' in the data. The iat_3^{ab} is considered as an aperiodic occurrence of 'ab' because $iat_3^{ab} \nleq per$.

Definition 2 (Period-support of pattern X.) Let $\widehat{IAT^X}$ be the set of all inter-arrival times in IAT^X that have $iat^X \leq PER(X)$. That is, $\widehat{IAT^X} \subseteq IAT^X$ such that if $\exists iat_k^X \in IAT^X : iat_k^X \leq PER(X)$, then $iat_k^X \in \widehat{IAT^X}$. The period-support of X, denoted as $PS(X) = |\widehat{IAT^X}|$.

Example 3 Continuing with the previous example, $\widehat{IAT^{ab}} = \{2, 2, 2, 1, 0\}$. Therefore, the *period-support* of '*ab*,' i.e. $PS(ab) = |\widehat{ab}| = |\{2, 2, 2, 1, 0\}| = 5$.

The *period-support*, as defined above, measures the number of cyclic repetitions of a pattern in the database. In other words, the proposed measure determines the interestingness of a pattern by taking into account both *support* and *inter-arrival times* in the data.

Definition 3 (Partial periodic pattern X.) A pattern X is a partial periodic pattern if $PS(X) \geq minPS$, where $minPS$ is the user-specified minimum period-support.

Example 4 Continuing with the previous example, if the user-specified $minPS = 2$, then '*ab*' is a partial periodic pattern because $PS(ab) \geq minPS$.

(Problem definition.) Given a temporal database (TDB), a set of items (I), *period* (*per*) and *minimum period-support* (*minPS*), the problem of finding partial periodic patterns involve discovering all patterns in TDB that have *period-support* no less than $minPS$.

The *support* of a pattern can be expressed in percentage of $|TDB|$. The *period-support* of a pattern can be expressed in percentage of $|TDB| - 1$, where $|TDB| - 1$ represents the maximum period-support a pattern can have in the database. The *inter-arrival times* of a pattern and the *period* can be expressed in percentage of $(t_{max} - t_{min})$. In this chapter, we will employ the above definitions of *support*, *period-support*, *inter-arrival times* and *period* for brevity.

The partial periodic patterns discovered by this model satisfy the *downward closure property* [1]. The correctness of this statement is based on Property 1 and shown in Lemma 1. In the next section, we describe the 3P-growth algorithm that employs this property to discover the complete set of partial periodic patterns from a temporal database.

Property 1 If $X \subset Y$, then $TS^X \supseteq TS^Y$. Therefore, $PS(X) \geq PS(Y)$.

Lemma 1 *Let X and Y be two patterns such that $X \subset Y$ and $X \neq \emptyset$. If Y is a partial periodic pattern, then X is a partial periodic pattern.*

Proof According to Definition 3, if Y is a partial periodic pattern, then $PS(Y) \geq minPS$. Based on Property 1, it turns out that $PS(X) \geq PS(Y) \geq minPS$. Henceforth, X is also a partial periodic pattern.

3 3P-Growth

In this section, we describe the 3P-growth algorithm to discover all partial periodic patterns in a temporal database. Our algorithm involves the following two steps: (i) compress the database into partial periodic pattern tree (3P-tree) and (ii) recursively mine the 3P-tree to find all partial periodic patterns. Before we discuss these two steps, we describe the 3P-tree structure.

Fig. 1 Conceptual structure of prefix-tree in 3P-tree. Dotted ellipse represents ordinary node, while other ellipse represents tail-node of sorted transactions with timestamps $ts_i, ts_j \in R$

3.1   The 3P-tree structure

A 3P-tree has two components: a 3P-list and a prefix-tree. The 3P-list consists of each distinct *item* (i), *period-support* (ps) and a pointer pointing to the first node in the prefix-tree carrying the item.

The prefix-tree in 3P-tree resembles the prefix-tree in FP-tree. However, to record the temporal occurrence information of the patterns, the nodes in 3P-tree explicitly maintain the occurrence information for each transaction by keeping an occurrence timestamp list, called a ts-**list**. To achieve memory efficiency, only the last node of every transaction maintains the ts-list. Hence, two types of nodes are maintained in a 3P-tree: **ordinary node** and **tail-node**. The former is a type of node similar to that used in an FP-tree [3], whereas the latter represents the last item of any sorted transaction. Therefore, the structure of a *tail*-node is $i[ts_p, ts_q, ..., ts_r], 1 \le p \le q \le r$, where i is the node's item name and $ts_i \in \mathbb{R}$ is the timestamp of a transaction containing the items from *root* up to the node i. The conceptual structure of 3P-tree is shown in Figure 1. Like an FP-tree, each node in 3P-tree maintains parent, children, and node traversal pointers. Please note that no node in 3P-tree maintains the support count as in an FP-tree. To facilitate a high degree of compactness, items in the prefix-tree are arranged in support-descending order.

One can assume that the structure of the prefix-tree in 3P-tree may not be memory efficient since it explicitly maintains timestamps of each transaction. However, it has been argued that such a tree can achieve memory efficiency by keeping transaction information only at the *tail*-nodes and avoiding the support count field at each node [6]. Furthermore, 3P-tree avoids the *complicated combinatorial explosion problem of candidate generation* as in Apriori-like algorithms [1]. Keeping the information pertaining to transactional identifiers in a tree can also be found in frequent pattern mining [9] and periodic-frequent pattern mining [6].

3.2   Construction of 3P-tree

Since partial periodic patterns satisfy the anti-monotonic property, partial periodic items (or 1-patterns) will play an important role in effective mining of these patterns. The set of partial periodic items in a temporal database for the user-defined *per* and

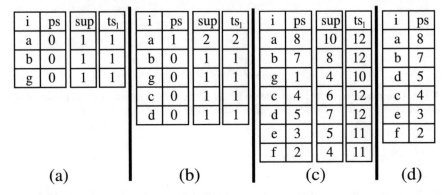

i	ps	sup	ts$_1$
a	0	1	1
b	0	1	1
g	0	1	1

i	ps	sup	ts$_1$
a	1	2	2
b	0	1	1
g	0	1	1
c	0	1	1
d	0	1	1

i	ps	sup	ts$_1$
a	8	10	12
b	7	8	12
g	1	4	10
c	4	6	12
d	5	7	12
e	3	5	11
f	2	4	11

i	ps
a	8
b	7
d	5
c	4
e	3
f	2

(a)   (b)   (c)   (d)

Fig. 2 Construction of 3P-List. **a** After scanning the first transaction. **b** After scanning the second transaction. **c** After scanning the entire database. **d** Final 3P-list containing the sorted list of items

$minPS$ can be discovered by populating the 3P-list with a scan on the database. Figure 2 shows the construction of 3P-list for Table 1 using Algorithm 1. Please note that the *per* and $minPS$ values have been set to 2 and 2, respectively.

Let '*sup*' and '*ts$_l$*' be two temporary lists that record the '*support*' and '*last occurrence of an item*' $i_j \in I$. The scan on the first transaction, "101 : 1 : *abg*," inserts the items '*a*', '*b*,' and '*g*' in the 3P-list and sets their *ps*, *sup*, and *ts$_l$* values to 0, 1, and 1, respectively (lines 5 and 6 in Algorithm 1). Figure 2a shows the 3P-list generated after scanning the first transaction. The scan on the second transaction, "102 : 1 : *acd*," inserts the items '*c*' and '*d*' in 3P-list by setting their *ps*, *s* and *ts$_l$* values to 0, 1, and 1, respectively. In addition, the *ps*, *s*, and *ts$_l$* values of an already existing item '*a*' are updated to 1, 2, and 1, respectively (lines 8–10 in Algorithm 1). Figure 2b shows the 3P-list generated after scanning the second transaction. Similar process is followed for remaining transactions in the database and 3P-list is updated accordingly. Figure 2c shows the 3P-list generated after scanning the entire database. As the partial periodic patterns satisfy the anti-monotonic property, the aperiodic items that have *ps* value less than $minPS$ are pruned from the 3P-list. The remaining items are sorted in descending order of their *support* (line 11 in Algorithm Algorithm 1). Figure 2d shows the final 3P-list containing sorted list of partial periodic items. Let CI denote this sorted list of partial periodic items.

After finding partial periodic items, we conduct another scan on the database and construct the prefix-tree of 3P-tree as in Algorithms 2 and 3. These algorithms are the same as those for constructing an FP-tree [3]. However, the major difference is that no node in 3P-tree maintains the *support* count as in an FP-tree.

A 3P-tree is constructed as follows. First, create the root node of the tree and labeled it as "*null*." Scan the database second time. The items in each transaction are processed in CI order (i.e., sorted according to descending support count) and a branch is created for each transaction such that only the tail-nodes record the timestamps of transactions. For example, the scan of the first transaction, "101 : 1 : *abg*," which contains two items (*a*, *b* in CI order), leads to the construction of

Algorithm 1 Construction of 3P-List(TDB: temporal database, I: set of items, $minPS$: minimum period-support, per: period)

1: The *timestamps* of the last occurring transactions of all items in the 3P-list are explicitly recorded for each item in a temporary array, called ts_l. Similarly, the *support* of all items in the 3P-list are explicitly recorded in another temporary array, called sup. (The 3P-tree will be constructed in the support descending order of items to achieve memory efficiency.) These two arrays can be ignored after finding partial periodic items (or 1-patterns).
2: Let $t = \{tid, ts_{cur}, X\}$ denote the current transaction with ts_{cur} and X representing the timestamp and a pattern, respectively.
3: **for** each transaction $t \in TDB$ **do**
4:     **for** each item $i \in X$ **do**
5:         **if** i does not exist in 3P-list **then**
6:             Add i to the 3P-list and set $ps(i) = 0$, $sup(i) = 1$ and $ts_l(i) = ts_{cur}$.
7:         **else**
8:             **if** $ts_{cur} - ts_l(i) \le per$ **then**
9:                 Set $ps(i) + +$.
10:            **end if**
11:            Set $ts_l = ts_{cur}$ and $sup(i) + +$.
12:        **end if**
13:    **end for**
14: **end for**
15: Prune all aperiodic-items from the 3P-list that have *period-support* less than $minPS$. Consider the remaining items in 3P-list as partial periodic items and sort them in descending order of their *support*. Let CI denote this sorted list of items.

the first branch of the tree with two nodes, $\langle a \rangle$ and $\langle b : 1 \rangle$, where a is linked as a child of the root and $b : 1$ is linked to a. The 3P-tree generated after scanning the first transaction is shown in Figure 3a. The scan on the second transaction, "101 : 1 : acd," containing the items a, d and c in CI order, would result in a branch where a is linked to the root, d is linked to a and $c : 1$ is linked to d. However, this branch would share a common prefix, a, with the existing path for first transaction. Therefore, we create two new nodes $\langle d \rangle$ and $\langle c : 1 \rangle$, and link d to a and c to d as shown in Figure 3b. A similar process is repeated for the remaining transactions and the tree is updated accordingly. Figure 3c shows the 3P-tree constructed after scanning the entire database. For simplicity, we do not show the node traversal pointers in trees; however, they are maintained like an FP-tree does.

Algorithm 2 3P-Tree(TDB, 3P-list)

1: Create the root of 3P-tree, T, and label it "*null.*"
2: **for** each transaction $t \in TDB$ **do**
3:     Set the timestamp of the corresponding transaction as ts_{cur}.
4:     Select and sort the partial periodic items in t according to the order of CI. Let the sorted candidate item list in t be $[p|P]$, where p is the first item and P is the remaining list.
5:     Call *insert_tree*($[p|P], ts_{cur}, T$).
6: **end for**

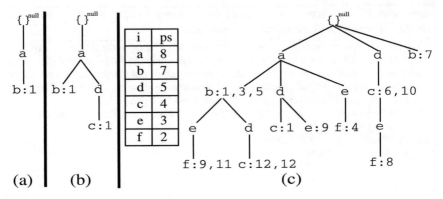

Fig. 3 Construction of 3P-tree. **a** After scanning the first transaction. **b** After scanning the second transaction. **c** Final 3P-tree generated after scanning the entire database

---

Algorithm 3 insert_tree($[p|P]$, ts_{cur}, T)

1: **while** P is non-empty **do**
2:   **if** T has a child N such that $p.itemName \neq N.itemName$ **then**
3:     Create a new node N. Let its parent link be linked to T. Let its node-link be linked to nodes with the same itemName via the node-link structure. Remove p from P.
4:   **end if**
5: **end while**
6: Add ts_{cur} to the leaf node.

The 3P-tree maintains the complete information of all partial periodic patterns in a database. The correctness is based on Property 2 and shown in Lemmas 2 and 3. For each transaction $t \in TDB$, $CI(t)$ is the set of all partial periodic items in t, i.e., $CI(t) = item(t) \cap CI$, and is called the partial periodic item projection of t (Fig. 4).

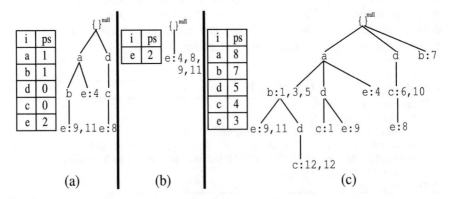

Fig. 4 Mining 3P-tree. **a** conditional pattern base of 'f.' **b** conditional 3P-tree of 'f'. **c** 3P-tree after pruning the item 'f'

Property 2 A 3P-tree maintains a complete set of partial periodic item projections for each transaction in a database only once.

Lemma 2 *Given a TDB and user-defined per and minPS values, the complete set of all partial periodic item projections of all transactions in the TDB can be derived from the 3P-tree.*

Proof Based on Property 2, each transaction $t \in TDB$ is mapped to only one path in the tree, and any path from the *root* up to a *tail* node maintains the complete projection for exactly n transactions, where n is the total number of entries in the ts-list of the *tail* node.

Lemma 3 *The size of 3P-tree (without the root node) on a TDB for the user-specified per and minPS values is bounded by* $\sum_{t \in TDB} |CI(t)|$.

Proof According to the 3P-tree construction process and Lemma 2, each transaction t contributes at most one path of size $|CI(t)|$ to 3P-tree. Therefore, the total size contribution of all transactions can be $\sum_{t \in TDB} |CI(t)|$ at best. However, since there are usually many common prefix patterns among the transactions, the size of 3P-tree is normally much smaller than $\sum_{t \in TDB} |CI(t)|$.

4 Finding Partial Periodic Patterns Using PAMI Software

In this section, we first describe the procedure to create a temporal database. Next, we describe the procedure to execute 3P-growth algorithm in PAMI software. Finally, we describe the procedure to utilize 3P-growth algorithm as a library in a python program.

4.1 Format to Create a Temporal Database

In order to discover periodic-frequent patterns from the data using PAMI software, one needs to construct a temporal database such that each transaction exists in the following format: ⟨*timestamp*⟩ ⟨*items appearing in a transaction*⟩. Please note that the *timestamp* and the items appearing in a transaction must be separated by space. Moreover, the *timestamp* must be provided in the relative context, and thus, its initial value must always be equal to 1. Multiple transactions can share the same timestamp and irregular time gaps may exist between two consecutive timestamps. An example of a transaction is as follows: "1 a b c d e," where 1 represents the timestamp of a transaction and a, b, c, d and e represents the items occurring in the corresponding transaction.

4.2   Executing 3P-Growth in PAMI

The step-by-step procedure to execute PFP-growth algorithm in PAMI software is as follows:

1. Clone the Github repository on your system by executing the following command on the terminal: git clone https://github.com/udayRage/PAMI.git.
2. The above command will create a directory, named PAMI, and will download all of the available algorithms.
3. Enter into 3P-growth directory by executing the following command: *cd PAMI/partialPeriodicFrequentPattern/basic.*
4. Execute the 3p-growth algorithm on the terminal using the following command: $python threePGrowth.py \langle input File \rangle \langle output File \rangle \langle minPS \rangle \langle per \rangle$
   E.g., *python ThreePGrowth.py sampleTDB.txt pattext 0.3 0.4*
   Please note that if you enter $minPS$ or per in float type, we calculate it as $len(Database) * minPS$ or $len(Database) * per$. If you enter $minSup$ or $maxPer$ in int we take it as it is.
5. The output file contains patterns and their respective *period-support* values. The format of the output file is $\langle Pattern : period\text{-}support \rangle$. The items within a pattern are separated by a tab space.

4.3   Utilising the 3P-Growth Algorithm in Your Python Program

Please download the recent version of PAMI library from the Python Package Index (PyPI) repository. This task can be easily achieved by executing the following command: *pip install PAMI*. Once the PAMI library was successfully installed on your machine, utilize the 3P-growth algorithm in your python code as shown in Algorithm 4. For more information, please refer to the documentation code provided within the 3P-growth algorithm.

Algorithm 4 Utilising code

```
1: from PAMI.partialPeriodicFrequentPattern.basic import threePGrowth as 3pgrowth
2: 3p = 3pgrowth.threePGrowth("inputFileName", "minPS", "per")
3: 3p.startMine()
4: partialPeriodicPatterns = 3p.getPatterns()
5: print("Total number of partial periodic Patterns:", len(partialPeriodicPatterns))
6: 3p.savePatterns("outpuFileName")
7: memUSS = 3p.getMemoryUSS()
8: print("Total Memory in USS:", memUSS)
9: memRSS = 3p.getMemoryRSS()
10: print("Total Memory in RSS", memRSS)
11: rTime = 3p.getRuntime()
12: print("Total ExecutionTime in seconds:", rTime)
```

References

1. R. Agrawal, T. Imieliński and A. Swami, Mining association rules between sets of items in large databases. Proceedings of the 1993 ACM SIGMOD International Conference on Management of Data (ACM, New York, 1993), pp. 207–216
2. Duy-Tai. Dinh, Bac Le, Philippe Fournier-Viger, Van-Nam. Huynh, An efficient algorithm for mining periodic high-utility sequential patterns. Appl. Intell. **48**(12), 4694–4714 (2018)
3. Jiawei Han, Jian Pei, Yiwen Yin, Runying Mao, Mining frequent patterns without candidate generation: A frequent-pattern tree approach. Data Min. Knowl. Discov. **8**(1), 53–87 (2004). (January)
4. R. Uday Kiran, H. Shang, M. Toyoda and M. Kitsuregawa, Discovering recurring patterns in time series. Proceedings of the 18th International Conference on Extending Database Technology
5. R. Uday Kiran, H. Shang, M. Toyoda and M. Kitsuregawa, Discovering partial periodic itemsets in temporal databases. In *Proceedings of the 29th International Conference on Scientific and Statistical Database Management (ACM, New York, 2017*, pp. 30:1–30:6
6. S.K. Tanbeer, C.F. Ahmed, B.S. Jeong and Y.K. Lee, Discovering periodic-frequent patterns in transactional databases. Pacific-Asia Conference on Knowledge Discovery and Data Mining (Springer, Berlin, 2009), pp. 242–253
7. S.K. Tanbeer, M.M. Hassan, A. Almogren, M. Zuair and B.-S. Jeong, Scalable regular pattern mining in evolving body sensor data. *Future Generation Comp. Syst.* **75**, 172–186 (2017)
8. J.N. Venkatesh, R. Uday Kiran, P. Krishna Reddy and M. Kitsuregawa, Discovering periodic-correlated patterns in temporal databases. T. Large-Scale Data- Knowl.-Centered Syst. **38**, 146–172 (2018)
9. M.J. Zaki and C.-J. Hsiao, Efficient algorithms for mining closed itemsets and their lattice structure. IEEE Trans. Knowl. Data Eng. **17**(4), 462–478 (2005)

Finding Periodic Patterns in Multiple Sequences

Philippe Fournier-Viger, Tin Truong Chi, Youxi Wu, Jun-Feng Qu, Jerry Chun-Wei Lin, and Zhitian Li

Abstract Discovering periodic patterns in data is an important data analysis task. A periodic pattern is a set of values that regularly appear together over time. Finding such patterns can be useful to understand the data and make predictions. However, most studies on periodic pattern mining have focused on identifying periodic patterns in a single discrete sequence. But for many applications, it is desirable to find patterns that are common to multiple sequences. For instance, it can be interesting to identify periodic behavior that is common to several customers in a store. This chapter gives an overview of the general problem of discovering periodic patterns in multiple sequences, and two special cases that are to find (1) frequent patterns and (2) rare correlated patterns in multiple sequences. Two algorithms are described, which can be applied in many domains where data can be modeled as discrete sequences such as to find periodic patterns in sequences of customer purchases, in sequences of words in a text document, and in sequences of treatments received by hospital patients. Research opportunities are also discussed.

P. Fournier-Viger (✉) · Z. Li
Harbin Institute of Technology (Shenzhen), Shenzhen, China
e-mail: philfv8@yahoo.com; philippe.fv@gmail.com

Z. Li
e-mail: tymonlee1@163.com

T. T. Chi
University of Dalat, Dalat, Vietnam
e-mail: tintc@dlu.edu.vn

Y. Wu
Hebei University of Technology, Tianjin, China
e-mail: wuc567@163.com

J.-F. Qu
School of Computer Engineering, Hubei University of Arts and Science, Xiangyang, China

J. C.-W. Lin
Department of Computing, Mathematics, and Physics, Western Norway University of Applied Sciences (HVL), Bergen, Norway
e-mail: jerrylin@ieee.org

© The Author(s), under exclusive license to Springer Nature Singapore Pte Ltd. 2021   81
R. Uday Kiran et al. (eds.), *Periodic Pattern Mining*,
https://doi.org/10.1007/978-981-16-3964-7_6

1 Introduction

Pattern mining is an important subfield of data mining that aims at finding interesting patterns in databases [1, 47]. A pattern can be described generally as some association between values appearing in data that can help understanding the data and/or support decision-making. Algorithms have been designed to find patterns in different types of data such as customer transaction databases [22, 26], graphs [37], trees [9, 70], process logs [8], trajectories [15, 71], sequences [25], spatial data [56] and ranking data [36]. Moreover, algorithms were proposed to find various types of patterns in data such as frequent patterns [2, 22, 25], rare patterns [40], and periodic patterns [4, 30, 61]

In the last decade, *periodic pattern mining* [4, 5, 30, 38, 39, 58, 61, 63] has emerged as an important data analysis task. It aims at finding patterns that regularly appear in the data. The traditional input of periodic pattern mining is a customer transaction database where each record contains a set of items purchased together at a given time by a customer [4, 30]. In that model, a periodic pattern can be, for example, that the customer regularly buys wine with cheese (e.g., every weekend). Finding such patterns can be useful for purposes such as marketing or product recommendation. For example, an online store could provide discounts on wine with cheese to a customer that regularly buys these items together.

Generally, a customer transaction database can be viewed as a sequence of discrete events or symbols that are sequentially ordered. Discrete sequences can be found in many domains besides shopping. For example, the sequence of words in a book can be viewed as a discrete sequence, as well as the sequence of locations visited by a person when driving a car. Hence, periodic pattern mining can be applied in many domains.

Several variations of the problem of periodic pattern mining have been studied but the majority of them focuses on analyzing a single discrete sequence. However, for many applications, it is desirable to analyze more than a single sequence at a time [20, 29]. For example, continuing the example of analyzing customer transactions, a business owner may want to analyze sequences of transactions made by several customers to discover periodic behavior that is common to several customers rather than analyzing each customer separately. This can provide insights such that many customers regularly buy bread and milk together.

Recently, a few algorithms have been designed to find periodic patterns common to multiple sequences such as MRCPPS [29], MPFPS [20]. These algorithms utilize different definitions of periodic patterns and also of sequences. This chapter gives an overview of the task of mining periodic patterns in a set of discrete sequences, the algorithms, and discuss applications and research opportunities.

The chapter is organized as follows: Section 2 briefly reviews the traditional problem of periodic pattern mining in sequence. Then, Sect. 3 presents the main model for discovering periodic patterns in multiple discrete sequences. Then, Sect. 4 briefly describes strategies used by the algorithms for this problem. Then, Sect. 5 discusses opportunities for research on this topic. Finally, Sect. 6 draws a conclusion.

2 Finding Periodic Patterns in a Single Discrete Sequence

The type of data considered in traditional periodic pattern mining studies is a discrete sequence [28], also called a sequence of transactions or transaction database [30].

Definition 1 (Discrete sequence) Let there be a set of items I (symbols or event types). An itemset X is a subset of I, that is, $X \subseteq I$. An itemset containing k items is said to be a k-itemset. A sequence s is an ordered list of itemsets $s = \langle T_1, T_2, \ldots T_m \rangle$, where $T_j \subseteq I$ ($1 \leq j \leq m$), j is the transaction identifier of T_j, and T_j is said to be a transaction.

Example 1 Let there be some products called a, b, c, d, and e that are sold in a retail store. This set of products is denoted as $I = \{a, b, c, d, e\}$. Then, a sequence of purchases made by a customer over time can be represented as shown in Fig. 1. This sequence indicates that a person has bought items a, b, and c at the same time, then purchased b and d together; then bought a, b, and e together; then purchased c; then bought a, b, d, and e together; and so on. This sequence contains eight itemsets. The itemset $\{a, b, c\}$ is a 3-itemsets.

Sequences of transactions can be used to represent data from multiple domains besides shopping data. For example, a sentence in a text can be viewed as a discrete sequence where each word is an item. Another example is a sequence of locations visited by a tourist in a city, where items represent locations. Another example of a discrete sequence is the genome of a virus such as $\langle \{A\}, \{T\}, \{A\}, \{A\}, \{C\}, \{A\}, \ldots \rangle$, where each item is a nucleotide denoted as A, C, G, and T [49]. Thus, as it can be observed by these examples, items in a discrete sequence can be ordered by time or other criteria. A discrete sequence where each transaction contains a single item is called a *string*, and a discrete sequence ordered by time can be called a *temporal database*, *temporal sequence*, *event sequence* [6, 32], or *event log* [18], depending on the context.

To find interesting patterns in a discrete sequence, it is necessary to define more clearly what we mean by periodic pattern and also to define some measures to select patterns that are periodic. In the traditional periodic pattern mining task, the type of patterns to be discovered is an itemset. The measures for identifying periodic itemsets are defined based on the following definitions.

Definition 2 (Subsequence) Let there be two sequences $s_a = \langle A_1, A_2, \ldots, A_k \rangle$ and $s_b = \langle B_1, B_2, \ldots, B_l \rangle$. The sequence s_a is a subsequence of s_b if and only if some integer numbers $1 \leq i1 < i2 < \ldots < ik \leq m$ can be found such that $A_1 \subseteq B_{i1}$, $A_2 \subseteq B_{i2}, \ldots, A_k \subseteq B_{il}$. The containment relationship between sequences is denoted as $s_a \sqsubseteq s_b$.

$$\langle \{a, b, c\}, \{b, d\}, \{a, b, e\}, \{c\}, \{a, b, d, e\}, \{a, c\}, \{b, c\}, \{b, e\} \rangle$$

Fig. 1 A discrete sequence of customer purchases

Example 2 The sequence $\langle \{a, b, c\}, \{b, e\} \rangle$ is a subsequence of the sequence illustrated in Fig. 1.

The traditional model for mining periodic patterns in a discrete sequence relies on the support and maximum periodicity measure to identify periodic patterns [4, 30].

Definition 3 (Support of an itemset in a sequence) Let there be an itemset X and a discrete sequence s. For an itemset X, the notation $TR(X, s)$ denotes the ordered set of transactions $TR(X, s) = \langle T_{g_1}, T_{g_2}, ..., T_{g_k} \rangle \sqsubseteq s$ where X appears in the sequence s. The support of X in s is the number of transactions containing X, that is, $sup(X, s) = |TR(X, s)|$.

Example 3 The itemset $\{a, b\}$ appears three times in the sequence of Fig. 1. Hence, the support of $\{a, b\}$ in that sequence is 3, which is denoted as $sup(\{a, b\}) = 3$.

Definition 4 (Maximum periodicity of an itemset in a sequence) Let there be a discrete sequence s. Two transactions T_x and T_y in s are said to be *consecutive with respect to* X if there does not exist a transaction $T_z \in s$ such that $x < z < y$ and $X \subseteq T_z$. The period of two consecutive transactions T_x and T_y for a pattern X is defined as $per(T_x, T_y) = y - x$. The periods of X in a sequence s are $pr(X, s) = \{per_1, per_2, ..., per_{k+1}\}$ where $per_1 = g_1 - g_0$, $per_2 = g_2 - g_1$, $... per_{k+1} = g_{k+1} - g_k$, and $g_0 = 0$ and $g_{k+1} = n$, respectively. The maximum periodicity of an itemset X in a sequence s is defined as $maxPr(X, s) = argmax(pr(X, s))$ [4].

Example 4 For example, consider the itemset $X = \{a, b\}$. That itemset appears in three transactions of the sequence of Fig. 1, namely, T_1, T_3, and T_5. Hence, it is said that $TR(X, s_1) = \{T_1, T_3, T_5\}$. The periods of X in that sequence are $pr(X, s) = \{1, 2, 2, 3\}$. The maximum periodicity of X in that sequence is $maxPr(X, s) = argmax(\{1, 2, 2, 3\}) = 3$.

In the traditional periodic pattern mining model, the goal is to find all the periodic itemsets in a sequence s. Let there be two user-specified thresholds called the maximum periodicity threshold $maxPer$ and the minimum support threshold $minSup$. An itemset X is deemed periodic in a sequence s if $maxPr(X, s) \leq maxPer$ and $sup(X, s) \geq minSup$.

Example 5 If $maxPr = 3$ and $minSup = 3$, the itemset $\{a, b\}$ is periodic in the sequence of Fig. 1 because its periods are $pr(\{a, b\}, s_1) = \{1, 2, 2, 3\}$, its maximum period is $maxPr(\{a, b\}, s) = max\{1, 2, 2, 3\} = 3 \leq maxPr$ and $sup(\{a, b\}, s) = 3 \geq minSup$.

In the above example, the sequence does not contain timestamps. But it should be noted that it is simple to extend the previous model to use timestamps. This can be done by replacing transaction identifiers by timestamps.

Several algorithms have been designed for the traditional problem of mining periodic patterns in a sequence [4, 30]. Moreover, in the last decades, several variations of

that problem have been proposed using alternative functions to evaluate the periodicity of each itemset X. Some functions that have been used to provide more flexibility are the minimum periodicity [30], average periodicity [30], and standard deviation of periods [20, 50, 51], respectively, defined as $minPr(X, s) = argmin(pr(X, s))$, $avgPr(X, s) = average(pr(X, s))$, and $stanDev(X, s) = \sqrt{\frac{\sum_{v \in pr(X,s)}(v - avgPr(X,s))^2}{avgPr(X,s)}}$
.

Example 6 Continuing the previous example, $minPr(X, s_1) = argmin(\{1, 2, 2, 3\}) = 1$, $avgPr(X, s_1) = average(\{1, 2, 2, 3\}) = \frac{1+2+2+3}{4} = 2$. Also, $stanDev$ $(X, s_1) = \sqrt{\frac{(1-2)^2+(2-2)^2+(2-2)^2+(3-2)^2}{4}} = \sqrt{0.5} \approx 0.71$.

Another function is the variance of periods [43, 54, 55]. In some other recent studies, a function to evaluate the periodic stability was introduced to find patterns that have a stable periodic behavior over time [27, 31]. The concept of periodicity has also been combined with various other functions to evaluate other aspects besides periodicity such as the utility (e.g., profit or importance) [12, 13, 23] and a statistical test [50]. However, all the above models are designed to find patterns in a single sequence rather than finding patterns that are common to multiple sequences. The next section discusses how periodic patterns can be discovered in multiple sequences.

3 Finding Periodic Patterns in Multiple Discrete Sequences

For many applications, the data is represented as a set of discrete sequences rather than a single sequence. This data format is called a *sequence database*, and is defined as follows.

Definition 5 (Sequence database) A sequence database D is a set of n sequences that is ordered, denoted as $D = \langle s_1, s_2, ..., s_n \rangle$. The i-th sequence of D is denoted as s_i, and its sequence identifier is said to be i.

Example 7 A small sequence database is shown in Table 1. This database contains four sequences that have the sequence identifiers 1, 2, 3, and 4. These four sequences could indicate the sequence of purchases made by four customers in a store. For example, sequence 1 indicates that a customer purchased items a, b, and e together;

Table 1 An example sequence database

Identifier	Sequence
1	$\langle \{a, b, e\}, \{a, b, e\}, \{a, d\}, \{a, e\}, \{a, b, c\} \rangle$
2	$\langle \{c\}, \{a, b, c, e\}, \{c, d\}, \{a, b, c, e\}, \{a, b, d\} \rangle$
3	$\langle \{b, c\}, \{a, b\}, \{a, c, d\}, \{a, c\}, \{a, b\} \rangle$
4	$\langle \{a, b, d, e\}, \{a, b, e\}, \{a, b, c\}, \{a, b, d, e\}, \{a, b\} \rangle$

followed by a, b, and e; followed by a and d; followed by a and e; and lastly followed by a, b, and c.

For such data, it can be desirable to find patterns that are periodic in multiple sequences rather than to consider each sequence separately. For this purpose, an evaluation function called the *sequence periodic ratio* was proposed [20, 29]. This function is defined as follows.

Definition 6 (Sequence periodic ratio) The number of sequences where an itemset X is periodic in a sequence database D is denoted and defined as $numSeq(X)$. The sequence periodic ratio of X in D is defined as $ra(X) = numSeq(X)/|D|$, where $|D|$ is the number of sequences in D.

In other words, the sequence periodic ratio represents the percentage of sequences from a database where an itemset X is periodic. Then, the problem of finding periodic patterns common to multiple sequences is defined as follows.

Definition 7 (General problem of mining periodic patterns common to multiple sequences) Let there be a sequence database, a minimum sequence periodic ratio threshold $minRa$ and a definition of what is a periodic pattern in a single sequence. An itemset X is a periodic pattern in D if $ra(X) \geq minRa$. The problem of *mining periodic patterns common to multiple sequences* is to find all periodic patterns in D.

It is to be noted that the above problem definition and sequence periodic ratio are defined in a general way such that various pattern evaluation functions could be used to evaluate if a pattern is periodic in each sequence. In the following two subsections, two instantiations of the general problem are presented to, respectively, find (1) periodic frequent patterns in multiple sequences and (2) rare correlated periodic patterns. For each of these problem instantiation, different evaluation functions are used to determine if a pattern is periodic in a sequence.

3.1   Finding Frequent Periodic Patterns

The first problem instantiation is designed to find periodic patterns that are frequent, that is, that appear in many sequences of a sequence database. For example, an application is to find periodic patterns that are common to many customers of a retail store. This problem is defined as follows [20].

Definition 8 (Problem of mining periodic frequent patterns common to multiple sequences) Let there be a sequence database D, and four user-defined thresholds, namely, the minimum support threshold $minSup$, maximum periodicity threshold $maxPr$, maximum standard deviation threshold $maxStd$, and minimum sequence periodic ratio threshold $minRa$. In that definition, an itemset X is periodic in a sequence s if $maxPer(X, s) \leq maxPr \wedge sup(X, s) \geq minSup \wedge stanDev(X, s) \leq maxStd\}|$. The *problem of mining frequent periodic patterns common to*

Table 2 Periodic Frequent Patterns found in the database of Fig. 1 for different threshold values

$minSup$	$maxPr$	$maxStd$	$minRa$	Periodic frequent patterns found
2	3	1.0	0.6	$\{a\}, \{e\}, \{a, e\}$
2	3	1.0	**0.4**	$\{a\}, \{b\}, \{c\}, \{e\}, \{a, b\}, \{a, e\},$ $\{b, e\}, \{a, b, e\}$
2	**1**	1.0	0.6	$\{a\}$
3	3	1.0	0.6	$\{a\}$
2	3	**1.5**	0.6	$\{a\}, \{b\}, \{e\}, \{a, b\}, \{a, e\},$

multiple sequences is to find all Periodic Frequent Patterns (PFP) [20]. An itemset X is a PFP in D if $ra(X) \geq minRa$.

Example 8 Consider the database D of Table 1, and that the user specifies some thresholds $minSup = 2$, $maxPr = 3$, $maxStd = 5.0$, and $minRa = 0.3$. Let $X = \{a, b\}$. The itemset X is periodic in sequence s_1, s_2, and s_4. Thus, $numSeq(X) = 3$. Furthermore, as the database contains four sequences, i.e., $|D| = 4$, the sequence periodic ratio of X is $ra(X) = 3/4 = 0.75$. Hence, X is a PFP in the database. The list of all PFP that can be found for different threshold values is shown in Table 2.

It is interesting to observe that the traditional problem of mining periodic frequent patterns in a single sequence can be viewed as a special case of the general problem of mining periodic frequent patterns common to multiple sequences such that the sequence database contains a single sequence and $minRa = 0$.

3.2 Finding Rare Correlated Periodic Patterns

The previous problem instantiation is designed to find periodic patterns that are frequent (appear in many sequences). Although these patterns may be interesting, rare patterns can also be valuable. To be able to find rare patterns, a second instantiation of the general problem of mining periodic patterns common to multiple sequence was defined to find rare correlated periodic patterns [29]. The motivation is to find patterns that are periodic, appear rarely but contain items that are strongly correlated in a sequence database. For the example of analyzing customer data, this can mean to find patterns that are periodic and rare (not purchased by many customers) but have a strong correlation.

To find rare patterns in data, several methods have been proposed in the literature [42, 46, 59, 60] using different definitions of what is a rare pattern and how to efficiently find these patterns. For an overview of studies on rare pattern mining, the reader can refer to a recent survey about finding rare patterns by Koh et al. [41]. In the following, a simple definition of what is a rare pattern is used. It is that an *itemset is rare* if its support is not greater than some maximum support threshold $maxSup$ set by the user.

But a problem with rare patterns is that because they seldomly appear, it is possible that some of them contains items that appear together by chance. To avoid this problem and find patterns representing items that are strongly correlated, several correlation measures have been used in itemset mining such as the bond [18, 21, 52, 69], affinity [3], all-confidence [52, 63], coherence, and mean [7, 57], each having different advantages and limitations [33]. In the problem instantiation presented in this section, the bond measure is used because it is a simple measure, it is easy to calculate, and also easy to interpret. The bond of an itemset is defined based on the following definitions.

Definition 9 (Disjunctive support) Let there be a sequence s and an itemset X. The disjunctive support of X in s is the number of transactions that contain one or more items from X. It is denoted as $dissup(X, s)$ and defined as $dissup(X, s) = max\{sup(X, s) | X \in s\}$.

Definition 10 (Bond) Let there be a sequence s and an itemset X. The bond of X in s is defined as $bond(X, s) = \frac{sup(X,s)}{dissup(X,s)}$.

The bond of an itemset can take a value in the [0, 1] interval such that a value of 0 indicates a low correlation and a value of 1 indicates the maximum correlation.

Example 9 The disjunctive support of the itemset $\{a, e\}$ in the first sequence (s_1) of the database of Table 3 is calculated as $dissup(\{a, e\}, s_1) = 5$, while the support of $\{a, e\}$ is $sup(\{a, e\}) = 3$. Hence, the bond of that itemset is $bond(\{a, e\}, s_1) = \frac{3}{5} = 0.6$.

Based on the concept of rare patterns and the bond measure for identifying correlated patterns, the second problem instantiation is defined.

Definition 11 (Rare correlated periodic pattern in a sequence) Let there be three thresholds, namely, the maximum support threshold ($maxSup$), minimum bond threshold ($minBond$), and maximum standard deviation threshold ($maxStd$). An itemset X is a rare correlated periodic pattern in a sequence s if $sup(X, s) \leq maxSup$, $stanDev(X, s) \leq maxStd$, and $bond(X, s) \geq minBond$.

Example 10 The periods of itemset $X = \{a\}$ in the sequence s_2 are $pr(X, s_2) = \{2, 2, 1, 0\}$. Hence, $sup(X, s_2) = 4$, $dissup(X, s_2) = 4$, $bond(\{a\}) = \frac{4}{4} = 1$,

Table 3 A second example sequence database

SID	Sequence
1	$\langle\{a, c, e\}, \{a, b, e\}, \{a, d\}, \{a, b, e\}, \{a, c\}\rangle$
2	$\langle\{c\}, \{a, b, c, e\}, \{c, d\}, \{a, b, c, e\}, \{a, b, d\}\rangle$
3	$\langle\{b, c\}, \{a, b\}, \{a, c, d\}, \{a, c\}, \{a, b\}\rangle$
4	$\langle\{a, b, d, e\}, \{a, b\}, \{a, c\}, \{a, b, d, e\}, \{a, d\}\rangle$

$max \, Pr(X, s_2) = 2$ and $avg \, Pr(X, s_2) = \frac{2+2+1+0}{4} = 1.25$. The standard deviation of periods is

$stan \, Dev(X, s_2) = \sqrt{\frac{(2-1.25)^2+(2-1.25)^2+(1-1.25)^2+(0-1.25)^2}{4}} = \sqrt{0.6875} \approx 0.83$. Thus, if $min \, Bond = 0.5$, $max \, Std = 0.9$, and $max \, Sup = 4$, the itemset X is a rare correlated periodic pattern in sequence s_2.

Definition 12 (Sequence periodic ratio) In the context of discovering rare correlated periodic patterns, the number of sequences where an itemset X is a rare correlated periodic pattern in a sequence database D is denoted and defined as $num \, Seq(X)$. Moreover, the sequence periodic ratio of X in D is defined as $ra(X) = num \, Seq(X)/|D|$, where $|D|$ is the number of sequences in D.

Definition 13 (Problem of mining periodic rare correlated periodic patterns common to multiple sequences) Consider a sequence database D and four user-specified thresholds: a maximum support threshold $(max \, Sup)$, a minimum bond threshold $(min \, Bond)$, a maximum standard deviation threshold $(max \, Std)$, and a minimum sequence periodic ratio threshold $(min \, Ra)$. The problem of mining RCPP common to multiple sequences is to find all the Rare Correlated Periodic Patterns (RCPP) in the database D, that is, each itemset X where $ra(X) \geq min \, Ra$ [29].

Example 11 Consider the database of Table 3 and that the thresholds are set by the user as $max \, Sup = 2$, $max \, Std = 1$, $min \, Bond = 0.6$, and $min \, Ra = 0.70$. The itemset $\{b, e\}$ is a rare correlated periodic pattern in sequence s_1, $sup(\{b, e\}, s_1) = 2$, $stan \, Dev(\{b, e\}, s_1) = 0.47$, and $bond(\{b, e\}, s_1) = 0.67$. Moreover, it can be found that $\{b, e\}$ is also a rare correlated periodic pattern in sequence s_2 and s_4. Hence, $num \, Seq(\{b, e\}) = 3$ and the sequence periodic ratio of $\{b,e\}$ is $ra(\{b, e\}) = 3/4 = 0.75$. Thus, $\{b, e\}$ is a RCPP.

The problem of discovering RCPP in a database is hard because the maximum support and standard deviation functions cannot be directly used to reduce the search space, as these functions are neither monotonic nor anti-monotonic. Thus, other ideas must be used to design efficient algorithms. Those will be discussed in the next section describing algorithms.

4 Two Algorithms

This section presents two efficient algorithms for mining periodic patterns in multiple sequence. The first subsection presents the MPFPS (mining periodic frequent patterns in multiple sequences) algorithm to solve the first problem instantiation from Sect. 3.1, while the following subsection presents the MRCPPS algorithm to solve the problem instantiation of Sect. 3.2.

4.1 The MPFPS Algorithm to Mine for Mining Frequent Periodic Patterns in Multiple Sequences

The key challenge to find periodic patterns in multiple sequences is how to reduce the search space to avoid looking at all the possible itemsets. In the field of pattern mining, several properties of pattern evaluation functions are used to reduce the search space. For the problem of finding periodic frequent patterns in multiple sequences, an upper bound on the sequence periodic ratio is used in the MPFPS algorithm for reducing the search space [20, 29], which is defined as follows.

Definition 14 Let there be an itemset X, a sequence s, and the $maxPer$ and $minSup$ thresholds. The itemset X is a *candidate* in a sequence s if $sup(X, s) \geq minSup$ and $maxPr(X, s) \leq maxPr$. The number of sequences where an itemset X is a candidate is denoted as $numCand(X)$ and defined as $numCand(X) = |\{s|maxPr(X, s) \leq maxPr \land sup(X, s) \geq minSup \land s \in D\}|$. The *boundRa* upper bound of X is denoted as $boundRa(X)$ and defined as $boundRa(X) = numCand(X)/|D|$.

Two properties of the *boundRa* upper bound are utilized by MPFPS to reduce the search space.

Theorem 1 *For any itemset X, the relationship $boundRa(X) \geq ra(X)$ holds [20].*

Proof We have an itemset X and:

$$
\begin{aligned}
ra(X) &= numSeq(X)/|D| \\
&= |\{s|maxPr(X, s) \leq maxPr \land sup(X, s) \\
&\quad \geq minSup \land stanDev(X, s) \leq maxStd \land s \in D\}|/|D| \\
&\leq |\{s|maxPr(X, s) \leq maxPr \land sup(X, s) \geq minSup \land s \in D\}|/|D| \\
&= numCand(X)/|D| \\
&= boundRa(X).
\end{aligned}
$$

$\square$

Theorem 2 *Let there be two itemsets X and X' such that $X' \subseteq X$. Then, $boundRa(X') \geq boundRa(X)$ [20].*

Theorem 3 *If $boundRa(X') < minRa$ for an itemset X', then X' and any superset $X \supset X'$ are not PFP [20].*

The proof of the above theorems can be found in the paper presenting the MPFPS algorithm [20].

The MPFPS algorithm searches for periodic patterns by first starting from 1-itemsets, and then it recursively adds an item to each itemset to search for larger itemsets. The items are appended to an itemset by following a total order on item called $\succ$. During the search for periodic itemsets, the MPFPS relies on the above

i-set	{a}		
sid-list	{0,1,2,3}		
tidlist-list	[{0,1,2,3,4}, {1,3,4},{1,2,3,4}, {0,1,2,3,4}]		

i-set	{e}	*i-set*	{a, e}
sid-list	{0,1,3}	*sid-list*	{0,1,3}
tidlist-list	[{0,1,3}, {1,3,4}, {0,1,3}]	*tidlist-list*	[{0,1,3}, {1,3}, {0,1,3}]

Fig. 2 The PFPS-lists of itemsets {a} (top-left), {e} (top-right), and {a, e} (bottom)

theorem to reduce the search space. If an itemset X has a *bound Ra* that is smaller than the *min Ra* threshold set by the user, all its supersets cannot be PFP and thus can be ignored to reduce the search space.

To be able to decide if an itemset is periodic, it is necessary to be able to calculate its support, periods, and its sequence periodic ratio. This is done using a special data structure named a PFPS-list. This data structure is constructed for each itemset that is considered by the algorithm during the search for PFP. It is defined based on the following definitions.

Definition 15 (Sequences containing an itemset) The ordered list of sequence containing an itemset X is denoted as $sequences(X)$ and defined as a list $sequences(X) = \{s | sup(X, s) > 0 \wedge s \in D\}$, ordered by increasing sequence identifiers.

Example 12 For Table 1, the list of sequences containing {e} is $sequences(\{e\}) = \{s_0, s_1, s_3\}$.

Definition 16 (The PFPS-list structure) Let there be an itemset X and a sequence database D. The PFPS-list LX of X is a table containing three rows (fields) called *i-set*, *sid-list*, and *tidlist-list*. The first row contains the itemset X, that is, $LX.i - set = X$. The second row stores the list of identifiers of sequences containing X, that is, $LX.sidlist = \{v_1, v_2 \ldots v_w\}$ where $w = |sequences(X)|$ and v_i ($1 \leq i \leq w$) is the sequence identifier of the i-th sequence in $sequences(X)$. The third row, *tidlist-list*, contains the identifiers of transactions containing X, for each sequence containing X. It is formally defined as a list $LX.tidlist - list = \{Z_1, Z_2 \ldots Z_w\}$ such that $w = |sequences(X)|$. Let s_i be the i-th sequence in $sequences(X)$ ($1 \leq i \leq w$). The value Z_i is $Z_i = \{z | X \subseteq T_z \wedge \langle T_z \rangle \sqsubseteq s_i\}$.

Example 13 The PFPS-lists of itemsets {a}, {e}, and {a, e} are presented in Figure 2 for the database of Table 1.

The PFPS-list of an itemset X indicates (1) the list of sequences where X appears (in the field $sid - list$) and (2) the list of transactions containing X for any sequence s (in the field $tidlist - list$). This latter information is all is needed to calculate $pr(X, s)$, to then derive $sup(X, s)$, $maxpr(X, s)$, and $stanDev(X, s)$, and thus check if X is a periodic frequent pattern. Thus, using the PFPS-list structure, it is possible to check if an itemset is periodic without scanning the database.

The MPFPS algorithm initially reads the database once to build the PFPS-list of each item. Then, MPFPS starts to search for larger itemsets. During this search, the PFPS-lists of each larger itemset is obtained by joining the PFPS-lists of two of its subsets. Thus, building the PFPS-lists of itemsets having more than 1 item does not require scanning the database again.

For an itemset P and some items x and y, the itemset $P \cup \{x\}$ is said to be an *extension* of P, which is briefly denoted as Px. The algorithm for constructing the PFPS-list of an itemset Pxy by joining the PFPS-lists of two itemsets Px and Py is shown in Algorithm 1 [20]. For instance, this algorithm can be applied to the PFPS-lists of $\{a\}$ and $\{e\}$ to build the PFPS-list of $\{a, e\}$, which are all depicted in Fig 2.

Algorithm 1: The Intersect procedure

---

    input : the PFPS-lists LPx and LPy of some itemsets Px and Py
    output: the PFPS-list $LPxy$ of itemset Pxy

1   $LPxy.i\text{-}set \leftarrow Px \cup \{y\}$; $LPxy.tidlist\text{-}list \leftarrow \emptyset$; $LPxy.sid\text{-}list \leftarrow \emptyset$;
2   **foreach** *sequence identifier* $sid \in LPx.sid\text{-}list$ *such that* $sid \in LPy.sid\text{-}list$ **do**
3      $tidListSidPx \leftarrow$ the tid list of sid in $LPx.tidlist\text{-}list$;
4      $tidListSidPy \leftarrow$ the tid list of sid in $LPy.tidlist\text{-}list$;
5      $tidListSidPxy \leftarrow tidListSiPx \cap tidListSiPy$;
6   **if** $tidListSidPxy \neq \emptyset$ **then**
7 | $LPxy.sid\text{-}list.append(sid)$; $LPxy.tidlist\text{-}list.append(tidListSidPxy)$;
8   **end**
9   **end**
10   **return** $LPxy$;

The MPFPS (Algorithm 2) takes as input a sequence database D and the user-defined thresholds $maxStd$, $minRa$, $maxPr$, and $minSup$. The algorithm explores the search space using a depth-first search, and returns the set of all PFP. MPFPS first reads the input database to calculate the following information for each item i and sequence s: $sup(\{i\}, s)$, $pr(\{i\}, s)$, $maxpr(\{i\}, s)$, and $stanDev(\{i\}, s)$. A loop is then done by MPFPS on each single item. An item i is considered to be periodic in a sequence s if $sup(\{i\}, s) \geq minSup$, $maxpr(\{i\}, s) \leq maxPr$, and $stanDev(\{i\}, s) < maxStd$. Then, the algorithm calculates the sequence periodic ratio of item i by dividing the number of sequences where i is periodic by the total number of sequences (line 3 to 4). If this value is not less than $minRa$, i is a PFP and it is output (line 5). Thereafter, the algorithm calculates the $boundRa$ upper bound of $\{i\}$ (line 6 to 7). After the loop is completed, for each item i such that $boundRa(\{i\}) \geq minRa$, the PFPS-list of $\{i\}$ is put in a list $boundPFPS$ (line 9), which is then sorted by the total order $\succ$ of increasing $boundRa$ values. After that, a *Search* procedure is invoked to recursively search for extensions of single items in $boundPFPS$. Other itemsets do not need to be explored according to Theorem 3. The *Search* procedure is initially called with $boundPFPS$, $minSup$, $maxPr$, $maxStd$, $minRa$, and D.

The *Search* procedure is described in Algorithm 3. The input of the procedure is the PFPS-lists of extensions of an itemset P. Recall that an extension of an itemset P with an item z refers to the itemset obtained by appending an item z to P, and is denoted as Pz. Moreover, the four thresholds are also received as parameters, as well as the input database D. When the *Search* procedure is first called, $P = \emptyset$ and all its extensions are 1-itemsets (items). The procedure executes two loops to join all pairs of extensions of P, denoted as Px and Py, where x, y are items and $y \succ x$, (line 1 to 12). The result of a join is an extension Pxy containing $|Px| + 1$ items. The PFPS-list of Pxy is obtained by calling Algorithm 1 with the PFPS-lists of Px and Py as parameters (line 3). Then, the procedure calculates $bound\,Ra(Pxy)$ and $numCand(Pxy)$ by reading Pxy (line 4 to 5). In the case, where the condition $bound\,Ra(Pxy) \geq minRa$ is not met, Pxy and all its supersets are not PFP and can be ignored from further processing according to Theorem 3 (line 6). If the condition is met, the PFPS-list of Pxy is stored into a variable $ExtensionsOfPx$ that contains all PFPS-lists of extensions of Px having a $bound\,Ra$ value that is no less than $minRa$ (line 7). Thereafter, the algorithm calculates the ratio $ra(Pxy)$ and if $ra(Pxy) \geq minRa$, then Pxy is output as a PFP (line 8 to 10). Finally, a recursive call of the *Search* procedure is made with $ExtensionsOfPx$ (PFPS-lists of itemsets extending Px that are PFP) to search for potential transitive extensions that are also PFP (line 13).

When the algorithm terminates, all PFP have been output. The proof that the algorithm is complete is omitted but it can be easily seen based on Theorem 3, as it guarantees that the algorithm only ignores itemsets that are non-PFP.

Algorithm 2: The MPFPS algorithm

input : D: a sequence database, $maxStd, minRa, maxPr, minSup$: the thresholds.
output: the set of periodic frequent patterns (PFPS).

1 Scan each sequence $s \in D$ to calculate $sup(\{i\}, s), pr(\{i\}, s), maxpr(\{i\}, s)$ and
   $stanDev(\{i\}, s)$ for each item $i \in I$;
2 **foreach** *item* $i \in I$ **do**
3     $numSeq(\{i\}) \leftarrow |\{s|maxpr(\{i\}, s) \leq maxPr \wedge stanDev(\{i\}, s) \leq$
       $maxStd \wedge sup(\{i\}, s) \geq minSup \wedge s \in D\}|$;
4     $ra(\{i\}) \leftarrow numSeq(\{i\})/|D|$;
5   **if** $ra(Px) \geq minRa$ **then** output Px
       $numCand(\{i\}) \leftarrow |\{s|maxpr(\{i\}, s) \leq maxPr \wedge sup(\{i\}, s) \geq minSup \wedge s \in D\}|$;
6     $bound\,Ra(\{i\}) \leftarrow numCand(\{i\})/|D|$;
7 **end**
8 $bound\,PFPS \leftarrow \{$PFPS-list of item $i|i \in I \wedge bound\,Ra(\{i\}) \geq minRa\}$;
9 Sort $bound\,PFPS$ by the order $\succ$ of ascending $bound\,Ra$ values;
10 Search ($bound\,PFPS, minSup, maxPr, maxStd, minRa, D$);

To illustrate how the algorithm is applied, a detailed example is next presented for mining PFP in the sequence of Table 1. Consider that the parameters are set as $minRa = 0.6, maxPr = 3, minSup = 2$, and $maxStd = 1.0$. The following steps are performed.

Algorithm 3: The *Search* procedure

---

input : *Extensions Of P*: a set of PFPS-lists of extensions of an itemset *P*,
       minSup, maxPr, maxStd, minRa: the thresholds, *D*: the database.
output: the set of periodic frequent patterns that extend *P*.

1 foreach *PFPS-list LPx $\in$ Extensions Of P and Px = LPx.i-set* **do**
2   **foreach** *PFPS-list LPy $\in$ Extensions Of P and Py = LPy.i-set such that y $\succ$ x* **do**
3   *LPxy* ← Intersect *(LPx, LPy)*;
4   *numCand(Pxy)* ← |{s|*maxpr(Pxy, s)* $\leq$ *maxPr* $\wedge$ *sup(Pxy, s)* $\geq$ *minSup* $\wedge$ *s* $\in$ *D*}|;
5   *boundRa(Pxy)* ← *numCand(Pxy)/|D|*;
6   **if** *boundRa(Pxy)* $\geq$ *minRa* **then**
7   *Extensions Of Px* ← *Extensions Of Px* $\cup$ {*LPxy*};
8   *numSeq(Pxy)* ← |{s|*maxpr(Pxy, s)* $\leq$ *maxPr* $\wedge$ *stanDev(Pxy, s)* $\leq$ *maxStd* $\wedge$ *sup(Pxy, s)* $\geq$ *minSup* $\wedge$ *s* $\in$ *LPxy.sid-list*}|;
9   *ra(Pxy)* ← *numSeq(Pxy)/|D|*;
10   **if** *ra(Pxy)* $\geq$ *minRa* **then** output *Pxy*
11   **end**
12   **end**
13   Search *(Extensions Of Px, minSup, maxPr, maxStd, minRa, D)*;
14 end

---

1. The algorithm reads the database and first checks single items. Consider the item a. The periods, maximum periodicity, and standard deviation of a are calculated for each sequence. The obtained values are shown in Table 4. Based on these values, it is found that a is periodic in the four sequences. For instance, a is periodic in s_2 because $maxpr(\{a\}, s_2) = 2 < maxPr$ and $stanDev(\{a\}, s_2) = 0.256 < maxStd$. The number of sequence where a is periodic is $numSeq(\{a\}) = 4$ and as there are four sequences in the database, $ra(\{a\}) = 4/4 = 1 \geq minRa$. Hence, $\{a\}$ is output as a PFP. The same process is repeated for all other items. It is found that only the items $\{a\}$ and $\{e\}$ are PFP.

2. Then, for each item having a *boundRa* value that is greater or equal to *minRa*, its PFPS-list is created and added to *boundPFPS*. In this example, only the PFPS-lists of $\{a\}$ and $\{e\}$ are built. The *boundRa* value of $\{a\}$ is 1.

3. After this, the list *boundPFPS* is sorted by increasing *boundRa* values and the algorithm invokes the *Search* procedure to find extensions of $\{a\}$ and $\{e\}$ that are PFP (Algorithm 3).

4. The *Search* procedure is a recursive method that extends itemsets having their PFPS-lists in *boundPFPS* by joining pairs of them. Initially, all PFPS-lists in *boundPFPS* are extensions of the empty set and the aim is to generate itemsets having two items.

5. The procedure first considers combining $\{a\}$ and $\{e\}$ to obtain $\{a, e\}$. These two itemsets can be joined because they have a common prefix (the empty set). To create the PFPS-list of $\{a, e\}$, the *Intersect* method is called with the PFPS-lists of $\{a\}$ and $\{e\}$ as parameters. The *sid-list* of $\{a, e\}$ is obtained by calculating the union of the *sid-lists* of $\{a\}$ and that of $\{e\}$ as $\{0, 1, 2, 3\} \bigcup \{0, 1, 3\} = \{0, 1, 3\}$.

Table 4 Periods, maximum periodicity, and standard deviation of itemset $\{a\}$ in sequences from Table 1

Sequence s_x	$pr(a, s_x)$	$maxPr(a, s_x)$	$stanDev(a, s_x)$	$\{a\}$ is periodic in s_x?
s_1	[1,1,1,1,1,0]	1	0.152	yes
s_2	[2,2,1,0]	2	0.256	yes
s_3	[2,1,1,1,0]	2	0.632	yes
s_4	[1,1,1,1,1,0]	1	0.152	yes

For sequence s_1, the tidlist of $\{a\}$ is $\{0, 1, 2, 3, 4\}$ while that of $\{e\}$ is $\{0, 1, 3\}$. Their intersection is $\{0, 1, 2, 3, 4\} \cap \{0, 1, 3\} = \{0, 1, 3\}$. Hence, the sequence identifier 1 is added to *sid-list* of $\{a, e\}$ and $\{0, 1, 3\}$ is stored in the *tidlist-list* of $\{a, e\}$. The same process is repeated to build the remaining part of the PFPS-list of $\{a, e\}$. The final PFPS-list of $\{a, e\}$ is shown at the bottom of Fig. 2.

6. Based on the PFPS-list of $\{a, e\}$, it is calculated that $boundRa(\{a, e\}) = 0.6$ which is no less than $minRa$. Hence, $\{a, e\}$ may be a PFP as well as its supersets.
7. The next step is to check in which sequences $\{a, e\}$ is periodic to obtain $ra(\{a, e\})$. As $numSeq(\{a, e\}) = 3$, we have $ra(\{a, e\}) = 0.75 \geq minRa$. Hence, the itemset $\{a, e\}$ is output as a PFP.
8. The *Search* procedure then applies this process again to generate other itemsets and recursively calls itself to find their extensions, in the same way.

It is to be noted that the MPFPS algorithm described in that subsection was originally called $MPFPS_{depth}$ [20]. There also exists a breadth-first search version of this algorithm called $MPFPS_{breadth}$ [20], which is not presented in this chapter due to space limitation. In terms of performance, both $MPFPS_{depth}$ and $MPFPS_{breadth}$ perform better on some datasets.

The MPFPS algorithm was applied to analyze text data where each sequences of a book can be viewed as a sequence. Then, the discovered PFP can reveal patterns that represent the writing styles of authors. The algorithm was also applied to find patterns in sequences of clicks on the FIFA website, where it was discovered that some web pages are periodically clicked by several users [20].

4.2 The MRCPPS Algorithm

The MRCPPS algorithm [29] was created by adapting the MPFPS algorithm to solve the second problem instantiation described in this chapter. Since both algorithms are designed to find periodic patterns in multiple sequences, they have many similarities. The key differences lie in the fact that these algorithms use different functions to select patterns. While MPFPS aims to find frequent periodic patterns, MRCPPS is designed to find rare correlated patterns using the bond measure.

The MRCPPS algorithm applies an upper bound on the *bond* measure to reduce the search space, which is called $upBondRa$, which is defined below.

Definition 17 (upBondRa) Let there be an itemset X. It is considered to be a candidate in a sequence s if $bond(X, s) \geq minBond$, where $minBond$ is a user-defined threshold. Then, the number of sequences where an itemset X is a candidate in a sequence database D is denoted as $numCand(X)$. The $upBondRa$ of X in D is defined as $upBondRa(X) = numCand(X)/|D|$.

The following theorem indicates that $upBondRa$ can be used to reduce the search space [29].

Theorem 4 *Let there be two itemsets $X \subset X'$. Then, the two following relationships hold: $upBondRa(X) \geq ra(X)$ and $upBondRa(X) \geq upBondRa(X')$.*

To calculate the *bond*, $maxPr$, and other evaluation functions for an itemset, the MRCPPS algorithm utilizes a novel data structure called RCPPS-list. This structure can be viewed as a more complex structure than the PFPS-list presented in the previous subsection. The reason is that a RCPPS-list stores additional information required to calculate the bond measure.

A RCPPS-list is created for each itemset X considered by MRCPPS. The RCPPS-list of X is a table providing information about an itemset X stored in three rows (fields): (1) *SIDlist*: is the list of identifiers of sequences containing X. (2) *list-conTIDlist*: contains the list of identifiers of transactions where X occurs (conTIDlist) for each sequence in SIDlist. (3) *list-disTIDlist*: is the list of identifiers of transactions containing at least one item from X (disTIDlist) for each sequence in SIDlist.

Example 14 The RCPPS-lists of $\{b\}$, $\{e\}$, and $\{b, e\}$ are shown in Fig. 3, for the database of Table 3.

Several useful information can be calculated directly from the RCPPS-list of an itemset. The field SIDlist provides the list of sequences containing the itemset. For instance, the field *SIDlist* of the RCPPS-list of $\{b, e\}$ indicates that it appears in sequences s_1, s_2, and s_4. The size of *conTIDlist* for a sequence indicates the support of the itemset in that sequence, while the *bond* can be calculated as the size of its *conTIDlist* divided by the size of its *disTIDlist* for the sequence. The *conTIDlist* and *disTIDlist* of itemset $\{b, e\}$ for sequence s_1 are [2, 4] and [1, 2, 4], respectively. As a result, the bond of $\{b, e\}$ in sequence s_1 is $\frac{2}{3}$.

The *stanDev* and periods of an itemset can also be calculated from its RCPPS-list, as well as the $upBondRa$ and ra values. The ra is used by the RCPP algorithm to determine if an itemset is a RCPP and the $upBondRa$ value is utilized to check if it should be extended to find larger RCPP based on Theorem 4.

The procedure $Construct$ for intersecting two RCPPS-lists of some itemsets Px and Py to obtain the RCPPS-list of an itemset Pxy is shown in Algorithm 4. As the principle of that procedure is similar to the $Intersect$ procedure of MPFPS, it is not described in more detail here.

i-set	{b}			
sid-list	{1,2,3,4}			
list-conTIDlist	{[2, 4], [2, 4, 5], [1, 2, 5], [1, 2, 4]}			
list-disTIDlist	{[2, 4], [2, 4, 5], [1, 2, 5], [1, 2, 4]}			

i-set	{e}		i-set	{b, e}
sid-list	{1,2,4}		sid-list	{1,2,4}
list-conTIDlist	{[1,2, 4], [2, 4], [1, 4]}		list-conTIDlist	{[2, 4], [2, 4], [1, 4]}
list-disTIDlist	{[2, 4], [2, 4], [1, 4]}		list-disTIDlist	{[2, 4], [2, 4], [1, 4]}

Fig. 3 The RCPPS-lists of itemsets {b} (top-left), {e} (top-right), and {b, e} (bottom) for the database of Table 3

Algorithm 4: The Construct procedure

---

input: LPx: the RCPPS-list of Px, LPy: the RCPPS-list of Py.
out   : the RCPPS-list of Pxy

1 $LPxy \leftarrow \emptyset$;
2 **foreach** i, j *where* $LPx.SIDlist(i) = LPy.SIDlist(j)$ **do**
3 | $conTIDlist \leftarrow LPx.list\text{-}conTIDlist(i) \cap LPy.list\text{-}conTIDlist(j)$;
4 | **if** $conTIDlist \neq \emptyset$ **then**
5 | | $disTIDlist \leftarrow LPx.list\text{-}disTIDlist(i) \cup LPy.list\text{-}disTIDlist(j)$;
6 | | $LPxy.SIDlist \leftarrow LPxy.SIDlist \cup LPx.SIDlist(i)$;
7 | | $LPxy.list\text{-}conTIDlist \leftarrow LPxy.list\text{-}conTIDlist \cup conTIDlist$;
8 | | $LPxy.list\text{-}disTIDlist \leftarrow LPxy.list\text{-}disTIDlist \cup disTIDlist$;
9 | **end**
10 **end**
11 **return** $LPxy$;

The pseudocode of MRCPPS is shown in Algorithm 5. The input is the thresholds $maxSup$, $maxStd$, $minBond$, and $minRa$, and a sequence database D. MRCPPS initially reads the sequence database to construct the RCPPS-lists of all items. Thereafter, the algorithm creates a set I^* to store 1-itemsets (single items) that have an $upBondRa$ value that is greater or equal to $minRa$. Items not in I^* will thereafter not be considered by the algorithm, based on Theorem 4. Then, MRCPPS sorts items in I^* according to the ascending order $\succ$ of $upBondRa$ values. After that, the recursive *Find* procedure (Algorithm 6) is called with the empty itemset $\emptyset$, the set of single items I^*, and the user-defined thresholds.

The *Find* procedure takes as input an itemset P, extensions of the form Pz such that $upBondRa(Pz) \geq minRa$, the sequence database D, $maxSup$, $maxStd$, $minBond$, and $minRa$. When the *Find* procedure is first called, $ExtensionsOfP$ contains single items and $P = \emptyset$. The procedure performs a loop to process each extension Px of P (line 1 to 15). The values $numSeq(Px)$ and $ra(Px)$ are first calculated based on the information stored in the RCPPS-list of Px (line 2 to 3). If the $ra(Px)$ value is no less than $minRa$, X is output as a RCPP (line 4). And in the case where $upBondRa(Px) \geq minRa$, the procedure will consider extending Px to generate larger itemsets. This is achieved by joining Px with each extension of the form Py where $y \succ x$ to obtain an itemset Pxy (line 7). The algorithm builds Pxy's RCPPS-list by calling the *Construct* procedure using the RCPPS-list of Px and Py (line 8). Each extension Pxy is added to a set $ExtensionsOfP$ of extensions that can be considered for further extensions if the $upBondRa$ value of Pxy is no less than $minRa$ (line 11 to 13). Extensions having an $upBondRa$ value smaller than $minRa$ are ignored to reduce the search space (by Theorem 4). Finally, a call to the *Search* procedure is made with Pxy as argument to calculate its ra value and explore its extensions(s) using a depth-first search (line 16). When the algorithm terminates all RCPP have been output.

Algorithm 5: The MRCPPS algorithm

---

   input : D: a sequence database,
                 $maxSup, maxStd, minBond, minRa$: the user-specified thresholds.
   output: the set of RCPPS

1 Scan D to calculate the RCPPS-lists of all items in I;
2 $I^* \leftarrow \emptyset$;
3 **foreach** *item* $i \in I$ **do**
4     $numCand(i) \leftarrow |\{s|bond(i, s) \geq minBond \wedge s \in D\}|$;
5     $upBondRa(i) \leftarrow numCand(i)/|D|$;
6   **if** $upBondRa(i) \neq 0 \wedge upBondRa(i) \geq minRa$ **then** $I^* \leftarrow I^* \cup \{i\}$
7 **end**
8 Sort I^* by the order $\succ$ of ascending $upBondRa$ values;
9 Find($\emptyset, I^*, D, minSup, maxStd, minBond, minRa$);

Algorithm 6: The Find procedure

---

   input : P: an itemset, $ExtensionsOfP$: a set of extensions of P, D: the sequence database
                 $maxSup, maxStd, minBond, minRa$: the user-specified thresholds.
   output: the set of RCPPS

1 **foreach** *itemset* $Px \in ExtensionsOfP$ **do**
2     $numSeq(Px) \leftarrow |\{s|sup(Px, s) \leq maxSup \wedge stanDev(Px, s) \leq$
      $maxStd \wedge bond(Px, s) \geq minBond \wedge s \in D\}|$;
3     $ra(Px) \leftarrow numSeq(Px)/|D|$;
4   **if** $ra(Px) \neq 0 \wedge ra(Px) \geq minRa$ **then** output Px $ExtensionsOfPx \leftarrow \emptyset$;
5   **foreach** *itemset* $Py \in ExtensionsOfP$ such that $y \succ x$ **do**
6        $Pxy \leftarrow Px \cup Py$;
7        $Pxy.RCPPS\text{-}list \leftarrow$ Construct($Px.RCPPS\text{-}list, Py.RCPPS\text{-}list$);
8        $numCand(Pxy) \leftarrow |\{s|bond(Pxy, s) \geq minBond \wedge s \in D\}|$;
9        $upBondRa(Pxy) \leftarrow numCand(Pxy)/|D|$;
10   **if** $upBondRa(Pxy) \neq 0 \wedge upBondRa(Pxy) \geq minRa$ **then**
11 | $ExtensionsOfPx \leftarrow ExtensionsOfPx \cup Pxy$;
12   **end**
13   **end**
14 **end**
15 Find($Px, ExtensionsOfPx, D, maxSup, maxStd, minBond, minRa$);

To illustrate how the MRCPPS algorithm is applied, a detailed example is next presented for mining the RCPP in the sequence database of Table 3. Consider that the parameters are set as $minSup = 2$, $maxStd = 1$, $minBond = 0.6$, and $minRa = 0.6$. The following steps are performed:

1. The algorithm reads the database and creates the RCPPS-lists of each item. For example, the RCPPS-lists of itemsets $\{b\}$ and $\{e\}$ are shown in Fig. 3.
2. The $upBondRa$ value of each item is calculated. For the item $\{b\}$, it is found that $upBondRa(b) = 1$, and thus its extensions should be considered. The items $\{a\}$, $\{c\}$, $\{d\}$, and $\{e\}$ also satisfy this condition. All these items are thus put in the variable I^* and the $Find$ procedure is called with them.
3. The $Find$ procedure first checks item b. Because $ra(\{b\}) = 0.25 \geq minRa$, the itemset $\{b\}$ is output as a RCPP.
4. Then, extensions of $\{b\}$ are considered such as $\{b, e\}$. The RCPPS-list of itemset $\{b, e\}$ is built by applying the $Construct$ procedure with the RCPPS-lists of $\{b\}$ and $\{e\}$. During this process, the list of sequences where both items are periodic and have an $upBondRa$ value no less than $minRa$ are identified. The obtained RCPPS-list of $\{b, e\}$ is shown in Fig. 3. Because, $upBondRa(\{b, e\}) = 0.75$, the $Find$ method is invoked to next search for extensions of $\{b, e\}$.
5. The $Find$ procedure finds that $ra(\{b, e\}) = 0.75 \geq minRa$ and thus the itemset $\{b, e\}$ is output as a RCPP. The search then continues in a similar way to process other itemsets until all RCPP have been found.

5 Research Opportunities

There are several research opportunities related to discovering periodic patterns in multiple sequences. A reason is that researchers working on pattern mining have up to now mostly focused on finding periodic patterns in a single sequence. A first research opportunity is to design more efficient algorithms in terms of runtime, memory usage, and scalability. This can be done using novel data structures, search strategies, and optimizations, but also by developing parallel versions of algorithms that can run on multi-thread, multi-core, GPU, or big data platforms. A second research opportunity is to adapt the definitions of periodic pattern to find other types of periodic patterns in multiple sequences. This can be done by using different pattern selection functions such as measures of stability [27, 31] but also by considering other types of patterns besides itemsets such as sequential patterns [14, 19, 25, 44, 62, 67, 68], episodes [6, 32], subgraphs [16, 17, 35, 37], trajectory patterns [71], and periodic patterns with gap constraints [65, 66]. A third research opportunity is to explore novel applications of periodic patterns. As discrete sequences are found in many domains, many applications can be considered such as smart homes [48], location prediction [64], sequence prediction [34], clustering [10, 11, 45], and privacy-preserving data mining [53].

6 Conclusion

This chapter has presented an overview of two recent algorithms for discovering periodic patterns in a set of discrete sequences, named MPFPS and MRCPPS. The pseudocodes of the algorithms as well as detailed examples were presented. Moreover, research opportunities were also discussed.

The original implementations and Java source code of the depth-first and breadth-first search version of MPFPS and MRCPPS, as well as the datasets that have been used for evaluating these algorithms can be found at http://www.philippe-fournier-viger.com/spmf/, as part of the SPMF data mining library [24].

References

1. C.C. Aggarwal, J. Han (eds.), *Frequent Pattern Mining* (Springer, Berlin, 2014)
2. R. Agrawal, R. Srikant, Fast algorithms for mining association rules in large databases. Proceedings of 20th International Conference on Very Large Data Bases (1994), pp. 487–499
3. C.F. Ahmed, S.K. Tanbeer, B. Jeong, H. Choi, A framework for mining interesting high utility patterns with a strong frequency affinity. Inf. Sci. **181**(21), 4878–4894 (2011). https://doi.org/10.1016/j.ins.2011.05.012
4. K. Amphawan, P. Lenca, A. Surarerks, Mining top-*K* periodic-frequent pattern from transactional databases without support threshold. Proceedings of the Third International Conference on Advanced in Information Technology (2009), pp. 18–29
5. K. Amphawan, P. Lenca, A. Surarerks, Mining top-*K* periodic-frequent pattern from transactional databases without support threshold. Proceedings of the Third International Conference on Advances in Information Technology (2009), pp. 18–29
6. X. Ao, H. Shi, J. Wang, L. Zuo, H. Li, Q. He, Large-scale frequent episode mining from complex event sequences with hierarchies. ACM Transactions on Intelligent Systems and Technology (TIST) **10**(4), 1–26 (2019)
7. M. Barsky, S. Kim, T. Weninger, J. Han, Mining flipping correlations from large datasets with taxonomies. Proc. VLDB Endow. **5**, 370–381 (2011)
8. A. Bogarín, R. Cerezo, C. Romero, A survey on educational process mining. Wiley Interdisciplinary Reviews: Data Mining and Knowledge Discovery **8**(1), e1230 (2018)
9. Y. Chi, R.R. Muntz, S. Nijssen, J.N. Kok, Frequent subtree mining-an overview. Fundamenta Informaticae **66**(1–2), 161–198 (2005)
10. D.T. Dinh, V.N. Huynh, k-PbC: an improved cluster center initialization for categorical data clustering. Applied Intelligence **50**(8), 1–23 (2020)
11. D.T. Dinh, V.N. Huynh, S. Songsak, Clustering mixed numerical and categorical data with missing values. Information Sciences **571**, 418–442 (2021)
12. D.T. Dinh, B. Le, P. Fournier-Viger, V.N. Huynh, An efficient algorithm for mining periodic high-utility sequential patterns. Applied Intelligence **48**(12), 4694–4714 (2018)
13. T. Dinh, V.N. Huynh, B. Le, Mining periodic high utility sequential patterns. Proceeding of the 2017 International Conference on Intelligent Information and Database Systems (Springer, Berlin, 2017), pp. 545–555
14. T. Dinh, M.N. Quang, B. Le, A novel approach for hiding high utility sequential patterns. Proceedings of the 6th International Symposium on Information and Communication Technology (2015), pp. 121–128
15. Z. Feng, Y. Zhu, A survey on trajectory data mining: Techniques and applications. IEEE Access **4**, 2056–2067 (2016)

16. P. Fournier-Viger, C. Cheng, J.C.W. Lin, U. Yun, R.U. Kiran, Tkg: Efficient mining of top-k frequent subgraphs. Proceedings of the 7th International Conference on Big Data Analytics (Springer, Berlin, 2019), pp. 209–226
17. P. Fournier-Viger, G. He, C. Cheng, J. Li, M. Zhou, J.C.W. Lin, U. Yun, A survey of pattern mining in dynamic graphs. Wiley Interdisciplinary Reviews: Data Mining and Knowledge Discovery 10(6), e1372 (2020)
18. P. Fournier-Viger, J. Li, J.C.W. Lin, T.T. Chi, R.U. Kiran, Mining cost-effective patterns in event logs. Knowledge-Based Systems 191, 105–241 (2020)
19. P. Fournier-Viger, J. Li, J.C.W. Lin, T. Truong, Discovering low-cost high utility patterns. Data Science and Pattern Recognition 4(2), 50–64 (2020)
20. P. Fournier-Viger, Z. Li, J.C. Lin, R.U. Kiran, H. Fujita, Discovering periodic patterns common to multiple sequences. Proceedings of the 20th International Conference on Big Data Analytics and Knowledge Discovery (2018), pp. 231–246
21. P. Fournier-Viger, J.C. Lin, T. Dinh, H.B. Le, Mining correlated high-utility itemsets using the bond measure. Proceedings of the 11th International Conference on Hybrid Artificial Intelligent Systems (2016), pp. 53–65
22. Fournier-Viger, P., Lin, J.C., Vo, B., Truong, T.C., Zhang, J., Le, H.B.: A survey of itemset mining. Wiley Interdiscip. Rev. Data Min. Knowl. Discov. 7(4), e1207 (2017)
23. P. Fournier-Viger, J.C.W. Lin, Q.H. Duong, T.L. Dam, Phm: mining periodic high-utility itemsets. Proceedings of the 16th Industrial Conference on Data Mining (Springer, Berlin, 2016), pp. 64–79
24. P. Fournier-Viger, J.C.W. Lin, A. Gomariz, T. Gueniche, A. Soltani, Z. Deng, H.T. Lam, The spmf open-source data mining library version 2. Joint European Conference on Machine Learning and Knowledge Discovery in Databases (Springer, Berlin, 2016), pp. 36–40
25. P. Fournier-Viger, J.C.W. Lin, U.R. Kiran, Y.S. Koh, A survey of sequential pattern mining. Data Science and Pattern Recognition 1(1), 54–77 (2017)
26. P. Fournier-Viger, J.C.W. Lin, T. Truong-Chi, R. Nkambou, A survey of high utility itemset mining, in High-Utility Pattern Mining (Springer, Berlin, 2019), pp. 1–45
27. Fournier-Viger, P., Wang, Y., Yang, P., Lin, J.C.W., Unil, Y.: Tspin: Mining top-k stable periodic patterns. Appl. Intell. (2021)
28. P. Fournier-Viger, P. Yang, R.U. Kiran, S. Ventura, J.M. Luna, Mining local periodic patterns in a discrete sequence. Information Sciences 544, 519–548 (2021)
29. P. Fournier-Viger, P. Yang, Z. Li, J.C.W. Lin, R.U. Kiran, Discovering rare correlated periodic patterns in multiple sequences. Data Knowl. Eng. 126, 101–733 (2020)
30. P. Fournier-Viger, P. Yang, J.C.W. Lin, Q.H. Duong, T. Dam, L. Sevcik, D. Uhrin, M. Voznak, Discovering periodic itemsets using novel periodicity measures. Advances in Electrical and Electronic Engineering 17(1), 33–44 (2019)
31. P. Fournier-Viger, P. Yang, J.C.W. Lin, R.U. Kiran, Discovering stable periodic-frequent patterns in transactional data. Proceedings of the 32nd International Conference on Industrial, Engineering and Other Applications of Applied Intelligent Systems (Springer, Berlin, 2019), pp. 230–244
32. P. Fournier-Viger, Y. Yang, P. Yang, J.C.W. Lin, U. Yun, Tke: Mining top-k frequent episodes. Proceedings of the 33rd International Conference on Industrial, Engineering and Other Applications of Applied Intelligent Systems (Springer, Berlin, 2020)
33. L. Geng, H.J. Hamilton, Interestingness measures for data mining: A survey. ACM Comput. Surv. 38, 9 (2006)
34. T. Gueniche, P. Fournier-Viger, R. Raman, V.S. Tseng, Cpt+: Decreasing the time/space complexity of the compact prediction tree. Pacific-Asia Conference on Knowledge Discovery and Data Mining (Springer, Berlin, 2015), pp. 625–636
35. S. Halder, M. Samiullah, Y.K. Lee, Supergraph based periodic pattern mining in dynamic social networks. Expert Systems with Applications 72, 430–442 (2017)
36. S. Henzgen, E. Hüllermeier, Mining rank data. Proceedings of the 17th International Conference on Discovery Science (Springer, Berlin, 2014), pp. 123–134

37. C. Jiang, F. Coenen, M. Zito, A survey of frequent subgraph mining algorithms. Knowledge Engineering Review **28**, 75–105 (2013)
38. R.U. Kiran, M. Kitsuregawa, P.K. Reddy, Efficient discovery of periodic-frequent patterns in very large databases. Journal of Systems and Software **112**, 110–121 (2016). https://doi.org/10.1016/j.jss.2015.10.035
39. R.U. Kiran, P.K. Reddy, Mining rare periodic-frequent patterns using multiple minimum supports. Proceedings of the 15th International Conference on Management of Data (2009)
40. Y.S. Koh, S.D. Ravana, Unsupervised rare pattern mining: A survey. ACM Transactions on Knowledge Discovery **10**(4), 45:1-45:29 (2016)
41. Y.S. Koh, S.D. Ravana, Unsupervised rare pattern mining: a survey. ACM Transactions on Knowledge Discovery from Data (TKDD) **10**(4), 1–29 (2016)
42. Y.S. Koh, N. Rountree, Finding sporadic rules using apriori-inverse. Proceedings of the 9th Pacific-Asia Conference, PAKDD 2005 (Springer, Berlin, 2005), pp. 97–106
43. V. Kumar, V. Kumari, Incremental mining for regular frequent patterns in vertical format. International Journal of Engineering and Technology **5**(2), 1506–1511 (2013)
44. B. Le, U. Huynh, D.T. Dinh, A pure array structure and parallel strategy for high-utility sequential pattern mining. Expert Systems with Applications **104**, 107–120 (2018)
45. N.V. Lu, T.N. Vuong, D.T. Dinh, Combining correlation-based feature and machine learning for sensory evaluation of Saigon beer. International Journal of Knowledge and Systems Science (IJKSS) **11**(2), 71–85 (2020)
46. Y. Lu, F. Richter, T. Seidl, Efficient infrequent pattern mining using negative itemset tree. Complex Pattern Mining (Springer, Berlin, 2020), pp. 1–16
47. J.M. Luna, P. Fournier-Viger, S. Ventura, Frequent itemset mining: A 25 years review. Wiley Interdisciplinary Reviews: Data Mining and Knowledge Discovery **9**(6), e1329 (2019)
48. I. Mukhlash, D. Yuanda, M. Iqbal, Mining fuzzy time interval periodic patterns in smart home data. International Journal of Electrical and Computer Engineering **8**(5), 3374 (2018)
49. S. Nawaz, P. Fournier-Viger, A. Shojaee, H. Fujita, Using artificial intelligence techniques for covid-19 genome analysis. Applied Intelligence **51**(5), 3086–3103 (2021)
50. V.M. Nofong, Discovering productive periodic frequent patterns in transactional databases. Annals of Data Science **3**(3), 235–249 (2016)
51. V.M. Nofong, Fast and memory efficient mining of periodic frequent patterns. Proceedings of the 10th Asian Conference on Intelligent Information and Database Systems (Springer, Berlin, 2018), pp. 223–232
52. E. Omiecinski, Alternative interest measures for mining associations in databases. IEEE Trans. Knowl. Data Eng. **15**(1), 57–69 (2003). https://doi.org/10.1109/TKDE.2003.1161582
53. M.N. Quang, T. Dinh, U. Huynh, B. Le, MHHUSP: An integrated algorithm for mining and Hiding High Utility Sequential Patterns. Proceedings of the 8th International Conference on Knowledge and Systems Engineering (2016), pp. 13–18
54. M.M. Rashid, I. Gondal, J. Kamruzzaman, Regularly frequent patterns mining from sensor data stream. Proceedings of the 20th International Conference on Neural Information Processing (Springer, Berlin, 2013), pp. 417–424
55. M.M. Rashid, M.R. Karim, B.S. Jeong, H.J. Choi, Efficient mining regularly frequent patterns in transactional databases. Proceedings of the 17th International Conference on Database Systems for Advanced Applications (Springer, Berlin, 2012), pp. 258–271
56. S. Shekhar, M.R. Evans, J.M. Kang, P. Mohan, Identifying patterns in spatial information: A survey of methods. Wiley Interdisciplinary Reviews: Data Mining and Knowledge Discovery **1**(3), 193–214 (2011)
57. A. Soulet, C. Raïssi, M. Plantevit, B. Crémilleux, Mining dominant patterns in the sky. Proceeding of the 11th IEEE International Conference on Data Mining (IEEE, New York, 2011), pp. 655–664
58. A. Surana, R.U. Kiran, P.K. Reddy, An efficient approach to mine periodic-frequent patterns in transactional databases. Proceedings of the 15th Pacific-Asia Conference on Knowledge Discovery and Data Mining (2011), pp. 254–266. https://doi.org/10.1007/978-3-642-28320-8_22

59. L. Szathmary, A. Napoli, P. Valtchev, Towards rare itemset mining. 19th IEEE International Conference on Tools with Artificial Intelligence, vol. 1 (IEEE, New York, 2007), pp. 305–312
60. L. Szathmary, P. Valtchev, A. Napoli, R. Godin, Efficient vertical mining of minimal rare itemsets. Proceedings of the 9th International Conference on Concept Lattices and Their Applications (2012), pp. 269–280
61. S.K. Tanbeer, C.F. Ahmed, B. Jeong, Y. Lee, Discovering periodic-frequent patterns in transactional databases. Proceedings of the 13th Pacific-Asia Conference on Knowledge Discovery and Data Mining (2009), pp. 242–253. https://doi.org/10.1007/978-3-642-01307-2_24
62. T. Truong, A. Tran, H. Duong, B. Le, P. Fournier-Viger, Ehusm: Mining high utility sequences with a pessimistic utility model. Data Science and Pattern Recognition 4(2), 65–83 (2020)
63. Venkatesh, J.N., Kiran, R.U., Reddy, P.K., Kitsuregawa, M.: Discovering periodic-frequent patterns in transactional databases using all-confidence and periodic-all-confidence. In: Proc. 27th International Conference on Database and Expert Systems Applications Part I, pp. 55–70 (2016). https://doi.org/10.1007/978-3-319-44403-1_4
64. M.H. Wong, V.S. Tseng, J.C. Tseng, S.W. Liu, C.H. Tsai, Long-term user location prediction using deep learning and periodic pattern mining. Proceedings of the 12th Conference on Advanced Data Mining and Applications (Springer, Berlin, 2017), pp. 582–594
65. Y. Wu, C. Shen, H. Jiang, X. Wu, Strict pattern matching under non-overlapping condition. Science China Information Sciences 60(1), 1–16 (2017)
66. Y. Wu, Y. Tong, X. Zhu, X. Wu, Nosep: Nonoverlapping sequence pattern mining with gap constraints. IEEE transactions on cybernetics 48(10), 2809–2822 (2017)
67. Y. Wu, L. Wang, J. Ren, W. Ding, X. Wu, Mining sequential patterns with periodic wildcard gaps. Applied intelligence 41(1), 99–116 (2014)
68. Y. Wu, C. Zhu, Y. Li, L. Guo, X. Wu, Netncsp: Nonoverlapping closed sequential pattern mining. Knowledge-based systems 196, 105–812 (2020)
69. N.B. Younes, T. Hamrouni, S.B. Yahia, Bridging conjunctive and disjunctive search spaces for mining a new concise and exact representation of correlated patterns. Proceedings of the 13th International Conference on Discovery Science (2010), pp. 189–204. https://doi.org/10.1007/978-3-642-16184-1_14
70. M.J. Zaki, Efficiently mining frequent trees in a forest: Algorithms and applications. IEEE transactions on knowledge and data engineering 17(8), 1021–1035 (2005)
71. D. Zhang, K. Lee, I. Lee, Hierarchical trajectory clustering for spatio-temporal periodic pattern mining. Expert Systems with Applications 92, 1–11 (2018)

Discovering Self-reliant Periodic Frequent Patterns

Vincent Mwintieru Nofong, Hamidu Abdel-Fatao, Michael Kofi Afriyie, and John Wondoh

Abstract Periodic frequent pattern discovery is a non-trivial task for analysing databases to reveal the recurring shapes of patterns' occurrences. Though significant strides have been made in their discovery for understanding large databases in decision-making, existing techniques still face a challenge of reporting a large number of periodic frequent patterns, most of which are often not useful as their periodic occurrences are either by random chance or can be inferred from the periodicities of other periodic frequent patterns. Reporting such periodic frequent patterns not only degrades the performance of existing algorithms but also could adversely affect decision-making. This study addresses these issues by proposing a novel algorithm named SRPFPM (Self-Reliant Periodic Frequent Pattern Miner) for mining and reporting the set of self-reliant periodic frequent patterns as those whose periodic occurrences have inherent item relationships and cannot be inferred from other periodic frequent patterns. Experimental analysis on benchmark datasets show that SRPFPM is efficient and effectively prunes periodic frequent patterns that are periodic due to random chance as well as those whose periodicities can be inferred from other periodic frequent patterns.

V. M. Nofong (✉) · H. Abdel-Fatao · M. K. Afriyie
University of Mines and Technology, P. O. Box 273, Tarkwa, Ghana
e-mail: vnofong@umat.edu.gh

H. Abdel-Fatao
e-mail: habdel-fatao@umat.edu.gh

M. K. Afriyie
e-mail: michaelkofiafriyie@yahoo.com

J. Wondoh
University of South Australia, Mawsonlakes Campus, Adelaide, SA, Australia
e-mail: John.Wondoh@unisa.edu.au

© The Author(s), under exclusive license to Springer Nature Singapore Pte Ltd. 2021   105
R. Uday Kiran et al. (eds.), *Periodic Pattern Mining*,
https://doi.org/10.1007/978-981-16-3964-7_7

1   Introduction

Frequent itemset (pattern) mining [3, 9, 27, 35, 37] is a fundamental data mining task (with a wide range of applications) that has been widely studied over the past years. The goal in frequent pattern mining is to identify all patterns that occur frequently in a given database. For any given database, a pattern is said to be frequent if its frequency of occurrence within the database is not less than a user-specified threshold. Over the past years, several techniques and approaches have been proposed for mining various categories of frequent patterns for domain-specific decision-making. Typical of such techniques include works that employ the: a priori candidate generation approach [3, 36]; vertical representation approach [30, 35, 37], frequent pattern growth approach [9, 27] and hierarchical approach [33].

Although traditional frequent pattern mining techniques are useful in revealing frequently occurring patterns in databases, they are incapable of revealing the occurrence shapes of patterns in databases. For instance, in market basket analysis, though frequent pattern mining algorithms will be able to reveal the frequent customer transactions, they will not be able to detect and report the periodic customer transactions. That is, they will fail to report the set of customer transactions which occur periodically. This drawback in frequent pattern mining algorithms is as a result of the frequency measure being the sole criteria in identifying interesting patterns in databases. Consequently, this make frequent pattern mining algorithms inapplicable in situations where the periodic occurrence of patterns in databases is relevant in decision-making. For example, in cases such as analysing the behaviour of website users or the purchase behaviour of customers, where the periodic occurrence of website visits or customer purchases are important, traditional frequent pattern mining algorithms will not be applicable.

To address this drawback in traditional frequent pattern mining and enable detect the periodic occurrences of patterns in databases for decision-making, periodic frequent pattern mining [6, 13, 20, 24, 28, 31, 32] emerged. Periodic frequent pattern mining is a non-trivial task for analysing databases to reveal recurring shapes of patterns' occurrences. The main objective in periodic frequent pattern mining is to find and report frequent patterns that occur periodically in databases. For any given database, a frequent pattern is periodic if its occurrence interval within the database is not more than a given periodicity threshold. Over the past years, several techniques and approaches have been proposed for mining categories of periodic frequent patterns in works such as [2, 4, 6–8, 12, 15–18, 20, 24, 32, 34].

Though periodic frequent pattern mining is essential in several applications, it is faced with some challenges. For example, existing periodic frequent pattern mining algorithms which employ the periodicity measure proposed in [32], in noisy databases will often report the noised maximal period of a pattern as its regular period since the periodicity measure in [32] is susceptible to noise. Further more, existing periodic frequent pattern mining algorithms often report a large number of periodic frequent patterns, most of which are periodic due to random chance or their periodicities can be inferred from other periodic frequent patterns. Reporting such periodic frequent patterns not only degrades the performance of existing periodic frequent pattern mining algorithms but also could adversely affect decision-making.

To address these issues, this study proposes a new framework for mining the set of self-reliant periodic frequent patterns (that is, periodic frequent patterns whose periodicities are not due to random chance and cannot be inferred from other periodic frequent patterns). To our best knowledge, this topic has not been explored so far. The contributions of this work are summarized as follows:

- We propose the self-reliant periodic frequent patterns as the set of periodic frequent patterns whose periodicities have inherent item relationships and cannot be inferred from other periodic frequent patterns.
- We further propose and develop a novel algorithm named SRPFPM (Self-Reliant Periodic Frequent Pattern Miner) for mining the complete set of self-reliant periodic frequent patterns from transaction databases.
- We perform an extensive experimental study on several real datasets to evaluate the performance of SRPFPM. Experimental results show that SRPFPM is efficient and effectively prunes periodic frequent patterns that are periodic due to random chance, as well as those whose periodic occurrences can be inferred.

The rest of this paper is organized as follows. Section 2 introduces the related concepts, definitions and related works. Section 3 presents the proposed definitions and the problem statement. Section 4 presents the proposed approach to mining the set of self-reliant periodic frequent patterns. Section 5 presents the experimental analysis while Sect. 6 presents our conclusions.

2 Background

2.1 Preliminaries

The problem of frequent itemset mining is as follow. Let $I = \{i_1, i_2,..., i_m\}$ be a set of literals, called items. A set $X_1 = \{i_a, \ldots, i_n\} \subseteq I$, where $a \leq n$ and $a, n \in [1, m]$, is called a pattern (or an itemset). A transaction database is a set of transactions $D = \{T_1, T_2, T_3, \ldots, T_k\}$ such that for each transaction T_a, $T_a \in I$ and T_a has a unique identifier a called its transaction ID (TID).

Example 1 Consider the transaction database in Table 1 (a sample customer transaction database - which will be used the running example), the set of items for this database is $I = \{a, b, c, d, e, f\}$. Transaction T_2 in Table 1 which has a transaction ID of 2 and three items $\{d, e, f\}$ is a length-3 itemset.

The *coverset* of an itemset, S in a database, D, denoted as $cov(S)$ is defined as $cov(S) = \{m | m \in D \land S \subseteq m\}$.

Example 2 In Table 1, given, $S = \{a, b\}$, then $cov(S) = \{1, 3, 5\}$ since $\{a, b\}$ appears in transactions 1, 3 and 5.

Table 1 Sample customer transactions

TID	Transaction	TID	Transaction	TID	Transaction	TID	Transaction
T_1	$\{a, b, c, e\}$	T_3	$\{a, b, c, d\}$	T_5	$\{a, b, c, e, f\}$	T_7	$\{c, d\}$
T_2	$\{d, e, f\}$	T_4	$\{c, d, e, f\}$	T_6	$\{a, d, e, f\}$	T_8	$\{e, f\}$

The *support count* of S in D is defined as $|cov(S)|$ and the *support* of S in D, denoted as $sup(S)$ is defined as

$$sup(S) = \frac{|cov(S)|}{|D|} \tag{1}$$

Example 3 In Table 1, given $S = \{a, b\}$, then $sup(S) = \frac{3}{8} = 0.375$ as $|cov(S)| = |\{1, 3, 5\}| = 3$ and $|D| = 8$.

Definition 1 (*Frequent itemset mining*). The problem of frequent itemset mining consists of discovering frequent itemsets [3]. An itemset S is a frequent itemset in a database D if its support, $sup(S)$, is not less than a user-specified minimum support threshold, *minsup*.

Example 4 Considering a *minsup* threshold of 0.5 on Table 1, the set of frequent itemsets and their respective supports in Table 1 will be $\{a\} : 0.5, \{c\} : 0.625, \{d\} : 0.625, \{e\} : 0.75, \{f\} : 0.625$, and $\{e, f\} : 0.625$.

To mine frequent patterns from databases, several algorithms have been proposed over the past years which employ various techniques (such as apriori candidate generation approach [3, 36]; vertical representation approach [30, 35, 37], frequent pattern growth approach [9, 27] and hierarchical approach [33]) in identifying frequent patterns. Notwithstanding the several propositions for mining frequently occurring patterns in databases for decision-making, as mentioned previously, these propositions are incapable of detecting and revealing the occurrence shapes of patterns in databases.

Inspired by the inability of frequent pattern mining algorithms in detecting and revealing the periodic shapes of patterns in databases, and the numerous applications of periodic frequent patterns in various domain-specific decision-making, several algorithms have been proposed to discover periodic frequent patterns in transaction databases in works such as [1, 2, 6, 10, 11, 16, 19, 23, 24, 28, 32, 38].

The concepts employed in periodic frequent pattern mining are presented as follows.

Definition 2 (*Consecutive transactions of an itemset*). Let $D = \{T_1, T_2, T_3, \ldots, T_u\}$ be a database with u transactions. Let the set of transactions in D containing an itemset S be denoted as $g(S) = \{T_{g_1}, T_{g_2}, T_{g_3}, \ldots, T_{g_{n-1}}, T_{g_n}\}$, where, $1 \leq g_1 < g_2 < \ldots g_{n-1} < g_n \leq u$. Two transactions $T_x \supset S$ and $T_y \supset S$ are said to be consecutive with respect to S if there does not exist a transaction $T_w \in g(S)$ such that $x < w < y$. The period of two consecutive transactions T_x and T_y in $g(S)$ is defined as $p(T_x, T_y) = (y - x)$, that is the number of transactions between T_x and T_y.

Example 5 Consider the itemset $\{c\}$ in Table 1 which appears in transactions T_1, T_3, T_4, T_5 and T_7. Transactions, T_1 and T_3, or T_3 and T_4, or T_4 and T_5 or T_5 and T_7 are its consecutive transactions. The period between consecutive transactions T_1 and T_3 thus become $p(T_1, T_3) = 3 - 1 = 2$.

Definition 3 (*Set of all periods of an itemset*). Let the set of transactions in a database D (with u transactions) containing an itemset S be denoted as $g(S) = \{T_{g_1}, T_{g_2}, T_{g_3}, \ldots, T_{g_{n-1}}, T_{g_n}\}$, such that $1 \leq g_1 < g_2 < \ldots g_{n-1} < g_n \leq u$. The coverset of S in D become, $cov(S) = \{g_1, g_2, g_3, \ldots, g_{n-1}, g_n\}$. The set of all periods of S in D denoted as $ps(S)$ is defined as $ps(S) = \{g_1 - g_0, g_2 - g_1, g_3 - g_2, \ldots, g_n - g_{n-1}, |D| - g_n\}$, where $g_0 = 0$ is a constant and $|D|$ is the size of the database.

Example 6 Consider the itemset $\{a, b\}$ in Table 1 (where $|D| = 8$) which appears in transactions T_1, T_3 and T_5. The coverset of $\{a, b\}$ thus become, $cov\{a, b\} = \{1, 3, 5\}$, hence, the set of all periods of $\{a, b\}$ in D based on Definition 3 will be evaluated as $ps(\{a, b\}) = \{1 - 0, 3 - 1, 5 - 3, 8 - 5\} = \{1, 2, 2, 3\}$.

Given the periods of an itemset, various definitions have been proposed for evaluating the periodicity of patterns and for mining periodic frequent patterns in transaction databases. These definitions and propositions are presented in Sect. 2.2.

2.2 Related Work

Given a transaction database D, a pattern S and its set of periods $ps(S)$ in D, Tanbeer et al. [32] introduced the concept to periodic frequent pattern mining by defining the periodicity of a pattern follows.

Definition 4 (*Periodicity of pattern* S [32]) For any given pattern S and its set of periods, $ps(S)$, its periodicity is defined as $Per(S) = max(ps(S))$.

Example 7 Considering the itemset $\{a, b\}$ in Table 1 having $ps(\{a, b\}) = \{1, 2, 2, 3\}$ as its set of periods, the periodicity of $\{a, b\}$ based on Definition 4 will be evaluated as $max(\{1, 2, 2, 3\}) = 3$.

Based on the proposed periodicity measure in Definition 4, Tanbeer et al. [32] defined a periodic frequent pattern as a frequent pattern whose periodicity is not greater than a user-given maximum periodicity threshold (*maxPer*).

Though the proposed periodicity measure and the defined periodic frequent patterns introduced in Tanbeer et al. [32] have been used mining periodic frequent patterns in transaction databases in works such as [12, 14, 31], Rashid et al. [28] argued that employing the periodicity measure proposed in [32] in mining periodic frequent pattern is inappropriate since it returns the maximum period for which a pattern does not appear in a database as its periodicity. Rashid et al. [28] further showed that, in noisy databases, the noised maximal interval will always be reported as the periodic interval if the periodicity measure proposed in [32] is employed in mining periodic frequent patterns.

To address this issue of reporting the possible "noised" maximal interval for which a pattern does not appear in the database as its periodicity, Rashid et al. [28] define the periodicity of a pattern under the name *patterns' regularity* as follows.

Definition 5 ((*Regularity of pattern* S [28]) For any given pattern S and its set of periods, $ps(S)$, its regularity $Reg(S)$, is defined as $Reg(S) = var(ps(S))$, where $var(ps(S))$ is the variance of $ps(S)$.

With the regularity (periodicity) measure in Definition 5, Rashid et al. [28] define a regular (periodic) frequent pattern as a frequent pattern whose regularity is not greater than a user-given maximum regularity threshold (*maxReg*). This concept proposed by Rashid et al. [28] has also been used in works such as [21, 29].

Nofong [24] with the aim of mining the set of periodic frequent patterns with similar periodic occurrences for decision-making, defined the periodicity of a pattern as follows.

Definition 6 ((*Periodicity of pattern* S [24]) Given a database D, a pattern S and its set of periods $ps(S)$ in D, the periodicity of S, $Prd(S)$, is defined as $Prd(S) = \bar{x}(ps(S))$, where $\bar{x}(ps(S))$ is the mean of $ps(S)$.

With the proposed periodicity measure in Definition 6, Nofong [24] define a periodic frequent pattern as a frequent pattern whose periodicity plus and minus the standard deviation (among the set of periods) are within the range $(p - p_1)$ to $(p + p_1)$, where p and p_1 are the user-given periodicity threshold and difference factor, respectively. This proposition have been used in mining periodic frequent patterns having similar periodicities in works such as [1, 2, 25, 26].

Fournier-Viger et al. [6] also introduced PFPM, an efficient algorithm having novel pruning techniques for discovering periodic frequent patterns in transaction databases. Unlike the techniques proposed in [24, 28, 32], PFPM discovers periodic frequent patterns using three measures (that is, the *minimum, maximum* and *average* periodicity measures) without requiring the minimum support threshold.

Notwithstanding the above propositions and works based on these propositions for mining categories of periodic frequent patterns for decision-making, existing periodic frequent pattern mining algorithms often still report a large number of periodic frequent patterns, most of which are often due to random chance or whose periodic occurrences can be inferred from other periodic frequent patterns. Reporting such periodic frequent patterns not only degrades the performance of existing algorithms, they could adversely affect decision-making if they happen to be false positively periodic.

This study addresses this issues by proposing the self-reliant periodic frequent patterns as the set of periodic frequent patterns whose periodic occurrences are not due to random chance and at the same time cannot be inferred from other periodic frequent patterns.

3   Definitions and Problem Statement

3.1   *Periodic Frequent Patterns*

This study adopts the periodic frequent pattern definition proposed in [6] as follows.

Definition 7 (*Periodic frequent pattern with novel periodicity measures* [6]) Let *minsup, minAvg, maxAvg, minPer* and *maxPer* be positive numbers, provided by the user. A pattern S in D is a periodic frequent pattern if and only if $sup(S) \geq minsup$, $minAvg \leq avgper(S) \leq maxAvg$, $minper(S) \geq minPer$ and $maxper(S) \leq maxPer$.

where $avgper(S), minper(S)$ and $maxper(S)$, respectively, refer to the average period, minimum period and maximum period in $ps(S)$

For any given dataset, Definition 7 will return the set of all periodic frequent patterns satisfying the user-given thresholds.

Example 8 Given $minsup = 0.25$, $minPer = 1$, $maxPer = 3$, $minAvg = 1$ and $maxAvg = 2$, and the database in Table 1, nine (9) periodic frequent patterns will be mined and returned as shown in Table 2.

For the given thresholds, the number of reported periodic frequent patterns (that is, nine) is more than the number of transactions in the database (that is, eight). Additionally, some of these reported periodic frequent patterns contain redundant information and their periodic occurrences can be inferred from other periodic frequent patterns. Typical examples of such periodic frequent patterns in Table 2 whose periodic occurrences can be inferred from other periodic frequent patterns are $\{a, b\}$, $\{b, c\}$ and $\{a, b, c\}$. Aside these periodic frequent patterns whose periodic occurrences can be inferred, some might be periodic due to random chance of occurrence without inherent item relationships.

Table 2 The set of periodic frequent patterns for the running example

PFP	sup(PFP)	Set of Periods	minper(PFP)	maxper(PFP)	avper(PFP)
$\{a\}$	0.5	$\{1, 2, 2, 1, 2\}$	1	2	1.60
$\{b\}$	0.375	$\{1, 2, 2, 3\}$	1	3	2.00
$\{c\}$	0.625	$\{1, 2, 1, 1, 2, 1\}$	1	2	1.33
$\{d\}$	0.625	$\{2, 1, 1, 2, 1, 1\}$	1	2	1.33
$\{a, b\}$	0.375	$\{1, 2, 2, 3\}$	1	3	2.00
$\{a, c\}$	0.375	$\{1, 2, 2, 3\}$	1	3	2.00
$\{b, c\}$	0.375	$\{1, 2, 2, 3\}$	1	3	2.00
$\{c, d\}$	0.375	$\{3, 1, 3, 1\}$	1	3	2.00
$\{a, b, c\}$	0.375	$\{1, 2, 2, 3\}$	1	3	2.00

To enable identify the set of periodic frequent patterns that have inherent item-relationships (that is, are not due to random chance) in decision-making, Nofong [24] proposed the set of productive periodic frequent patterns as follows.

Definition 8 (*Productive periodic frequent pattern* [24]) A periodic frequent pattern, S in D, is a productive periodic frequent pattern if and only if, for all S_1, S_2 such that, $S_1 \subset S, S_2 \subset S, S_1 \cup S_2 = S$ and $S_1 \cap S_2 = \phi$, then, $sup(S) > sup(S_1) \times sup(S_2)$.

The condition (for all S_1, S_2 such that $S_1 \subset S, S_2 \subset S, S_1 \cup S_2 = S$ and $S_1 \cap S_2 = \phi$, then, $sup(S) > sup(S_1) \times sup(S_2)$) specifies that the frequency of the periodic frequent pattern must be greater than that which would be expected under any assumption of independence between any partition of the pattern S into two independent itemsets. That is, a periodic frequent pattern is productive if and only if every rule that can be formed from it is productive.

Example 9 Applying the productiveness criterion in mining the set of productive periodic frequent patterns with same thresholds (that is, $minsup = 0.25, minPer = 1, maxPer = 3, minAvg = 1$, and $maxAvg = 2$) on the database in Table 1, eight (8) productive periodic frequent patterns will be reported as $\{a\}, \{b\}, \{c\}, \{d\}, \{a, b\}, \{a, c\}, \{b, c\}$, and $\{a, b, c\}$ with same properties as shown in Table 2. The periodic frequent pattern $\{c, d\}$ reported in Table 2 will not be reported since it does not satisfy the productiveness criterion in Definition 8 and hence will be pruned during the discovery process.

Afriyie et al. [1, 2] with the aim of identifying periodic frequent patterns whose periodic occurrences cannot be inferred from other periodic frequent patterns and additionally are more preferable in decision-making employed the concept of frequent generators [22] in introducing the non-redundant periodic frequent patterns as follows.

Definition 9 (*Non-redundant periodic frequent pattern* [1]) Given a database D and the set of periodic frequent patterns from D as $Per_D = \{S_1, S_2, \ldots S_j\}$, a periodic frequent pattern, S_n, is a non-redundant periodic frequent pattern if $\nexists S_u \in Per_D$ such that, $S_u \subset S_n$ and, $sup(S_n) = sup(S_u)$.

The defined non-redundant periodic frequent patterns in Definition 9 based on frequent generators differ from close periodic frequent patterns in the sense that, all proper subsets of the non-redundant periodic frequent patterns will be non-redundant unlike closed periodic frequent patterns (which may have some proper subsets not closed). Additionally, unlike closed periodic frequent patterns, non-redundant periodic frequent patterns will be more preferable since some closed periodic frequent patterns might be contain redundant information. Example 10 illustrates why non-redundant periodic frequent patterns will be preferred in decision-making over closed periodic frequent patterns.

Example 10 Assuming {*male, prostate cancer, old age, obese*} is a closed periodic frequent pattern, it is more preferable to keep its non-redundant periodic subsets (generators) such as {*prostate cancer, obese, old age*} for decision-making over {*male, prostate cancer, old age, obese*} since {*male*} subsumes {*prostate cancer*} - that is, only males can have prostate cancer. This subsumption thus make the closed periodic frequent pattern {*male, prostate cancer, old age, obese*} uninteresting since the information about being a male when you have prostate cancer is redundant - hence the need to keep the minimal information contained in the non-redundant subset (generator) {*prostate cancer, obese, old age*}.

Example 11 Applying Definition 9 in mining the set of non-redundant periodic frequent patterns with same thresholds (that is, $minsup = 0.25$, $minPer = 1$, $maxPer = 3$, $minAvg = 1$, and $maxAvg = 2$) on the database in Table 1, six (6) non-redundant periodic frequent patterns will be reported as $\{a\}$, $\{b\}$, $\{c\}$, $\{d\}$, $\{a, c\}$, and $\{c, d\}$ with same properties as shown in Table 2. The periodic frequent patterns $\{a, b\}$, $\{b, c\}$ and $\{a, b, c\}$ that are reported in Table 2 will pruned since their periodic occurrence can be inferred from other periodic frequent patterns.

As can be observed from the results based on Definitions 7, 8 and 9, Definition 7 will report a set of periodic frequent patterns that contain those without inherent item-relationships as well as those whose periodic occurrences can be inferred from other periodic frequent patterns. Though Definition 8 will report the set of periodic frequent patterns not due to random chance, they might report periodic frequent patterns that contain redundant information. For example, $\{a, b\}$, $\{b, c\}$ and $\{a, b, c\}$ which are reported in Definition 8 are identified as containing redundant information in Definition 9. Also, though Definition 9 will report the set of periodic frequent patterns whose periodic occurrences cannot be inferred from other periodic frequent patterns, it might report those that are periodic due to random chance. For instance, $\{c, d\}$ that is reported in Definition 9 is identified in Definition 8 as non-productive.

To addresses this challenge and ensure periodic frequent patterns that have inherent item relationships and whose periodic occurrences cannot be inferred from other periodic frequent patterns are mined for decision-making, we propose and introduce the self-reliant periodic frequent patterns as the set of periodic frequent patterns that are periodic due to inherent item relationship and whose periodic occurrences cannot be inferred from other periodic frequent patterns.

3.2 Self-reliant Periodic Frequent Patterns

To enable identify the set of periodic frequent patterns whose periodic occurrences are due to inherent item-relationship and at the same time cannot be inferred from other periodic frequent patterns, we combine the concepts of productive periodic frequent patterns ([24]) and the non-redundant periodic frequent patterns ([1, 2]) and define the our proposed self-reliant periodic frequent patterns as follows.

Definition 10 (*Self-reliant periodic frequent pattern*) Given a database D and the set of periodic frequent patterns from D as $Per_D = \{S_1, S_2, \ldots S_j\}$, a periodic frequent pattern, S_n, is a self-reliant periodic frequent pattern if and only if: $\nexists S_u \in Per_D | S_u \subset S_n \wedge sup(S_n) = sup(S_u)$, and, for all S_1, S_2 such that, $S_1 \subset S_n, S_2 \subset S_n, S_1 \cup S_2 = S_n \wedge S_1 \cap S_2 = \phi$, then, $sup(S_n) > sup(S_1) \times sup(S_2)$.

Definition 10 basically states that for any periodic frequent pattern to be self-reliant, it must be both productive and non-redundant among the set of discovered periodic frequent patterns. That is, its periodic occurrence in the given database should not be easily inferred from other periodic frequent patterns and the items within it must have inherent item relationship.

Example 12 Applying Definition 10 in mining the set of self-reliant periodic frequent patterns with, $minsup = 0.25$, $minPer = 1, maxPer = 3, minAvg = 1$, and $maxAvg = 2$ on the database in Table 1, five (5) self-reliant periodic frequent patterns will be reported as $\{a\}, \{b\}, \{c\}, \{d\}$, and $\{a, c\}$ (having same properties as shown in Table 2). The periodic frequent patterns $\{a, b\}, \{b, c\}, \{c, d\}$ and $\{a, b, c\}$ that are reported in Table 2 will pruned since they are either non-productive or redundant.

With the self-reliant periodic frequent patterns defined in Definition 10, the problem statement can be described as follows

Definition 11 (***Problem definition***) Given a transaction database D, let minsup, minAvg, maxAvg, minPer and maxPer be positive numbers, provided by the user, mine and report the set of self-reliant periodic frequent patterns from D where for any reported periodic frequent pattern S in D, $sup(S) \geq minsup$, $minAvg \leq avgper(S) \leq maxAvg$, $minper(S) \geq minPer$ and $maxper(S) \leq maxPer$.

4 Mining Self-reliant Periodic Frequent Patterns

To identify the self-reliant periodic frequent patterns among all periodic frequent patterns in a databases, we propose a Self-RelianceTest() function which is employed during the periodic frequent pattern mining process in pruning those that are not self-reliant. The following subsections details the proposed self-reliance test function and the algorithms employed in mining the set of self-reliant periodic frequent patterns.

4.1 Self-reliance Test

For a given a periodic frequent pattern S mined from a transaction database D, we use the Self-RelianceTest() function to check if S is a self-reliant periodic frequent pattern or not. The Self-RelianceTest() function takes as input the periodic frequent pattern (S) to be tested and the set of productive periodic frequent patterns ($pPer_D$). The self-reliance of a periodic frequent pattern S is then tested as follows.

Function Self-RelianceTest(S, $pPer_D$)

Input: Periodic Frequent Pattern S, HashMap of Productive PFPs, $pPer_D$
Output: S.class as either Self-Reliant or not Self-Reliant
1 Create S.class $= null$
2 **if** S is a length-1 pattern **then**
3   **if** $sup(S) = sup(\emptyset)$ **then**
4   S.class = Not Self-Reliant
5   **else**
6   S.class = Self-Reliant; Add S to $pPer_D$
7 **else if** S is a length-2 pattern **then**
8   Let S_1 be the first item in S and S_2 the second item in S
9   **if** $sup(S) > sup(S_1) \times sup(S_2) \wedge sup(S_1) \neq sup(S) \wedge sup(S_2) \neq sup(S)$ **then**
10   S.class = Self-Reliant
11   Add S to $pPer_D$ /* Add S to set of productive PFPs */
12   **else**
13   S.class = Not Self-Reliant
14 **else**
15   Generate powerset of S as Γ, remove S and the empty set ($\emptyset$) from Γ
16   ProductiveTest = 0
17   **for** *each* subset S_i in Γ **do**
18   **for** *each* subset S_j in Γ **do**
19   **if** $S_i = S_j$ **then**
20   continue
21   **else**
22   **if** $S_i \cap S_j = \emptyset \wedge S_i \cup S_j = S \wedge sup(S) > sup(S_i) \times sup(S_j)$ **then**
23   continue
24   **else**
25   ProductiveTest+=1
26   **if** *ProductiveTest=0* **then**
27   Add S to $pPer_D$ /* Add S to set of productive PFPs */
28   RedundanceTest=0
29   **for** *each* subset S_k in Γ **do**
30   **if** $sup(S) = sup(S_k)$ **then**
31   RedundanceTest+=1
32   break
33   **else**
34   continue
35   **if** $RedundanceTest = 0 \wedge pPer_D$ contains S **then**
36   S.class = Self-Reliant; Add S to $pPer_D$
37   **else**
38   S.class = Not Self-Reliant
39 **return** S.class

Given a periodic frequent pattern S and the set which contains all productive periodic frequent patterns from D (note that this set can be empty), Line 1 of the Self-RelianceTest() function creates the class of S as null. If S is a length-1 pattern, Line 6 assigns its class as *Self-Reliant* provided the support of S is not equal to that of an empty set (an empty set occurs with a support of 1.0 in every database). S is then added to the set of productive periodic frequent patterns ($pPer_D$) in Line 6. If the support of S is equal to that of the empty set, its class is assigned as *Not Self-Reliant* in Line 4. This support comparison is to test if the pattern is non-redundant since all length-1 patterns are automatically productive. As such all length-1 periodic frequent patterns that are non-redundant will be classified as self-reliant since they are automatically productive.

Example 13 Using our running example in Table 1, all the length-1 periodic frequent patterns as shown in Table 2, that is, $\{a\}, \{b\}, \{c\}, \{d\}$) will be classified as self-reliant and will all be added to $pPer_D$.

If the periodic frequent pattern S is a length-2 pattern, Line 8 assigns S_1 and S_2 as the first and second items in S respectively. Line 9 in the Self-RelianceTest() function simultaneously tests if S is productive and non-redundant. If S is both productive and non-redundant, its class is assigned as *Self-Reliant* in Line 10 and S added to the set of all productive periodic frequent patterns $pPer_D$, otherwise the class of S is assigned as *Not Self-Reliant* in Line 13.

Example 14 For our running example, given $S = \{a, b\}$ from Table 2, Line 8 will assign $S_1 = \{a\}$ and $S_2 = \{b\}$. S is then tested for self-reliance in Line 9. Though $sup(\{a, b\}) = 0.375$ is greater than $sup(\{a\}) \times sup(\{b\})$ (that is, 0.188) since the support of S is same as that of S_2, S will be classified as *Not Self-Reliant* in Line 13. For the case of $S = \{a, c\}$ in our running example, Line 8 will assign $S_1 = \{a\}$ and $S_2 = \{c\}$. S will then tested for self-reliance in Line 9. Since $sup(\{a, c\}) = 0.375$ is greater than $sup(\{a\}) \times sup(\{c\})$ (that is, 0.313) and the support of S is not same as that of S_1 or S_2, S will be classified as *Self-Reliant* in Line 10 and added to $pPer_D$ in Line 11.

If the periodic frequent pattern S has a length of more than 2 (that is, length-3 and above), Lines 14 to 38 of the Self-RelianceTest() function tests for its self-reliance as follows. Line 15 generates the powerset of S as Γ and subsequently removes S and the empty set in Γ - this leaves only the proper subsets of S in Γ.

Example 15 For our running example, given $S = \{a, b, c\}$, Γ in Line 15 initially will be $\{\{\emptyset\}, \{a\}, \{b\}, \{c\}, \{a, b\}, \{a, c\}, \{b, c\}, \{a, b, c\}\}$, but will finally become $\Gamma = \{\{a\}, \{b\}, \{c\}, \{a, b\}, \{a, c\}, \{b, c\}\}$ after the empty set and S are removed.

Line 16 creates a variable ProductiveTest whose value is assigned 0. Lines 17 to 25 then iteratively pick any two subsets in Γ to test if all possible subset combinations satisfy productiveness criteria in Line 22. If any two possible subset combinations do not satisfy the criteria in Line 22, the ProductiveTest variable is incremented by 1. After all possible subset combinations are tested for productiveness, S will be added to the set of productive periodic frequent patterns $pPer_D$ if value of the ProductiveTest variable is still 0.

Line 28 creates the RedundanceTest variable and assigns the value 0 to it. Lines 29 to 34 iteratively pick all subsets of S in Γ one after the other to test if S is redundant or non-redundant. If S is a generator periodic frequent pattern (that is, all subsets of S do not satisfy the criteria in Line 30) and S is in the set of productive periodic frequent patterns, the class of S is assigned as *Self-Reliant* in Line 36 and subsequently added to $pPer_D$, else its class is assigned as *Not Self-Reliant* in Line 38.

Table 3 h_n: unsorted length-1 items with their coversets from Table 1

Item	Coverset (TIDs)	Item	Coverset (TIDs)	Item	Coverset (TIDs)
{b}	{1, 3, 5}	{d}	{2, 3, 4, 6, 7}	{e}	{1, 2, 4, 5, 6, 8}
{f}	{2, 4, 5, 6, 8}	{a}	{1, 3, 5, 6}	{c}	{1, 3, 4, 7}

4.2 The Proposed Method: SRPFPM Algorithm

To mine the set of self-reliant periodic frequent patterns from transactional databases for decision-making, we propose the Self-Reliant Periodic Frequent Pattern Miner (SRPFPM) algorithm shown in Algorithm 1. The proposed SRPFPM algorithm employs the Apriori-like candidate generation technique in [3]. However, unlike the approach proposed in [3], SRPFPM scans the database once to store the transaction IDs of each item to avoid repeated scanning of the database during the mining process. The proposed SRPFPM algorithm as shown in Algorithm 1 takes as input the database and the user-desired thresholds - that is, the minimum support ($minsup$), minimum average period ($minAvg$), maximum average period ($maxAvg$), the minimum period ($minPer$), the maximum period ($maxPer$) and returns the set of self-reliant periodic frequent patterns ($sPer_D$) based on the user-given thresholds.

Given a transaction database D and the desired user thresholds ($minsup, minAvg$, $maxAvg, minPer \ maxPer$), SRPFM mines the self-reliant periodic frequent patterns from the database as follows.

Line 1 of Algorithm 1 creates a hashmap h_n to store the set of all length-items and all their transaction IDs in D while Line 2 creates the hashmaps $sPer_D$, $pPer_D$ and L to store the self-reliant periodic frequent patterns, productive periodic frequent patterns and length-1 frequent items respectively. Lines 3 to 11 scan the database to identify and store the unique length-1 items with all their transaction IDs as follows. For each length-1 item a_y in each transaction T in the database, if that item is not in h_n, Line 6 creates its coverset and adds the transaction ID to the coverset of a_y. The item together with the coverset (containing its transaction ID) is then added to h_n in Line 7. If the length-1 item a_y is in h_n, its coverset is obtained from h_n and updated to contain the new transaction ID in Line 10. h_n is then updated in Line 11 with the new coverset of a_y. After all transactions in D are scanned, h_n will contain all length-1 items with their all their respective transaction IDs in D.

Example 16 For our sample customer transaction database shown in Table 1, Lines 1 to 11 will return h_n (the set of all length-items and their coversets) as shown in Table 3.

With all length-1 items and their coversets (transaction IDs) in h_n, Lines 12 to 19 of Algorithm 1 identifies the set of length-1 self-reliant periodic frequent patterns as follows. For each item a_y in h_n, its coverset is obtained from h_n and its set of periods subsequently derived in Line 14 as $ps(a_y)$. The given item a_y is then tested against the user-given thresholds for frequency and periodicity in Line 15. If a_y is frequent

---

Algorithm 1: SRPFPM($D, minsup, minAvg, maxAvg, minPer, maxPer$)

Input: Dataset $D, minsup, minAvg, maxAvg, minPer, maxPer$
Output: Set of Self-Reliant Periodic Frequent Patterns, $sPer_D$
1 Create HashMap h_n /* to store all length-1 items in D */
2 Create HashMaps: $sPer_D, pPer_D, L$
3 **for** *each* transaction $T \in D$ **do**
4   **for** *each* length-1 item $a_y \in T$ **do**
5   **if** $a_y \notin h_n$ **then**
6   Create $cov(a_y) = \{$ TID of $a_y\}$ /* TID = Transaction ID */
7   Add $(a_y, cov(a_y))$ to h_n
8   **else**
9   Let $(a_y, cov(a_y)) = h_n(a_y)$
10   Udate $cov(a_y)$ as $cov(a_y) = cov(a_y) \cup$ TID of a_y
11   Update h_n with $(a_y, cov(a_y))$
12 **for** *each* item $a_y \in h_n$ **do**
13   Let $(a_y, cov(a_y)) = h_n(a_y)$
14   Get $ps(a_y)$
15   **if** $sup(a_y) \geq \varepsilon \wedge minAvg \leq avper \leq maxAvg \wedge minper \geq minPer \wedge maxper \leq maxPer$ **then**
16   Self-RelianceTest($a_y, pPer_D$)   /* calls the self-reliance test on a_y */
17   **if** $a_y.class = $ *Self-Reliant* **then**
18   Add $(a_y, ps(a_y))$ to $sPer_D$
19   Add $(a_y, cov(a_y))$ to L
20 Sort L in descending order of items
21 MinePFPs($L, minsup, minAvg, maxAvg, minPer, maxPer, sPer_D, pPer_D$)
22 **return** $sPer_D$

Table 4 L: sorted length-1 periodic frequent items

Item	Coverset (TIDs)	Item	Coverset (TIDs)
$\{a\}$	$\{1, 3, 5, 6\}$	$\{c\}$	$\{1, 3, 4, 7\}$
$\{b\}$	$\{1, 3, 5\}$	$\{d\}$	$\{2, 3, 4, 6, 7\}$

and periodic based on the user-given thresholds, it is tested for self-reliance in Line 16 by calling the Self-RelianceTest function on a_y as Self-RelianceTest($a_y, pPer_D$). If the Self-RelianceTest function returns the class of a_y as *Self-Reliant*, a_y with its set of periods is added to the set of self-reliant periodic frequent patterns $sPer_D$ in Line 18 while a_y with its coverset is added to the hashmap L in Line 19. The set L is then sorted in descending order of length-1 items in Line 20.

Example 17 For our running example, given $minsup = 0.25$, $minPer = 1$, $maxPer = 3$, $minAvg = 1$, and $maxAvg = 2$, Lines 12 to 20 will produce L from h_n as shown in Table 4. All items shown in Table 4 and their set of periods will be added to the set of self-reliant periodic frequent patterns ($sPer_D$) in Line 18 of Algorithm 1 and the set of productive periodic frequent patterns ($pPer_D$) in Line 27 of the Self-RelianceTest function.

The remaining self-reliant periodic frequent patterns are then mined from L by calling Algorithm 2 on L as MinePFPs($L, minsup, minAvg, maxAvg, minPer,$

$maxPer, sPer_D, pPer_D$). When Algorithm 2 exits after mining the remaining self-reliant periodic frequent patterns, Line 22 of Algorithm 1 returns the set of all self-reliant periodic frequent patterns ($sPer_D$) in D.

Algorithm 2 mines the remaining self-reliant periodic frequent patterns as follows. Given L which initially is supposed to contain the set of length-1 self-reliant periodic frequent patterns, Line 1 of Algorithm 2 creates a hashmap TempL for temporary storage. If there are no length-1 self-reliant periodic frequent patterns in L, as shown in Lines 3 and 4, Algorithm 2 terminates and the self-reliant periodic frequent pattern mining process ends. While L contains length-1 items, Algorithm 2 mines the self-reliant periodic frequent patterns from L using a nested for-loops (from Lines 7 to 21 of Algorithm 2) as follows.

Algorithm 2: MinePFPs($L, minsup, minAvg, maxAvg, minPer, maxPer,$ $sPer_D, pPer_D$)

Input: $L, minsup, minAvg, maxAvg, minPer, maxPer, sPer_D, pPer_D$

1 Create Hashmap TempL
2 Let $P_{a_n}[0, b]$ be the the length-b prefix of a_n
3 **if** $|L| = 0$ **then**
4 | End
5 **else**
6 | **while** $|L| > 1$ **do**
7 | | **for** $k = 0$ to $|L|$-1 **do**
8 | | | Let $(a_k, cov(a_k)) = L[k]$
9 | | | **for** $l = (k + 1)$ to $|L|$-1 **do**
10 | | | | Let $(a_l, cov(a_l)) = L[l]$
11 | | | | **if** $P_{a_k}[0, |a_k|$-$1] = P_{a_l}[0, |a_l|$-$1]$ **then**
12 | | | | | Create $S = a_k \cup a_l$, and $cov(S) = cov(a_k) \cap cov(a_l)$
13 | | | | **if** $sup(S) \geq minsup$ **then**
14 | | | | | Get $ps(S)$
15 | | | | | **if** S *is periodic* **then**
16 | | | | | | Self-RelianceTest($S, pPer_D$)
17 | | | | | | **if** $S.class = Self$-$Reliant$ **then**
18 | | | | | | | Add $(S, cov(S))$ to TempL
19 | | | | | | | Add $(S, ps(S))$ to $sPer_D$
20 | | $L = $ TempL
21 | | TempL.clear()

In the first for-loop within L (from index $k = 0$ to $|L| - 1$), the tuple at the kth-index, that is, $(a_k, cov(a_k))$ is obtained in Line 8 as $L[k]$. While still at the kth-index, the second for-loop within L (from index $l = (k + 1)$ to $|L| - 1$) starts in Line 9 as follows. Each tuple in the lth-index, that is, $(a_l, cov(a_l))$ is obtained in Line 10 as $L[l]$. If a_k and a_l have common length-$(|a_k| - 1)$ prefixes, that is, $P_{a_k}[0, |a_k| - 1] = P_{a_l}[0, |a_l| - 1]$, a candidate frequent pattern, S, is created in Line 12 as $S = a_k \cup a_l$ and $cov(S) = cov(a_k) \cap cov(a_l)$. If S is frequent in D, its set of periods $ps(S)$ is obtained in Line 14. If S is periodic based on the user-desired periodicity measures, it is tested for self-reliance in Line 16. If S is self-reliant,

S and its coverset is added to $TempL$ in Line 18. This ensures only frequent and self-reliant periodic patterns are kept and used in the next iteration to generate the candidate periodic frequent patterns. This serves as a pruning mechanism. Given S is self-reliant, S with its set of periods ($ps(S)$) is added to the set of self-reliant periodic frequent patterns.

Example 18 For our running example, given L in Table 4, during the first loop within L, $(\{a\}, \{1, 3, 5, 6\})$ will be obtained in Line 8 as the kth-index while $(\{b\}, \{1, 3, 5\})$ will be obtained in Line 10 as the lth-index. Since $P_{\{a\}}[0, |\{a\}| - 1]$ will be same as $P_{\{b\}}[0, |\{b\}| - 1]$ in Line 11, candidate periodic frequent pattern S will be created as $S = \{a\} \cup \{b\} = \{a, b\}$ and $cov(S) = \{1, 3, 5, 6\} \cap \{1, 3, 5\} = \{1, 3, 5\}$ in Line 12. Given the minimum support threshold of 0.25, $S = \{a, b\}$ will be frequent in Line 13 hence, its set of periods will be obtained in Line 14 as $ps(\{a, b\}) = \{1, 2, 2, 3\}$. Based on the given periodicity thresholds ($minPer = 1, maxPer = 3, minAvg = 1$, and $maxAvg = 2$), $\{a, b\}$ will be periodic in Line 15. The pattern $\{a, b\}$ will then be tested for self-reliance using the Self-RelianceTest() function in Line 16. Though $\{a, b\}$ is productive, its class will be assigned in Line 13 of the Self-RelianceTest() function as *Not Self-Reliant* as it is not a generator pattern, that is, $\{a, b\}$ is redundant, hence $\{a, b\}$ will be pruned and not added to TempL.

For each kth-index in the first for-loop, the second for-loop repeats till all indexes in L are iterated in the second for-loop. When both nested loops are complete, L is recreated in Line 20 from $TempL$ and the content of $TempL$ is cleared in Line 21. The size of L is checked and the nested looping repeats on L until $|L| = 1$ at which point the self-reliant periodic frequent pattern mining process terminates. Line 22 of Algorithm 1 then returns the set of self-reliant periodic frequent patterns discovered in D based on the user-desired thresholds.

Example 19 Continuing with the running example, after the first nested for-loops, L (which will be recreated in Line 20 from TempL) will comprise of only $\{a, c\}$ and its coverset. The pattern $\{a, c\}$ which will be generated in Line 12 in the first iteration will be self-reliant in Line 17 hence it will be added to $sPer_D$. Since $L = 1$ after the first nested for-loops, self-reliant periodic frequent pattern mining process terminates. In our running example, Line 22 of Algorithm 1 then returns $sPer_D = \{\{a\}, \{b\}, \{c\}, \{d\}\{a, c\}\}$ and their respective set of periods.

5 Experimental Analysis

To evaluate the performance of the proposed SRPFPM algorithm, we compared its performance with our implementations[1] of PPFP [24], NPFPM [1, 2] and PFP*. PPFP and NPFPM are existing periodic frequent pattern mining algorithms for discovering productive and non-redundant periodic frequent patterns respectively while

---

[1] These implementations are based on novel periodicity measures in Definition 7 and not the measures in periodicity measures employed in the original implementations found in [1, 2, 24]

PFP* discovers the set of all periodic frequent patterns. The compared algorithms (SRPFPM, PPFP, NPFPM and PFP*) are all implemented in Python. The experimental analysis was carried out on a tenth-generation 64 bit Core i7 processor running Windows 10, and equipped with 16 GB of RAM. Four benchmark datasets as described in Table 5 (commonly used in frequent pattern mining) were utilized.

Table 5 Dataset characteristics

Dataset	Unique Items	Total Transactions	Nature
Retail	16,470	88,162	Sparse
Mushroom	119	8,416	Dense, long transactions
Chainstore	46,086	1,112,949	Sparse
Kosarak25K	41,270	25,000	Partly dense

The experiment consisted of running the algorithms on each dataset with fixed *minPer* and *minAvg* values, while varying the *maxAvg* and *maxPer* parameters. Execution times, memory consumption and number of patterns found were measured for each algorithm. The execution times and memory consumption are average values of each algorithm run ten times. All memory measurements were done using the Python API except that of the original implementation of PFPM [5] which was done using the Java API. Note that results for varying the *minPer* and *minAvg* values are not shown because they have less influence on the number of periodic frequent patterns reported compared to the other parameters.

5.1 Number of Reported Patterns

The number of reported periodic frequent patterns of the compared algorithms are shown in Tables 6, 7, 8 and 9 for the Chainstore, Kosarak25K, Retail and Mushroom datasets respectively. It can be observed that at a fixed *minimum support count, minPer, minAvg* and *maxAvg*, the four algorithms report a larger number of periodic frequent patterns as the *maxPer* increases (see Tables 6, 7 and 9).

Similarly, keeping the *minimum support count, minPer, minAvg* and *maxPer* constant, the four algorithms likewise report a large number of periodic frequent pattern mining process (see Table 8). It can be observed in Tables 7, 8 and 9 that the self-reliance test is able to prune periodic frequent patterns that do not have inherent item relationships as well as those whose periodic occurrences can be inferred from other periodic frequent patterns.

Table 6 Reported periodic frequent patterns in Chainstore dataset with *minsup count = 1, minPer = 1, minAvg = 1, maxAvg = 2000*

	Number of Periodic Frequent Patterns		
	maxPer = 5000	*maxPer = 3000*	*maxPer = 1000*
PFP*	123	50	3
PPFP	123	50	3
NPFPM	123	50	3
SRPFPM	123	50	3

Table 7 Reported periodic frequent patterns in Kosarak25K dataset with *minsup count = 50, minPer = 1, minAvg = 5, maxAvg = 500*

	Reported Periodic Frequent Patterns		
	maxPer = 2000	*maxPer = 1000*	*maxPer = 1*
PFP*	2384	447	0
PPFP	2364	444	0
NPFPM	2381	447	0
SRPFPM	2364	444	0

Table 8 Reported periodic frequent patterns in Retail dataset with *minsup count = 1, minPer = 1, minAvg = 5, maxPer = 1500*

	Reported Periodic Frequent Patterns		
	maxAvg = 2000	*maxAvg = 1000*	*maxAvg = 1*
PFP*	313	255	0
PPFP	309	253	0
NPFPM	196	168	0
SRPFPM	192	166	0

Table 9 Reported periodic frequent patterns in Mushroom dataset with *minsup count = 1, minPer = 1, minAvg = 5, maxAvg = 2000*

	Reported Periodic Frequent Patterns		
	maxPer = 5000	*maxPer = 4000*	*maxPer = 3000*
PFP*	754	80	14
PPFP	689	75	13
NPFPM	728	80	14
SRPFPM	663	75	13

5.2 Memory Usage Evaluation

The memory usage of the compared algorithms are shown in Tables 10, 11 and 12 for the Chainstore, Kosarak25K and Retail datasets respectively. It can be observed that at a fixed *minimum support count, minPer, minAvg* and *maxAvg*, the four algorithms consume more memory in discovering the set of periodic frequent patterns as the *maxPer* increases (see Tables 10 and 11). Similarly, keeping the *minimum support count, minPer, minAvg* and *maxPer* constant, the four algorithms likewise consume more memory during the periodic frequent pattern mining process (see Table 12).

It can be observed that the amount memory consumed by all algorithms is dependent on the database size.

Table 10 Memory usage: periodic frequent pattern discovery in Chainstore dataset with *minsup count = 1, minPer = 1, minAvg = 1, maxAvg = 2000*

	Memory (MB)		
	maxPer = 3000	*maxPer = 2000*	*maxPer = 1000*
PFP*	545.2	541.6	533.5
PPFP	544.6	540.8	532.4
NPFPM	545.4	542.1	532.7
SRPFPM	543.5	542.2	531.8

Table 11 Memory usage: periodic frequent pattern discovery in Kosarak25K dataset with *minsup count = 50, minPer = 1, minAvg = 5, maxAvg = 500*

	Memory (MB)		
	maxPer = 1000	*maxPer = 500*	*maxPer = 1*
PFP*	21.7	18.9	17.8
PPFP	20.9	18.5	17.8
NPFPM	21.0	18.7	17.8
SRPFPM	21.0	18.6	17.8

Table 12 Memory usage: periodic frequent pattern discovery in Retail dataset with *minsup count = 1, minPer = 1, minAvg = 5, maxPer = 1500*

	Memory (MB)		
	maxAvg = 3000	*maxAvg = 500*	*maxAvg = 1*
PFP*	65.2	64.3	60.3
PPFP	64.9	64.1	60.4
NPFPM	62.7	61.8	59.9
SRPFPM	62.5	61.6	60.1

5.3 Runtime Analysis

The runtimes of the compared algorithms are shown in Figs. 1, 2, 3 and 4 for the Chainstore, Kosarak25K, Mushroom and Retail datasets respectively. It can be observed that at a fixed *minimum support count, minPer, minAvg* and *maxAvg*, the four algorithms take longer times to discover the set of periodic frequent patterns as the *maxPer* increases (see Figs. 1, 2 and 3). Similarly, keeping the *minimum support count, minPer, minAvg* and *maxPer* constant, the four algorithms also take a longer time to discover the set of periodic frequent patterns as *maxAvg* increases (see Fig. 4).

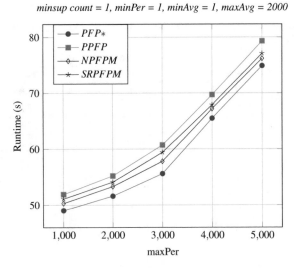

minsup count = 1, minPer = 1, minAvg = 1, maxAvg = 2000

Fig. 1 Runtime: periodic frequent pattern discovery in Chainstore dataset

It can also be observed that in the four datasets, though the runtimes of all four algorithms are quite similar, except for the Retail dataset (see Fig. 4), PFP* and NPFPM are slightly faster than PPFP and SRPFPM. This is because the productive-

minsup count = 50, minPer = 1, minAvg = 5, maxAvg = 500

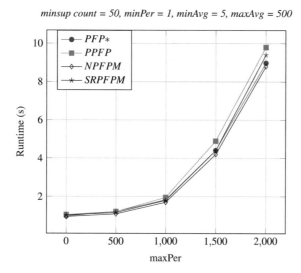

Fig. 2 Runtime: periodic frequent pattern discovery in Kosarak25K dataset

ness test in PPFP and SRPFPM is more time consuming than the redundance test in NRPFPM. In the Chainstore dataset for example, all four algorithms are reporting same number of periodic frequent patterns (see Table 6). As such the productiveness test, non-redundance test and self-reliance test will adversely affect the runtimes of PPFP, NPFPM and SRPFPM.

minsup count = 1, minPer = 1, minAvg = 1, maxAvg = 2000

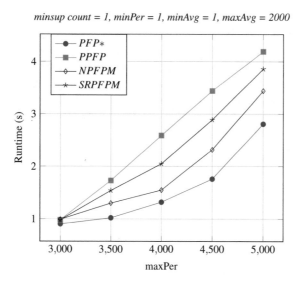

Fig. 3 Runtime: periodic frequent pattern discovery in Mushroom dataset

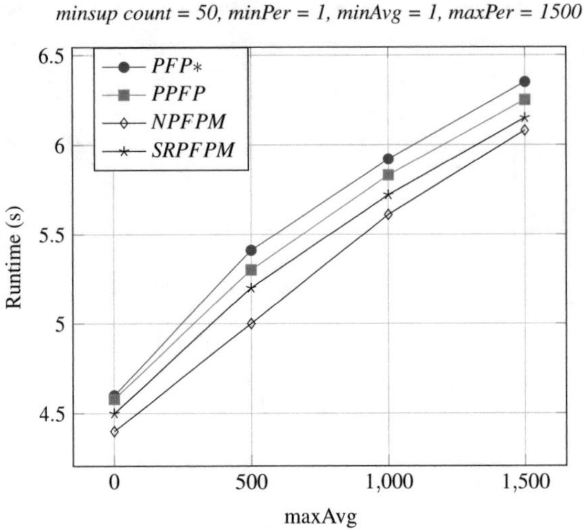

Fig. 4 Runtime: periodic frequent pattern discovery in Retail dataset

5.4 Scalability

For our scalability test, we compare SRPFPM with the original implementation of PFPM [5] with regards to runtime and memory usage in periodic frequent pattern mining. Though PFPM takes less time in discovering periodic frequent patterns compared to SRPFPM (see Figs. 5, 6 and 7), it consumes more memory during the discovery process compared to SRPFPM (see Tables 13, 14 and 15).

Though PFPM is efficient with regards to time, SRPFPM is efficient with regards to memory usage. It is also worth noting that PFPM will always report a larger number of periodic frequent patterns for decision-making, most of which will be periodic due random occurrence chance (that is, without inherent item relationship) or their periodicities can be inferred from other periodic frequent patterns. Employing such

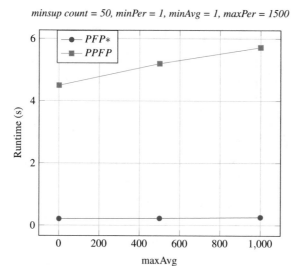

Fig. 5 Runtime: periodic frequent pattern discovery in Retail dataset

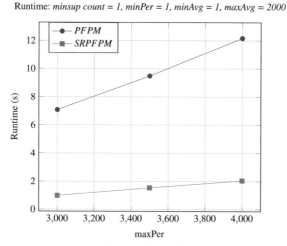

Fig. 6 Runtime: periodic frequent pattern discovery in Mushroom dataset

patterns reported in PFPM in decision-making could be detrimental if they happen to be false positively periodic. SRPFPM will however prune such patterns and report a smaller set of self-reliant periodic frequent patterns that will be more preferable in decision-making.

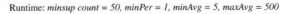

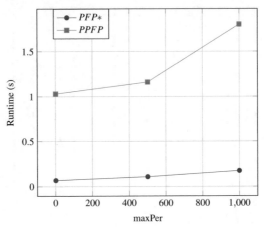

Fig. 7 Runtime: periodic frequent pattern discovery in Kosarak25K dataset

Table 13 Memory usage: periodic frequent pattern discovery in Chainstore dataset with *minsup count = 1, minPer = 1, minAvg = 1, maxAvg = 2000*

	Memory (MB)		
	maxPer = 3000	*maxPer = 2000*	*maxPer = 1000*
PFPM	813.4	799.5	506.2
SRPFPM	543.5	542.2	531.8

Table 14 Memory usage: periodic frequent pattern discovery in Kosarak25K dataset with *minsup count = 50, minPer = 1, minAvg = 5, maxAvg = 500*

	Memory (MB)		
	maxPer = 1000	*maxPer = 500*	*maxPer = 1*
PFPM	545.2	507.8	455.3
SRPFPM	21.0	18.6	17.8

Table 15 Memory usage: periodic frequent pattern discovery in Retail dataset with *minsup count = 1, minPer = 1, minAvg = 5, maxPer = 2000*

	Memory (MB)		
	maxAvg = 300	*maxAvg = 200*	*maxAvg = 100*
PFPM	1153.2	995.6	829.3
SRPFPM	24.4	19.8	19.0

6 Conclusions

In this paper, we have proposed the use of a self-reliance measure in mining our proposed set of self-reliant periodic frequent patterns. Using this measure provides the advantages of reporting a smaller number of periodic frequent patterns whose periodic occurrences have inherent item relationships as well as contain non-redundant information. We subsequently propose and develop an efficient algorithm named SRPFPM to efficiently discover only periodic frequent patterns that are self-reliant using the self-reliant measure. Experimental analysis on benchmark datasets show that SRPFPM is efficient and effectively prunes periodic frequent patterns that are periodic due to random chance, as well as those whose periodicities can be inferred from other periodic frequent patterns.

References

1. M.K. Afriyie, V.M. Nofong, J. Wondoh and H. Abdel-Fatao, Mining Non-redundant Periodic Frequent Patterns, in *Intelligent Information and Database Systems. LNCS*, ed. by N.T. Nguyen, K. Jearanaitanakij, A. Selamat, B. Trawiński and S. Chittayasothorn, vol. 12033 (Springer, Berlin, 2020), pp. 321-331
2. Afriyie, M. K, Nofong, V. M, Wondoh, J. & Abdel-Fatao, H.:Efficient Mining of Non-redundant Periodic Frequent Patterns. Vietnam Journal of Computer Science, 8(4) 1-15 (2021)
3. Agrawal, R., Imieliński, T. & Swami, A.: Mining Association Rules between Sets of Items in Large Databases. SIGMOD Rec. 22(2) 207–216. ACM (1993)
4. Amphawan K, Surarerks A, Lenca P: Mining Periodic-Frequent Itemsets with Approximate Periodicity using Interval Transaction-ids List Tree. In: Proceedings of the 3rd International Conference on Knowledge Discovery and Data Mining. pp 245-248 (2010)
5. Fournier-Viger, P., Lin, J. C. W., Gomariz, A., Gueniche, T., Soltani, A., Deng, Z., & Lam, H. T.: The SPMF Open-Source Data Mining Library Version 2. In: Proceedings of European Conference on Machine Learning and Knowledge Discovery in Databases, pp. 36-40, Springer, Cham (2016)
6. Fournier-Viger, P., Lin, CW., Duong, QH., Dam, TL., Ševčík, L., Uhrin, D., & Voznak, M.: PFPM: Discovering Periodic Frequent Patterns with Novel Periodicity Measures. In: Proceedings of the 2nd Czech-China Scientific Conference, InTech (2017)
7. Fournier-Viger P., Li Z., Lin, J.C.W., Kiran R.U., Fujita H.: Discovering Periodic Patterns Common to Multiple Sequences. In: Ordonez C., Bellatreche L. (eds) Big Data Analytics and Knowledge Discovery. LNCS, vol 11031, pp 231-246 (2018)
8. Fournier-Viger P., Yang P., Lin J.C.W., Kiran R.U. (2019) Discovering Stable Periodic-Frequent Patterns in Transactional Data. In: Wotawa F., Friedrich G., Pill I., Koitz-Hristov R., Ali M. (eds) Advances and Trends in Artificial Intelligence. From Theory to Practice. LNCS, vol 11606, pp 230-244 (2019)
9. Han, J., Pei, J., & Yin, Y.: Mining Frequent Patterns without Candidate Generation. In: ACM SIGMOD Rec. 29(2) 1-12. ACM (2000)
10. W.N. Ismail, M.M. Hassan, H.A. Alsalamah, Mining of Productive Periodic-Frequent Patterns for IoT Data Analytics. Future Generation Computer Systems 88, 512–523 (2018)
11. W.N. Ismail, M.M. Hassan, H.A. Alsalamah, G. Fortino, Mining Productive-Periodic Frequent Patterns in Tele-Health Systems. Journal of Network and Computer Applications 115, 33–47 (2018)

12. Kiran, R. U., & Kitsuregawa, M.: Novel Techniques to Reduce Search Space in Periodic-Frequent Pattern Mining. In: Bhowmick, S. S., Dyreson, C. E., Jensen, C. S., Lee, M. L., Muliantara, A., Thalheim, B. (eds) DASFAA 2014. LNCS, vol 8422, pp. 377-391. Springer International Publishing (2014)
13. Kiran, R. U., & Kitsuregawa, M.: Discovering Quasi-Periodic-Frequent Patterns in Transactional Databases. In: Bhatnagar, V., Srinivasa, S. (eds) BDA 2013. LNCS, vol. 8302, pp. 97-115. Springer International Publishing (2013)
14. R.U. Kiran, P.K. Reddy, Towards Efficient Mining of Periodic-Frequent Patterns in Transactional Databases, in *DASFAA 2010*, ed. by P.G. Bringas, A. Hameurlain, G. Quirchmayr. LNCS, vol. 6262 (Springer, Heidelberg, 2010), pp. 194–208
15. R.U. Kiran, P.K. Reddy, An Alternative Interestingness Measure for Mining Periodic-Frequent Patterns, in *DASFAA 2011, LNCS*, vol. 6587, ed. by J.X. Yu, M.H. Kim, R. Unland (Springer, Heidelberg, 2011), pp. 183–192
16. R.U. Kiran, M. Kitsuregawa, P.K. Reddy, Efficient Discovery of Periodic-Frequent Patterns in Very Large Databases. Journal of Systems and Software **112**, 110–121 (2016)
17. Kiran R.U., Venkatesh J.N., Fournier-Viger P., Toyoda M., Reddy P.K., Kitsuregawa M.: Discovering Periodic Patterns in Non-uniform Temporal Databases. In: Kim J., Shim K., Cao L., Lee JG., Lin X., Moon YS. (eds) Advances in Knowledge Discovery and Data Mining. LNCS, vol 10235, pp 604-617 (2017.)
18. R.U. Kiran, A. Anirudh, C. Saideep, M. Toyoda, P.K. Reddy, M. Kitsuregawa, Finding Periodic-Frequent Patterns in Temporal Databases using Periodic Summaries. Data Science and Pattern Recognition **3**(2), 24–46 (2019)
19. Kiran, R.U., Saideep, C., Zettsu, K., Toyoda, M., Kitsuregawa, M., Reddy, P.K.: Discovering Partial Periodic Spatial Patterns in Spatiotemporal Databases. In: Proceedings of the 2019 IEEE International Conference on Big Data. IEEE, pp 233-238 (2019)
20. Kiran R.U.& Reddy P.K.: Mining Rare Periodic-Frequent Patterns using Multiple Minimum Supports. In: Proceedings of the 15th International Conference on Management of Data. pp 7-8 (2010)
21. V. Kumar, V. Valli-Kumari, Incremental Mining for Regular Frequent Patterns in Vertical Format. Int. J. Eng & Tech. **5**(2), 1506–1511 (2013)
22. Li, J., Li, H., Wong, L., Pei, J., & Dong, G. Minimum Description Length Principle: Generators Are Preferable to Closed Patterns. In: Proceedings of the 21st National Conference on Artificial Intelligence, pp. 409–414 (2006)
23. J.C.W. Lin, J. Zhang, P. Fournier-Viger, T.P. Hong, J. Zhang, A Two-Phase Approach to Mine Short-Period High-Utility Itemsets in Transactional Databases. Adv. Eng. Inf. **33**, 29–43 (2017)
24. V.M. Nofong, Discovering Productive Periodic Frequent Patterns in Transactional Databases. Annals of Data Science **3**(3), 235–249 (2016)
25. V.M. Nofong, J. Wondoh, Towards Fast and Memory Efficient Discovery of Periodic Frequent Patterns. Journal of Information and Telecommunication **3**(4), 480–493 (2019)
26. Nofong, V. M. (2018). Fast and Memory Efficient Mining of Periodic Frequent Patterns. In: Sieminski, A., Kozierkiewicz, A., Nunez, M., Ha, Q. T. (eds) Modern Approaches for Intelligent Information and Database Systems, SCI, vol. 769, pp. 223-232, Springer, Cham (2018)
27. Pei, J., Han, J., Lu, H., Nishio, S., Tang, S., & Yang, D.: H-mine: Hyper-Structure Mining of Frequent Patterns in Large Databases. In: Proceedings IEEE International Conference on Data Mining, pp. 441-448, IEEE (2001)
28. M.M. Rashid, M.R. Karim, B.S. Jeong, H.J. Choi, Efficient Mining Regularly Frequent Patterns in Transactional Databases, in *DASFAA 2012, LNCS*, vol. 7238, ed. by S. Lee, Z. Peng, X. Zhou, Y. Moon, R. Unland, J. Yoo (Springer, Heidelberg, 2012), pp. 258–271
29. M.M. Rashid, I. Gondal, J. Kamruzzaman, Regularly Frequent Patterns Mining from Sensor Data Stream, in *NIP 2013, LNCS*, vol. 8227, ed. by M. Lee, A. Hirose, Z.G. Hou, R. Kil (Springer, Berlin Heidelberg, 2013), pp. 417–424
30. Shenoy, P., Haritsa, J. R., Sudarshan, S., Bhalotia, G., Bawa, M., & Shah, D.: Turbo-charging Vertical Mining of Large Databases. In ACM SIGMOD Record 29(2), 22-33. ACM (2000)

31. A. Surana, R.U. Kiran, P.K. Reddy, An Efficient Approach to Mine Periodic-Frequent Patterns in Transactional Databases, in *LNAI*, vol. 7104, ed. by L. Cao, J.Z. Huang, J. Bailey, Y.S. Koh, J. Luo (Springer, Heidelberg, 2012), pp. 254–266
32. S.K. Tanbeer, C.F. Ahmed, B.S. Jeong, Y.K. Lee, Discovering Periodic-Frequent Patterns in Transactional Databases, in *PAKDD 2009, LNAI*, vol. 5476, ed. by T. Theeramunkong, B. Kijsirikul, N. Cercone, T. Ho (Springer, Heidelberg, 2009), pp. 242–253
33. F.C. Tseng, Mining Frequent Itemsets in Large Databases: The Hierarchical Partitioning Approach. Expert Systems with Applications **40**(5), 1654–1661 (2013)
34. Venkatesh J.N., Uday Kiran R., Krishna Reddy P., & Kitsuregawa M.: Discovering Periodic-Correlated Patterns in Temporal Databases. In: Hameurlain A., Wagner R., Hartmann S., Ma H. (eds) Transactions on Large-Scale Data- and Knowledge-Centered Systems XXXVIII. LNCS, vol 11250, pp. 146-172 (2018)
35. Zaki, M. J. (2000). Scalable Algorithms for Association Mining. IEEE Transactions on Knowledge and Data Engineering, 12(3), 372-390 (2000)
36. M.J. Zaki, S. Parthasarathy, M. Ogihara, W. Li, Parallel Algorithms for Discovery of Association Rules. Data Mining and Knowledge Discovery **1**(4), 343–373 (1997)
37. Zaki, M. J., & Gouda, K.: Fast Vertical Mining using Diffsets. In: Proceedings of the 9th ACM SIGKDD International Conference on Knowledge Discovery and Data Mining, pp. 326–335 (2003)
38. D. Zhang, K. Lee, I. Lee, Mining hierarchical semantic periodic patterns from GPS-collected spatio-temporal trajectories. Expert Systems with Applications. **122**, 85–101 (2019)

Discovering Periodic High Utility Itemsets in a Discrete Sequence

Philippe Fournier-Viger, Youxi Wu, Duy-Tai Dinh, Wei Song,
and Jerry Chun-Wei Lin

Abstract Periodic itemset mining is the task of finding all the sets of items (events or symbols) that regularly appear in a sequence. One of the most important applications is customer behavior analysis, where a periodic itemset found in a sequence of customer transactions indicates that the customer regularly buys some items together. Using this information, marketing strategies can be tailored and product recommendation can be done. However, a major limitation of traditional periodic itemset mining is that the relative importance of each item is not taken into account and that each item cannot appear more than once at each time step of the sequence. But in real life, not all items are equally important (e.g., selling a cake yields less profit than selling a computer) and a customer may buy multiple units of the same item at the same time (e.g., many cakes). To address these factors, the task of periodic frequent itemset mining was generalized as that of periodic high utility itemset mining, where the goal is to find the sets of items that not only periodically appear in a sequence but also have a high importance (e.g., yield a high profit). This chapter provides an overview of this task and presents an algorithm to solve this problem. Moreover, a

P. Fournier-Viger (✉)
Harbin Institute of Technology (Shenzhen), Shenzhen, China
e-mail: philfv8@yahoo.com; philippe.fv@gmail.com

Y. Wu
Hebei University of Technology, Tianjin, China
e-mail: wuc567@163.com

D.-T. Dinh
Japan Advanced Institute of Science and Technology, Nomi, Japan
e-mail: taidinh@jaist.ac.jp

W. Song
North China University of Technology, Beijing, China
e-mail: songwei@ncut.edu.cn

J. C.-W. Lin
Department of Computing, Mathematics, and Physics, Western Norway University of Applied
Sciences (HVL), Bergen, Norway
e-mail: jerrylin@ieee.org

© The Author(s), under exclusive license to Springer Nature Singapore Pte Ltd. 2021   133
R. Uday Kiran et al. (eds.), *Periodic Pattern Mining*,
https://doi.org/10.1007/978-981-16-3964-7_8

variation of this task that consisting of discovering irregular high utility itemsets is also discussed. Finally, some research opportunities are listed.

1 Introduction

In recent decades, more and more data has been collected and stored in databases. Making sense of this data has become challenging as analyzing large volumes of data by hand is prone to error and time consuming. As a solution, data mining has emerged as an important task. It consists of applying algorithms to (semi)-automatically analyze data. Various types of data can be analyzed using data mining techniques such as spatial data [65], sequences [26], trajectories [16, 80], databases [24, 27], and graphs [42]. Data mining algorithms can generally be viewed as designed to build predictive models from data (to predict the future) or to find interesting patterns that can help to understand the data (to describe the data or understand the past).

The task of finding interesting patterns in data is known as *pattern mining*. The goal is to find sets of values that appear together in the data and meet some criteria set by the user. The most fundamental problem in pattern mining is known as *frequent itemset mining* [1, 24, 52]. The input is a database of transactions (records), as well as a parameter called the minimum support threshold *minsup*. The output is the frequent itemsets, that is the sets of value that appear at least in *minsup* records of the input database. For example, frequent itemset mining can be applied on a database of transactions made by a customer to reveal information such that the customer often purchased the items *cake* and *milk* together.

Frequent itemset mining is useful and has many applications. However, it considers that records are unordered. But for several domains such as analyzing customer transactions, the records (transactions) are ordered sequentially (e.g., by time). To find patterns that regularly appear in a sequence of events (e.g., transactions), the problem of frequent itemset mining was redefined as *frequent periodic itemset mining* [3, 4, 32]. A periodic pattern is a set of values that regularly appear in a sequence. For instance, a pattern found in transaction data from a customer may indicate that the customer regularly buys milk with cake.

Although periodic frequent pattern mining is useful, a drawback is that the relative importance of each item is not taken into account. In other words, it is assumed that all items are equally important. Moreover, another limitation is that the quantities of items are assumed to be binary (an item either appears or not in a record). However, in real life, items may appear more than once at each time step (e.g., a customer may buy five apples or just one), and some items are more important than others (e.g., selling a smartphone yield more profit than selling a pen). As a result, algorithms for periodic frequent itemset mining may find a huge amount of periodic patterns that are not important (e.g., yield a low profit) and miss many rare periodic patterns that have a high importance (e.g., yield a high profit). To address these issues, the task of periodic pattern mining was generalized as that of *periodic high utility itemset mining* (PHUIM) [23]. The goal is to find the sets of items that not only periodically

Table 1 A sequence of transactions (transaction database)

Transaction ID	Transaction
T_1	a, c
T_2	e
T_3	a, b, c, d, e
T_4	b, c, d, e
T_5	a, c, d
T_6	a, c, e
T_7	b, c, e

appear in a sequence but also have a high importance (e.g., yield a high profit). This chapter describes the problem of PHUIM.

The chapter is organized as follows. Section 2 briefly reviews related work on frequent periodic itemset mining. Then, Section 3 describes the problem of periodic high utility itemset mining and an efficient algorithm named PHM (Periodic High utility itemset Miner) [23] to solve this problem. Thereafter, Section 4 discusses a problem variation to find irregular high utility itemsets using an algorithm called $PHM_{irregular}$. Finally, Section 5 discusses some other variations [11] and research opportunities, and Section 6 draws a conclusion.

2 Preliminaries about Periodic Itemset Mining

This section briefly reviews concepts from periodic itemset mining that are useful for this chapter. Let there be a finite set I of items. An item is a symbol or event type. An itemset X is a set of items ($X \subseteq I$). An itemset X that has k items is called a k-itemset. The input in periodic itemset mining is a sequence of transactions, also called a transaction database.

Definition 1 (Transaction database) A transaction database is a sequence of transactions $s = \langle T_1, T_2, \ldots T_m \rangle$, where $T_j \subseteq I$. The integer w used to denote a transaction T_w is called its transaction identifier and is unique.

Example 1 Consider the database of Table 1. This database is a sequence of seven transactions made by a customer, denoted as $T_1, T_2, \ldots, T_7$. Each transaction is a set of items that the customer purchased at a given time from the sets of products $I = \{a, b, c, d, e\}$. For instance, the transaction T_5 indicates that the customer has bought items a, c and d together. In this case, transactions are ordered by purchase times. An itemset such as $\{a, b, c, d, e\}$ is a 5-itemset because it contains five items.

It is to be noted that although a sequence of transactions is often ordered by time, it can be ordered according to other criteria. For example, a sequence of nucleotides in the genome of a virus such as $\langle \{A\}, \{A\}, \{C\}, \{G\}, \{T\}, \{A\}, \ldots \rangle$ can be viewed as a sequence of transactions where the order does not depend on time [54]. Another example is to consider a text document as a sequence of transactions, where each transaction is the set of words (items) appearing in a sentence. In that case, the sequential ordering represents the order between sentences rather than time. If a transaction database is ordered by time or contains timestamps, it is sometimes called a temporal database or temporal transaction database.

To find periodic patterns in a sequence of transactions, the task of frequent periodic itemset mining has been proposed, which aims at finding itemsets that regularly appear over time [3, 32]. This task is defined based on the following definitions.

Definition 2 (Sequence containment) Consider a sequence $s_v = \langle V_1, V_2, \ldots, V_k \rangle$ and another sequence $s_w = \langle W_1, W_2, \ldots, W_l \rangle$. It is said that the sequence s_v is a subsequence of s_w (denoted as $s_v \sqsubseteq s_w$) iff some integers $1 \leq b1 < b2 < \ldots < bk \leq m$ exist such that $V_1 \subseteq W_{b1}, V_2 \subseteq W_{b2}, \ldots, V_k \subseteq W_{bl}$.

Example 2 The sequence $s_v = \langle (a, e), (b, d) \rangle$ is a subsequence of the sequence of transactions depicted in Table 1 since $\{a, e\}$ is a subset of transaction T_3 and it is followed by $\{b, d\}$ in T_4. Another example is that $\langle (a, e), (b, d) \rangle \sqsubseteq \langle (a, e), (b)(b, c, d) \rangle$.

To check if an itemset is periodic in a sequence of transactions, it is necessary to define what it means to be periodic using some criteria [3, 32]. This is done by using some evaluation functions that measure how periodic a pattern is. The basic evaluation functions used in frequent periodic itemset mining to select periodic patterns are the support and the maximum periodicity, which are based on the following definitions.

Definition 3 (Support function) The support of an itemset X in a sequence of transactions s is the number of transactions from s that contain X. The support is formally defined as $sup(X, s) = |TR(X, s)|$ and simply denoted as $sup(X)$ without s when the context is clear.

Example 3 Consider the itemset $X = \{a, e\}$ and the sequence s of Table 1. The itemset X has a support of 2 since X appears in two transactions (T_3 and T_6). This is denoted as $sup(X, s) = 2$.

Definition 4 (Consecutive transactions with respect to an itemset) For a sequence of transactions s and an itemset $X \sqsubseteq s$, the ordered list of transactions (from s) containing X is denoted and defined as $TR(X, s) = \langle T_{g_1}, T_{g_2}, \ldots, T_{g_k} \rangle \sqsubseteq s$. Two transactions T_x and T_y in $TR(X, s)$ are said to be *consecutive with respect to* X if there is no transaction $T_z \in s$ such that $x < z < y$ and $X \subseteq T_z$.

Example 4 Consider the sequence s of Table 1 and the itemset $X = \{b, c\}$. The list of transactions from s containing X is $TR(X, s_1) = \{T_3, T_4, T_7\}$. The transactions T_3 and T_4 are consecutive transactions with respect to X. The transactions T_4 and T_7 are also consecutive transactions with respect to X.

Definition 5 (Periods of an itemset) For an itemset X, the period of two consecutive transactions T_x and T_y containing X is defined as $per(T_x, T_y) = y - x$. The periods of an itemset X in a sequence s are denoted and defined as $pr(X, s) = \{per_1, per_2, ..., per_{k+1}\}$ where $per_1 = g_1 - g_0$, $per_2 = g_2 - g_1$, $...per_{k+1} = g_{k+1} - g_k$, and $g_0 = 0$ and $g_{k+1} = n$, respectively.

Example 5 Continuing the previous example, the periods of X in s are $pr(X, s) = \{3, 1, 3, 0\}$. Consider another itemset $Y = \{a, c\}$. The periods of Y in s are $pr(Y, s) = \{1, 2, 2, 1, 1\}$.

Definition 6 (Maximum periodicity function) A common evaluation function to determine if an itemset X is periodic in a sequence s is called the maximum periodicity and is defined as $maxPr(X, s) = argmax(pr(X, s))$ [3].

Example 6 The maximum periodicity of $X = \{a, e\}$ and $Y = \{a, c\}$ in s are respectively $maxPr(X, s) = argmax(\{3, 1, 3, 0\}) = 3$ and $maxPr(Y, s) = argmax(\{1, 2, 2, 1, 1\}) = 2$.

Traditionally, an itemset X is called a *frequent periodic itemset* if it has a support that is no less than a $minSup$ threshold set by the user ($sup(X, s) \geq minsup$), and if the maximum periodicity of X is not greater than a user-specified threshold $maxPer$ ($maxPr(X, s) \leq maxPer$) [3, 32]. For instance, if the user sets $minSup = 2$ and $maxPer = 2$, the itemset $\{a, c\}$ is a periodic itemset in the sequence of Table 1.

Though finding frequent periodic itemsets can be useful, a problem with this definition of periodic itemset is that it is too strict since if an itemset has a single period that is greater than $maxPer$, it is deemed non-periodic. For instance, if $maxPer = 7$ days and a customer bought bread every day except during 8 consecutive days, then the itemset $\{bread\}$ will not be periodic. Several alternative evaluations functions have been proposed to address this problem. Some functions that have been studied are

- the *average periodicity* [32] defined as $avgPr(X, s) = average(pr(X, s))$,
- the *minimum periodicity* [32] defined as $minPr(X, s) = argmin(pr(X, s))$,
- the *standard deviation of periods* [21, 55, 56], defined as $stanDev(X, s) = stanDev(pr(X, s))$,
- and the *variance of periods* [44, 61, 62].

Besides, some measures were proposed to evaluate whether an itemset is periodically stable over time (consecutive periods remain more or less the same) [28, 33], and statistical testing has also been employed to find statistically significant periodic patterns [55].

In the remaining of this chapter, the following definition of periodic itemsets is used, as it is more general than the traditional definition [23, 32].

Definition 7 (Periodic itemset) An itemset X is considered to be a *periodic itemset* if it satisfies three conditions: (1) $minAvg \leq avgper(X) \leq maxAvg$, (2) $minper(X) \geq minPer$, and (3) $maxper(X) \leq maxPer$, where $minAvg$, $maxAvg$, $minPer$ and $maxPer$ are positive numbers.

This definition is quite flexible as it let the user specifies on average how large the periods of an itemset should be (by condition 1), and two constraints on the minimum and maximum size of periods (by conditions 2 and 3). It is to be noted that some conditions can also be omitted if needed. For example, if only condition 3 is used, the traditional definition of periodic pattern is obtained.

Example 7 Consider that $minAvg = 1$, $maxAvg = 2$, $minPer = 1$, and $maxPer = 3$. The itemset $X = \{b, c, e\}$ is a periodic itemset since $avgper(X) = 1.75 \leq maxAvg$, $avgper(X) = 1.75 \geq minAvg$, $minper(X) = 1 \geq minPer$, $maxper(X) = 3 \leq maxPer$.

It can be observed that the definition of periodic itemset does not consider timestamps. However, it can be easily generalized to be used for transactions that have timestamps. In that case, the transaction identifiers can be replaced by timestamps for the calculation of the periods of an itemset. In the case where two transactions have the same timestamp, an arbitrary order between them can be established.

To develop efficient algorithms for finding periodic itemsets using the above definition, a few important properties of the evaluation functions are the following [3, 23, 32].

Property 1 (The average, minimum, and maximum periodicity are monotonic) Consider two itemsets X and Y satisfying the relationship $X \subset Y$. It follows that $avgper(Y) \geq avgper(X)$. Moreover, it can proved that $minper(Y) \geq minper(X)$ and
$maxper(Y) \geq maxper(X)$.

Property 2 (Eliminating non-periodic itemsets using the maximum periodicity) Consider an itemset X and a sequence of transaction s. The itemset X and its supersets are not periodic itemsets if $maxper(X) > maxPer$ [3].

Property 3 (Eliminating non-periodic itemsets using the average periodicity) An itemset X is not a periodic itemset in a sequence of transactions s if $avgper(X) > maxAvg$. This condition can be rewritten as $|sup(X, s)| < (|s|/maxAvg) - 1$ [23, 32].

As it will be shown, the Property 2 and 3 can be used to avoid exploring the whole search space of itemsets, and thus to improve efficiency of algorithms for mining periodic itemsets.

3 Periodic High Utility Itemset Mining

Though discovering periodic frequent itemsets is useful and can reveal interesting patterns in data [3], it has two important limitations inherited from frequent itemset mining [1, 24, 40, 79]. They are that (1) an item cannot appear more than once in each transaction and (2) all items are considered to have the same importance.

However, these assumptions are not true for several applications. For instance, in the context of customer transaction data analysis, a client may buy more than one bottle of milk in a transaction and some items may be viewed as more important than others because their sale yields a higher profit. Ignoring this information may lead to finding many uninteresting patterns (e.g., patterns that yield a low profit for a store).

To find patterns that have a high importance (e.g., a high profit) rather than only frequent patterns, the problem of *frequent itemset mining* was generalized as *high utility itemset mining* [27]. Then, inspired by this, frequent periodic itemset mining was generalized as periodic high utility itemset mining [23]. The next subsection first briefly reviews concepts from high utility itemset mining and presents the problem of periodic high utility itemset mining that is a generalization of periodic frequent itemset mining. Then, the next subection presents an efficient algorithm, named PHM (Periodic High utility itemset Miner) [23].

3.1   The Problem

The input data in high utility itemset mining is a quantitative transaction database, and the goal is to find *high utility itemsets* (itemsets that have a high importance such as that yield a high profit) [27]. The concept of a quantitative transaction database is defined next.

Definition 8 (Quantitative transaction database) A quantitative transaction database s (also called sequence of quantitative transactions) is a transaction database where a positive number $p(i)$ named external utility is given for each item i, which represents its relative importance in the database. Moreover, a positive number $q(i, T_c)$ called internal utility is provided to indicate the importance of each item i in each transaction T_c.

Example 8 Table 2 lists seven transactions of a quantitative transaction database that are the transactions made by a customer over time. There are five items $I = \{a, b, c, d, e\}$. In each transaction, the relative importantce of each item represents its purchase quantity. For example, in transaction T_6, 2 units of item a is purchased ($q(a, T_5) = 2$), 6 units of item c is purchased ($q(c, T_5) = 6$), and 2 units of item e is purchased ($q(e, T_5) = 2$). The relative importance of items (the external utility) is given in the last line of Table 2. For instance, it is indicated that selling one unit of item d yields a profit of $p(d) = 2\$$ while the sale of a unit of a gives a $p(a) = 5\$$ profit.

Transactions in a quantitative transaction database can be ordered by time or according to other criteria and may represent shopping data or other types of symbolic data. In the case where all external utility values and internal utility values are either 0 or 1, a quantitative transaction database is a transaction database, as defined in the previous section.

Table 2 A quantitative transaction database

Transaction ID	Quantitative transaction
T_1	$(a, 1), (c, 1)$
T_2	$(e, 1)$
T_3	$(a, 1), (b, 5), (c, 1), (d, 3), (e, 1)$
T_4	$(b, 4), (c, 3), (d, 3), (e, 1)$
T_5	$(a, 1), (c, 1), (d, 1)$
T_6	$(a, 2), (c, 6), (e, 2)$
T_7	$(b, 2), (c, 2), (e, 1)$
External utility values of items	
$p(a) = 5, p(b) = 2, p(c) = 1, p(d) = 2, p(e) = 3$	

In high utility itemset mining, patterns are selected based on their utility (e.g., importance or profit), an evaluation function defined as follows.

Definition 9 (Utility of items and itemset) Let there be a transaction T, an itemset X, and an item i. The utility of i in T is the product of its internal and external utility, that is $u(i, T) = p(i) \times q(i, T)$. For an itemset $X \subseteq T$, the utility of X in T is the sum of the product of the internal and external utility of each item in X, that is : $u(X, T) = \sum_{i \in X} u(i, T)$. For an itemset $X \nsubseteq T$, the utility of X in T is 0, that is $u(X, T) = 0$. The utility of X in the input database s is the sum of its utility for all transactions that is: $u(X) = \sum_{X \subseteq T \in D} u(X, T)$.

Example 9 For the database of Table 2, the utility of b in T_7 is $u(b, T_7) = q(b, T_7) \times p(b) = 2 \times 2 = 4$. The utility of c in T_7 is $u(c, T_7) = q(c, T_7) \times p(c) = 2 \times 1 = 2$. The utility of $\{b, c\}$ in T_7 is $u(\{b, c\}, T_7) = u(b, T_7) + u(c, T_7) = 4 + 2 = 6$. The utility of $\{b, c\}$ in the database is $u(\{b, c\}) = u(\{b, c\}, T_3) + u(\{b, c\}, T_4) + u(\{b, c\}, T_7) = (10 + 1) + (8 + 3) + (4 + 2) = 28$.

In the problem of high utility itemset mining, the goal is to find all high utility itemsets. The problem is defined as follows.

Definition 10 (High utility itemset mining) Let there be an itemset X, a database s, and a minimum utility threshold $minUtil$ set by the user (a positive number. If $u(X) \geq minUtil$ then X is a high utility itemset. Otherwise, it is a low utility itemset. The goal of *high utility itemset mining* is to find all high utility itemsets.

Example 10 Continuing the running example, assume that $minUtil = 35$. Then, there are three high utility itemsets, which are $\{b, c, d, e\}$ with a utility of 40, $\{b, c, e\}$ with a utility of 37, and $\{b, d, e\}$ with a utility of 36.

To find high utility itemsets, several efficient algorithms have been designed such as Two-phase [50], HUP-Growth [48], FHM [30], ULB-Miner [14], mHUIMiner [58],

Table 3 The set of PHUIs in the running example

Itemset	u(X)	sup(X)	minper(X)	maxper(X)	avgper(X)
{b, e}	31	3	1	3	1.75
{b, c, e}	37	3	1	3	1.75
{b, c}	28	3	1	3	1.75
{a, c}	34	4	1	2	1.4

EFIM [81], and HUI-Miner [49]. A good overview of techniques and algorithms for high utility itemset mining can be found in a recent survey [27].

The concept of high utility itemset has combined with that of periodic patterns to find patterns that are not only periodic but also have a high importance (yield a large profit). This problem called periodic high utility itemset mining [23] is defined as follows.

Definition 11 (Periodic high utility itemset mining) An itemset X is considered to be a *periodic high utility itemset* (PHUI) if it satisfies four conditions: (1) $minAvg \leq avgper(X) \leq maxAvg$, (2) $minper(X) \geq minPer$, (3) $maxper(X) \leq maxPer$, and (4) $u(X) \geq minUtil$, where $minAvg$, $maxAvg$, $minPer$, $maxPer$ and $minUtil$ are positive numbers.

Example 11 Consider again the sequence of Table 2 as example. Let $minPer = 1$, $maxPer = 3$, $minAvg = 1$, $maxAvg = 2$, and $minUtil = 28$. The periodic high utility itemsets discovered in this data are listed in Table 3.

It can be shown that the traditional problem of periodic frequent itemset mining is a special case of that problem where all internal and external utility values are set to 0 or 1 and only condition (3) is used.

3.2 The PHM Algorithm

This subsection describes the PHM algorithm to discover all periodic high utility itemsets in a quantitative transaction database, where transactions are ordered by time or other criteria. The PHM algorithm is an extension of the FHM [30] algorithm, a popular, simple, and efficient algorithm for mining high utility itemsets, which is an improved version of the HUI-Miner algorithm [49].

The search space of periodic high utility itemset mining contains $2^{|I|} - 1$ possible itemsets. For instance, in the running example, $I = \{a, b, c, d, e\}$. Hence, there are $2^{|5|} - 1 = 31$ possible itemsets such as $\{a\}, \{b\}, \dots \{a, b\}, \{a, c\} \dots \{b, c, d\}, \{b, c, e\}$ $\dots \{a, b, c, d, e\}$. A naive algorithm to find all periodic high utility itemsets would scan the database and calculate the utility and periods of all possible itemsets to find the solution. But this is inefficient as the search space can be very large. The PHM algorithm adopts a different approach to avoid looking at all possibilities. The next paragraphs introduce the key ideas of this algorithm, and then the pseudocode is presented.

For the purpose of processing, the PHM algorithm supposes that there exists a *total order* $\succ$ on the items from I. This order $\succ$ can be any order such as the alphabetical order. This order is used by PHM to search for itemsets in a systematic way, that is to not look at the same itemset more than once.

To eliminate many itemsets that are not high utility itemsets from the search space, PHM uses the *Transaction-Weighted Utilization (TWU)* measure, which was proposed by Liu et al. [50].

Definition 12 (Transaction utility, TWU) Let there be an itemset X and a transaction T. The notation $TU(T)$ denotes the *transaction utility* of transaction T, which is defined as $TU(T) = \sum_{x \in T} u(x, T)$. The notation $TWU(X)$ denotes the TWU of X, which is defined as $TWU(X) = \sum_{X \subseteq T \in D)} TU(T)$.

Example 12 Following the previous example, the transactions $T_1, T_2, \ldots T_7$, respectively have transaction utility values of $6, 3, 25, 20, 8, 22$ and 9. The TWU of the itemset $\{b, c\}$ is $TWU(\{b, c\}) = TU(T_3) + TU(T_4) + TU(T_7) = 25 + 20 + 9 = 54$.

A powerful property for reducing the search space using the TWU is the following [50].

Theorem 1 *(Pruning search space using the TWU) For an itemset X, if $TWU(X) < minUtil$, then $u(X) < minUtil$, and for any superset $Y \supset X$, we have $u(Y) < minUtil$. In other words, both X and Y are not high utility itemsets and can be ignored.*

To be able to calculate the utility and periods of itemsets, the PHM algorithms utilize a structure called *utility list* [49], which is defined as follows. Initially, PHM reads the database to create the utility list of each itemset containing one item. Then, PHM generates utility lists of larger itemsets by joining utility lists of smaller itemsets. The advantage of this approach is that all itemsets can be explored without having to repeatedly read the database. The utility list structure is defined as follows.

Definition 13 (Utility list) Recall that $\succ$ is a total order defined on items in I. Let there be an itemset X and a database s. The utility list of X is denoted as $ul(X)$ and is defined as a list of tuples where there is a tuple $(tid, iutil, rutil)$ for each transaction T where $X \subseteq T$. The *iutil* element of a tuple stores the value $u(X, T)$. The *rutil* element of a tuple stores the value $\sum_{i \in T \wedge i \succ x \forall x \in X} u(i, T)$ called *remaining utility*.

Example 13 For instance, assume that $\succ$ is the alphabetical order. The utility list of $\{a\}$ contains four tuples: $\{(T_1, 5, 1), T_3, 5, 20), (T_5, 5, 3), (T_6, 10, 12)\}$. The utility list of $\{d\}$ has three tuples: $\{(T_3, 6, 3), (T_4, 6, 3), (T_5, 2, 0)\}$. And the utility list of $\{a, d\}$ contain two tuples: $\{(T_3, 11, 3), (T_5, 7, 0)\}$.

The utility lists of itemsets having single items are constructed by reading the database. For itemsets having more than one item, the following join operation is used to build the utility list [49]. Let there be two items $x \succ y$ such that $x, y \in I$. The utility list of itemset $\{x, y\}$, denoted as $ul(\{x, y\})$ can be created by adding a tuple $(ex.tid, ex.iutil + ey.iutil, ey.rutil)$ to $ul(\{x, y\})$ for each pair of tuples $ex \in ul(\{x\})$ and $ey \in ul(\{y\})$ such that $ex.tid = ey.tid$. For itemsets containing more than two items, the join operation is done as follows. Let there be two itemsets $P \cup \{x\}$ and $P \cup \{y\}$ such that $x \succ y$ and $P \subset I$. The utility list of itemset $P \cup \{x, y\}$, denoted as $ul(P \cup \{x, y\})$ is created by adding a tuple $(ex.tid, ex.iutil + ey.iutil - ep.iutil, ey.rutil)$ to $ul(P \cup \{x, y\})$ for each set of tuples $ex \in ul(\{x\})$, $ey \in ul(\{y\})$, $ep \in ul(P)$ such that $ex.tid = ey.tid = ep.tid$.

The utility list of an itemset is useful as it allows to directly obtain its utility and to reduce the search space. The utility of an itemset X is simply the sum of the $iutil$ values in $ul(X)$. To reduce the search space using the utility, the following property is used [49]:

Theorem 2 *(Search space reduction with utility list) For an itemset X, an extension of X is an itemset that is obtained by appending an item y to X such that $y \succ i$, $\forall i \in X$. If the sum of $iutil$ and $rutil$ values in $ul(X)$ is less than $minUtil$, X and its extensions are low utility itemsets and can be ignored.*

The utility list structure can also be used to directly calculate the periods of an itemset X, and thus determine if it is a periodic itemset. This can be done by looking at the *tid* elements of the utility list of X.

Pseudocode. The pseudocode of the main procedure of the PHM algorithm is shown in Algorithm 1. The input is a quantitative transaction database as well as the five thresholds $minUtil, minAvg, maxAvg, minPer$ and $maxPer$. The algorithm outputs the set of all periodic high utility itemsets. PHM initially reads the input database to calculate the following values for each item $i \in I$: $sup(\{i\}), TWU(\{i\})$, $minper(\{i\})$ and $maxper(\{i\})$. Moreover, PHM calculates $\gamma = (|s|/maxAvg) - 1$, which is needed by Property 3 to reduce the search space. Afterward, a set of items I^* is created by PHM, which contains each item i that has a TWU value that is greater or equal to $minUtil$, a maximum periodicity that is no less than $maxPer$, and where i appears in at least γ transactions (as per the Property 3). Thereafter, the total order $\succ$ on items is established as in the HUI-Miner algorithm [49] as the order of ascending TWU values and the alphabetical order when two items have the same TWU. The next action done by PHM is to read the database again and sort in memory each transaction according to $\succ$. At the same time, a utility list is built for each item $i \in I^*$. Then, PHM starts to recursively search for periodic high utility itemsets by calling the *Search* procedure with the following parameters: the empty itemset $\emptyset$, the set $I^*, \gamma, minUtil, minAvg, minPer, maxPer$ and $|s|$.

Algorithm 2 presents the *Search* procedure. It receives an itemset P, some extensions of P that have the form Pz (which was obtained by adding an item z to P), as well as the original input parameters and the input sequence length $|s|$. The first step done by this procedure is to iterate over the extensions in P. For each such extension

---

Algorithm 1: The PHM algorithm

---

input : s: a transaction database,
         $minUtil$, $minAvg$, $maxAvg$, $minPer$ and $maxPer$: the thresholds
output: the set of periodic high utility itemsets

1 Scan s once to calculate $TWU(\{i\})$, $minper(\{i\})$, $maxper(\{i\})$, and $|g(\{i\})|$ for each item
   $i \in I$;
2 $\gamma \leftarrow (|s|/maxAvg) - 1$;
3 $I^* \leftarrow$ each item i such that $\text{TWU}(i) \geq minUtil$, $sup(\{i\}) \geq \gamma$ and
   $maxper(\{i\}) \leq maxPer$;
4 Establish the total order $\succ$ on I^* as that of TWU ascending values;
5 Search $(\emptyset, I^*, \gamma, minUtil, minAvg, minPer, maxPer, |s|)$;

Px, the procedure first calculates the average periodicity of Px as the ratio of $|s|$ to the number of elements in the utility list of Px plus one. Thereafter, the procedure checks if the following conditions are met: (1) the average periodicity of Px is no less than $minAvg$ and no greater than $maxAvg$, (2) the sum of the *iutil* values of the utility list of Px is no less than $minUtil$ (by Theorem 2), and (3) according to the utility list of Px, the minimum (maximum) periodicity of Px is no less (not greater) than $minPer$ ($maxPer$). If these conditions are met, then Px is output as a periodic high utility itemset. Afterward, if the number of elements in the utility list of Px is at least γ, the sum of *iutil* and *rutil* values in the utility list of Px is no less than the minimum utility treshold, and $maxper(Px)$ is no greater than $maxPer$, extensions of Px will be considered as potential periodic high utility itemsets (based on Theorem 2 and Properties 2 and 3). To explore these extensions, a loop is done where each extension Py of P is merged with Px to produce a new extension Pxy having $|Px|+1$ items. The *Construct* procedure of FHM [30] is applied to combine the utility lists of P, Px and Py to generate the utility list of Pxy, with some minor modifications. The difference is that periods are calculated for Pxy while building its utility list so that $maxPer(Pxy)$ and $minPer(Pxy)$ can be obtained. Then, the *Search* procedure is recursively invoked with Px and all extensions of the form Pxy to explore the search space and find all periodic high utility itemsets that are transitive extensions of Px.

When the PHM algorithm terminates all periodic high utility itemsets have been output. This can be demonstrated by observing that the algorithm can recursively explore the whole search space and only applies Theorem 2, Properties 2 and 3, to eliminate non-periodic high utility itemsets.

Optimizations. Note that for a fast implementation of PHM, there are several possible optimizations. Three optimizations proposed in PHM are called (1) Estimated-Utility Co-occurrence Pruning, (2) Estimated Average Periodicity Pruning, and (3) Abandoning List Construction early. Details about these optimizations can be found in the PHM paper [23]. Besides, several other optimizations used in other HUI-Miner or FHM-based algorithms could be integrated into PHM to further enhance its performance such as the memory-buffering technique of ULB-Miner [14].

Algorithm 2: The *Search* procedure

input : P: an itemset, *ExtensionsOfP*: a set of extensions of P, γ, $minUtil$, $minAvg$, $minPer$, $maxPer$, $|s|$

output: the set of periodic high utility itemsets

1 **foreach** *itemset* $Px \in$ *ExtensionsOfP* **do**
2      $avgperPx \leftarrow |s|/(|Px.utilitylist| + 1)$;
3   **if** $SUM(Pxy.utilitylist.iutils) \geq minUtil \wedge minAvg \leq avgperPx \leq$
     $maxAvg \wedge Px.utilitylist.minp \geq minPer \wedge Px.utilitylist.maxp \leq maxPer$ **then**
     output Px **if** $SUM(Px.utilitylist.iutils)+SUM(Px.utilitylist.rutils) \geq minUtil \wedge$
     $avgperPx \geq \gamma$ *and* $Px.utilitylist.maxp \leq maxPer$ **then**
4   *ExtensionsOfPx* $\leftarrow \emptyset$;
5   **foreach** *itemset* $Py \in$ *ExtensionsOfP* *such that* $y \succ x$ **do**
6              $Pxy \leftarrow Px \cup Py$;
7              $Pxy.utilitylist \leftarrow$ Construct (P, Px, Py);
8   *ExtensionsOfPx* $\leftarrow$ *ExtensionsOfPx* $\cup \{Pxy\}$;
9   **end**
10   Search $(Px, ExtensionsOfPx, \gamma, minUtil, minAvg, minPer, maxPer, |s|)$;
11   **end**
12 **end**

Implementation and datasets. The original Java implementation of PHM with all optimizations, and datasets are offered in the open-source SPMF data mining library [25] at http://www.philippe-fournier-viger.com/spmf/.

4 Irregular High Utility Itemset Mining

The problem of mining periodic high utility itemsets reviewed in the previous section is interesting as it can reveal itemsets that periodically appear in a sequence of quantitative transactions and also have a high importance (e.g., yield a high profit). This section presents a related problem, which is that of discovering *irregular high utility itemsets*. Intuitively, an irregular itemset is an itemset that typically has a long time delay between each of its consecutive occurrences. The problem of irregular itemset mining [45] is defined as follows.

Definition 14 (Regularity) The regularity of an itemset X in a sequence of quantitative transactions s, denoted as $reg(X)$, is the smallest period among the periods in $pr(X, s)$, when the first and last periods are excluded.

Example 14 Consider the sequence s of Table 1 and the itemsets $X = \{b, c\}$ and $Y = \{a, c\}$. The periods of X in s are $pr(X, s) = \{3, 1, 3, 0\}$ while that of Y in s are $pr(Y, s) = \{1, 2, 2, 1, 1\}$. Hence, the regularity of X and Y in s are, respectively, $reg(X) = 1$ and $reg(Y) = 1$.

Definition 15 (Irregular high utility itemset mining) An itemset X is considered to be an *irregular high utility itemset* (IHUI) if it satisfies two conditions: (1)

$reg(X) \geq minReg$, and (2) $u(X) \geq minUtil$, where $minReg$ and $minUtil$ are positive numbers.

The PHM_irregular Algorithm. The problem of discovering irregular high utility itemsets can be solved using the PHM algorithm presented in the previous section by simply setting $minPer = minReg$, $maxPer = \infty$, $minAvg = 0$ and $maxAvg = \infty$. The original implementation of that variation of PHM is called $PHM\_irregular$ and is available in the open-source SPMF pattern mining library [25] at http://www.philippe-fournier-viger.com/spmf/.

5 Other Variations and Research Opportunities

The previous section has presented the basic problem of periodic high utility itemset mining and the variation of irregular high utility itemset mining. Some other variations have been proposed such as (1) mining periodic high utility sequential patterns [9, 11–13, 46, 47, 59, 60] where each transaction is a quantitative sequence, (2) productive-associated periodic high utility itemsets mining [41], where a statistical test is used to filter spurious patterns, and (3) partial periodic high utility itemsets [63], which utilizes a different measure of periodicity.

There are several possibilities for future work. A few of them are:

- Designing faster and more memory-efficient algorithms for periodic high utility itemset mining by taking advantage of the numerous work on high utility itemset mining [27], or other optimizations in periodic itemset mining such as approximate periodicity calculations [5].
- Proposing new problems by drawing inspiration from variations of the high utility itemset mining problem such as considering multiple minimum utility thresholds [64], using the average utility function [68], discovering the top-k high itemsets having the highest utility [15, 66], mining high utility itemsets in a stream or incrementally updated data [77, 78], discovering on-shelf high utility itemsets [37], finding a summary of all high utility itemsets [29], and considering negative utility values [17].
- Using different measures of periodicity or other measures related to time such as the stability [28, 33].
- Applying periodic high utility itemset mining in new applications such as for smart homes [53], intelligent systems [43], location prediction [71] and sequence prediction [38].
- Designing measures to find periodic high utility itemsets common to multiple sequences similarly to studies on mining periodic patterns common to multiple sequences [21, 31],
- Integrating various correlation measures in the mining process to filter spurious patterns besides the one that was used in productive-associated periodic high utility itemset mining [41], such as the bond [22, 36, 57, 76], affinity [2], all-confidence [57, 70], coherence and mean [7, 67].

- Combining the concept of high utility periodic patterns with other concepts such as episode patterns [6, 34, 35], subgraphs [18, 19, 39, 42], sequential patterns [20, 26, 69, 74, 75], trajectory patterns [80], and periodic patterns with gap constraints [72, 73], and clustering [8, 10, 51].

6 Conclusion

This chapter has presented an overview of how to discover periodic and irregular high utility itemsets in a sequence of quantitative transactions (also called a transaction database). The problems have been described and two algorithms have been explained, namely, PHM and PHM_irregular. There are several possible research opportunities on this topic. Some of them have been listed in this chapter.

References

1. R. Agrawal, R. Srikant, Fast algorithms for mining association rules in large databases. Proceedings of 20th International Conference on Very Large Data Bases (1994), pp. 487–499
2. C.F. Ahmed, S.K. Tanbeer, B. Jeong, H. Choi, A framework for mining interesting high utility patterns with a strong frequency affinity. Inf. Sci. **181**(21), 4878–4894 (2011)
3. K. Amphawan, P. Lenca, A. Surarerks, Mining top-K periodic-frequent pattern from transactional databases without support threshold. Proceedings of the Third International Conference on Advanced in Information Technology (2009), pp. 18–29
4. K. Amphawan, P. Lenca, A. Surarerks, Mining top-K periodic-frequent pattern from transactional databases without support threshold. Proceedings of the Third International Conference on Advances in Information Technology (2009), pp. 18–29. 10.1007/978-3-642-10392-6_3
5. K. Amphawan, A. Surarerks, P. Lenca, Mining periodic-frequent itemsets with approximate periodicity using interval transaction-ids list tree. 2010 Third International Conference on Knowledge Discovery and Data Mining (IEEE, New York, 2010), pp. 245–248
6. X. Ao, H. Shi, J. Wang, L. Zuo, H. Li, Q. He, Large-scale frequent episode mining from complex event sequences with hierarchies. ACM Transactions on Intelligent Systems and Technology (TIST) **10**(4), 1–26 (2019)
7. M. Barsky, S. Kim, T. Weninger, J. Han, Mining flipping correlations from large datasets with taxonomies. Proc. VLDB Endow. **5**, 370–381 (2011)
8. D.T. Dinh, V.N. Huynh, k-pbc: an improved cluster center initialization for categorical data clustering. Appl. Intell. **50**, 1–23 (2020)
9. D.T. Dinh, V.N. Huynh, B. Le, P. Fournier-Viger, U. Huynh, Q.M. Nguyen, A survey of privacy preserving utility mining. High-Utility Pattern Mining (2019), pp. 207–232
10. D.T. Dinh, V.N. Huynh, S. Songsak, Clustering mixed numerical and categorical data with missing values. Inf. Sci. **571**, 418–442 (2021)
11. D.T. Dinh, B. Le, P. Fournier-Viger, V.N. Huynh, An efficient algorithm for mining periodic high-utility sequential patterns. Appl. Intell. **48**(12), 4694–4714 (2018)
12. T. Dinh, V.N. Huynh, B. Le, Mining periodic high utility sequential patterns. Proceedings of the 2017 International Conference on Intelligent Information and Database Systems (Springer, Berlin, 2017), pp. 545–555
13. T. Dinh, M.N. Quang, B. Le, A novel approach for hiding high utility sequential patterns. Proceedings of the 6th International Symposium on Information and Communication Technology (2015), pp. 121–128

14. Q.H. Duong, P. Fournier-Viger, H. Ramampiaro, K. Nørvåg, T.L. Dam, Efficient high utility itemset mining using buffered utility-lists. Appl. Intell. **48**(7), 1859–1877 (2018)
15. Q.H. Duong, B. Liao, P. Fournier-Viger, T.L. Dam, An efficient algorithm for mining the top-k high utility itemsets, using novel threshold raising and pruning strategies. Knowl.-Based Syst. **104**, 106–122 (2016)
16. Z. Feng, Y. Zhu, A survey on trajectory data mining: Techniques and applications. IEEE Access **4**, 2056–2067 (2016)
17. P. Fournier-Viger, Fhn: efficient mining of high-utility itemsets with negative unit profits. International Conference on Advanced Data Mining and Applications (Springer, Berlin, 2014), pp. 16–29
18. P. Fournier-Viger, C. Cheng, J.C.W. Lin, U. Yun, R.U. Kiran, Tkg: Efficient mining of top-k frequent subgraphs. Proceedings of the 7th International Conference on Big Data Analytics (Springer, Berlin, 2019), pp. 209–226
19. P. Fournier-Viger, G. He, C. Cheng, J. Li, M. Zhou, J.C.W. Lin, U. Yun, A survey of pattern mining in dynamic graphs. Wiley Interdisciplinary Reviews: Data Mining and Knowledge Discovery **10**(6), e1372 (2020)
20. P. Fournier-Viger, J. Li, J.C.W. Lin, T. Truong, Discovering low-cost high utility patterns. Data Science and Pattern Recognition **4**(2), 50–64 (2020)
21. P. Fournier-Viger, Z. Li, J.C. Lin, R.U. Kiran, H. Fujita, Discovering periodic patterns common to multiple sequences. Proceedings of the 20th International Conference on Big Data Analytics and Knowledge Discovery (2018), pp. 231–246. https://doi.org/10.1007/978-3-319-98539-8_18
22. P. Fournier-Viger, J.C., Lin, T. Dinh, H.B. Le, Mining correlated high-utility itemsets using the bond measure. Proceedings of the 11th International Conference on Hybrid Artificial Intelligent Systems (2016), pp. 53–65. https://doi.org/10.1007/978-3-319-32034-2_5
23. P. Fournier-Viger, J.C. Lin, Q. Duong, T. Dam, PHM: mining periodic high-utility itemsets. Proceedings of the 16th Industrial Conference, ICDM 2016, ed. by P. Perner (Springer, New York, 2016), pp. 64–79. https://doi.org/10.1007/978-3-319-41561-1_6
24. P. Fournier-Viger, J.C. Lin, B. Vo, T.C. Truong, J. Zhang, H.B. Le, A survey of itemset mining. Wiley Interdiscip. Rev. Data Min. Knowl. Discov. **7**(4), e1207 (2017)
25. P. Fournier-Viger, J.C.W. Lin, A. Gomariz, T. Gueniche, A. Soltani, Z. Deng, H.T. Lam, The spmf open-source data mining library version 2. Joint European Conference on Machine Learning and Knowledge Discovery in Databases (Springer, Berlin, 2016), pp. 36–40
26. P. Fournier-Viger, J.C.W. Lin, U.R. Kiran, Y.S. Koh, A survey of sequential pattern mining. Data Science and Pattern Recognition **1**(1), 54–77 (2017)
27. P. Fournier-Viger, J.C.W. Lin, T. Truong-Chi, R. Nkambou, A survey of high utility itemset mining. High-Utility Pattern Mining (Springer, Berlin, 2019), pp. 1–45
28. P. Fournier-Viger, Y. Wang, P. Yang, J.C.W. Lin, Y. Unil, A survey of sequential pattern mining. Appl. Intell. Data Sci. Pattern Recog. **1**(1), 54–77 (2021)
29. P. Fournier-Viger, C.W. Wu and V.S. Tseng, Novel concise representations of high utility itemsets using generator patterns. International Conference on Advanced Data Mining and Applications (Springer, Berlin, 2014), pp. 30–43
30. P. Fournier-Viger, C.W. Wu, S. Zida, V.S. Tseng, Fhm: Faster high-utility itemset mining using estimated utility co-occurrence pruning. International Symposium on Methodologies for Intelligent Systems (Springer, Berlin, 2014), pp. 83–92
31. P. Fournier-Viger, P. Yang, Z. Li, J.C.W. Lin, R.U. Kiran, Discovering rare correlated periodic patterns in multiple sequences. Data Knowl. Eng. **126**, 101–733 (2020). https://doi.org/10.1016/j.datak.2019.101733
32. P. Fournier-Viger, P. Yang, J.C.W. Lin, Q.H. Duong, T. Dam, L. Sevcik, D. Uhrin, M. Voznak, Discovering periodic itemsets using novel periodicity measures. Advances in Electrical and Electronic Engineering **17**(1), 33–44 (2019)
33. P. Fournier-Viger, P. Yang, J.C.W. Lin and R.U. Kiran, Discovering stable periodic-frequent patterns in transactional data. Proceedings of the 32nd International Conference on Industrial, Engineering and Other Applications of Applied Intelligent Systems (Springer, Berlin, 2019), pp. 230–244

34. P. Fournier-Viger, P. Yang, J.C.W. Lin and U. Yun, Hue-span: Fast high utility episode mining. Proceedings of the 14th International Conference on Advanced Data Mining and Applications (Springer, Berlin, 2019), pp. 169–184
35. P. Fournier-Viger, Y. Yang, P. Yang, J.C.W. Lin, U. Yun, Tke: Mining top-k frequent episodes. Proceedings of the 33rd International Conference on Industrial, Engineering and Other Applications of Applied Intelligent Systems (Springer, Berlin, 2020)
36. P. Fournier-Viger, Y. Zhang, J.C.W. Lin, D.T. Dinh, H. Le Bac, Mining correlated high-utility itemsets using various measures. Logic Journal of the IGPL **28**(1), 19–32 (2020)
37. P. Fournier-Viger, S. Zida, Foshu: faster on-shelf high utility itemset mining–with or without negative unit profit. Proceedings of the 30th Annual ACM Symposium on Applied Computing (ACM, New York, 2015), pp. 857–864
38. T. Gueniche, P. Fournier-Viger, R. Raman, V.S. Tseng, Cpt+: Decreasing the time/space complexity of the compact prediction tree. Pacific-Asia Conference on Knowledge Discovery and Data Mining (Springer, Berlin, 2015), pp. 625–636
39. S. Halder, M. Samiullah, Y.K. Lee, Supergraph based periodic pattern mining in dynamic social networks. Expert Syst. Appl. **72**, 430–442 (2017)
40. J. Han, J. Pei, Y. Yin, R. Mao, Mining frequent patterns without candidate generation: A frequent-pattern tree approach. Data Min. Knowl. Discov. **8**(1), 53–87 (2004)
41. W. Ismail, M.M. Hassan, G. Fortino, Productive-associated periodic high-utility itemsets mining. 2017 IEEE 14th International Conference on Networking, Sensing and Control (ICNSC) (IEEE, New York, 2017), pp. 637–642
42. C. Jiang, F. Coenen, M. Zito, A survey of frequent subgraph mining algorithms. Knowl. Eng. Rev. **28**, 75–105 (2013)
43. Kim, H., Yun, U., Vo, B., Lin, J.C.W., Pedrycz, W.: Periodicity-oriented data analytics on time-series data for intelligence system. IEEE Systems Journal (2020)
44. V. Kumar, V. Kumari, Incremental mining for regular frequent patterns in vertical format. International Journal of Engineering and Technology **5**(2), 1506–1511 (2013)
45. S. Laoviboon, K. Amphawan, Mining high-utility irregular itemsets. In: High-Utility Pattern Mining (Springer, Berlin, 2019), pp. 175–205
46. B. Le, D.T. Dinh, V.N. Huynh, Q.M. Nguyen, P. Fournier-Viger, An efficient algorithm for hiding high utility sequential patterns. Int. J. Approximate Reasoning **95**, 77–92 (2018)
47. B. Le, U. Huynh, D.T. Dinh, A pure array structure and parallel strategy for high-utility sequential pattern mining. Expert Syst. Appl. **104**, 107–120 (2018)
48. C.W. Lin, T.P. Hong, W.H. Lu, An effective tree structure for mining high utility itemsets. Expert Syst. Appl. **38**(6), 7419–7424 (2011)
49. M. Liu, J. Qu, Mining high utility itemsets without candidate generation. Proceedings of the 21st ACM international conference on Information and Knowledge Management (ACM, New York, 2012), pp. 55–64
50. Y. Liu, W.K. Liao, A. Choudhary, A two-phase algorithm for fast discovery of high utility itemsets. Pacific-Asia Conference on Knowledge Discovery and Data Mining (Springer, Berlin, 2005), pp. 689–695
51. N.V. Lu, T.N. Vuong, D.T. Dinh, Combining correlation-based feature and machine learning for sensory evaluation of Saigon beer. International Journal of Knowledge and Systems Science (IJKSS) **11**(2), 71–85 (2020)
52. J.M. Luna, P. Fournier-Viger, S. Ventura, Frequent itemset mining: A 25 years review. Wiley Interdisciplinary Reviews: Data Mining and Knowledge Discovery **9**(6), e1329 (2019)
53. I. Mukhlash, D. Yuanda, M. Iqbal, Mining fuzzy time interval periodic patterns in smart home data. International Journal of Electrical and Computer Engineering **8**(5), 3374 (2018)
54. S. Nawaz, P. Fournier-Viger, A. Shojaee, H. Fujita, Using artificial intelligence techniques for covid-19 genome analysis. Appl. Intell. **51**(5), 3086–3103 (2021)
55. V.M. Nofong, Discovering productive periodic frequent patterns in transactional databases. Annals of Data Science **3**(3), 235–249 (2016)
56. V.M. Nofong, Fast and memory efficient mining of periodic frequent patterns. Proceedings of the 10th Asian Conference on Intelligent Information and Database Systems (Springer, Berlin, 2018), pp. 223–232

57. E. Omiecinski, Alternative interest measures for mining associations in databases. IEEE Trans. Knowl. Data Eng. **15**(1), 57–69 (2003). https://doi.org/10.1109/TKDE.2003.1161582
58. A.Y. Peng, Y.S. Koh and P. Riddle, mhuiminer: A fast high utility itemset mining algorithm for sparse datasets. Pacific-Asia Conference on Knowledge Discovery and Data Mining (Springer, Berlin, 2017), pp. 196–207
59. M.N. Quang, T. Dinh, U. Huynh and B. Le, MHHUSP: An integrated algorithm for mining and Hiding High Utility Sequential Patterns. Proceedings of the 8th International Conference on Knowledge and Systems Engineering (IEEE, New York, 2016), pp. 13–18
60. M.N. Quang, U. Huynh, T. Dinh, N.H. Le and B. Le, An Approach to Decrease Execution Time and Difference for Hiding High Utility Sequential Patterns. Proceedings of the 5th International Symposium on Integrated Uncertainty in Knowledge Modelling and Decision Making (Springer, Berlin, 2016), pp. 435–446
61. M.M. Rashid, I. Gondal, and J. Kamruzzaman, Regularly frequent patterns mining from sensor data stream. Proceedings of the 20th International Conference on Neural Information Processing (Springer, Berlin, 2013), pp. 417–424
62. M.M. Rashid, M.R. Karim, B.S. Jeong and H.J. Choi, Efficient mining regularly frequent patterns in transactional databases. Proceedings of the 17th International Conference on Database Systems for Advanced Applications (Springer, Berlin, 2012), pp. 258–271
63. T.Y. Reddy, R.U. Kiran, M. Toyoda, M., P.K. Reddy and M. Kitsuregawa, Discovering partial periodic high utility itemsets in temporal databases. International Conference on Database and Expert Systems Applications (Springer, Berlin, 2019), pp. 351–361
64. H. Ryang, U. Yun, K.H. Ryu, Discovering high utility itemsets with multiple minimum supports. Intelligent data analysis **18**(6), 1027–1047 (2014)
65. S. Shekhar, M.R. Evans, J.M. Kang, P. Mohan, Identifying patterns in spatial information: A survey of methods. Wiley Interdisciplinary Reviews: Data Mining and Knowledge Discovery **1**(3), 193–214 (2011)
66. W. Song, L. Liu and C. Huang, Tku-ce: Cross-entropy method for mining top-k high utility itemsets. International Conference on Industrial, Engineering and Other Applications of Applied Intelligent Systems (Springer, Berlin, 2020), pp. 846–857
67. A. Soulet, C. Raïssi, M. Plantevit and B. Crémilleux, Mining dominant patterns in the sky. Proceedings of the 11th IEEE International Conference on Data Mining (IEEE, New York, 2011), pp. 655–664
68. T. Truong, H. Duong, B. Le, P. Fournier-Viger, Efficient vertical mining of high average-utility itemsets based on novel upper-bounds. IEEE Trans. Knowl. Data Eng. **31**(2), 301–314 (2018)
69. T. Truong, A. Tran, H. Duong, B. Le, P. Fournier-Viger, Ehusm: Mining high utility sequences with a pessimistic utility model. Data Science and Pattern Recognition **4**(2), 65–83 (2020)
70. J.N. Venkatesh, R.U. Kiran, P.K. Reddy and M. Kitsuregawa, Discovering periodic-frequent patterns in transactional databases using all-confidence and periodic-all-confidence. Proceedings of the 27th International Conference on Database and Expert Systems Applications Part I (Springer, Berlin, 2016), pp. 55–70
71. M.H. Wong, V.S. Tseng, J.C. Tseng, S.W. Liu and C.H. Tsai, Long-term user location prediction using deep learning and periodic pattern mining. Proceedings of the 12th Conference on Advanced Data Mining and Applications (Springer, Berlin, 2017), pp. 582–594
72. Y. Wu, C. Shen, H. Jiang, X. Wu, Strict pattern matching under non-overlapping condition. SCIENCE CHINA Inf. Sci. **60**(1), 1–16 (2017)
73. Y. Wu, Y. Tong, X. Zhu, X. Wu, Nosep: Nonoverlapping sequence pattern mining with gap constraints. IEEE transactions on cybernetics **48**(10), 2809–2822 (2017)
74. Y. Wu, L. Wang, J. Ren, W. Ding, X. Wu, Mining sequential patterns with periodic wildcard gaps. Appl. Intell. **41**(1), 99–116 (2014)
75. Y. Wu, C. Zhu, Y. Li, L. Guo, X. Wu, Netncsp: Nonoverlapping closed sequential pattern mining. Knowl.-Based Syst. **196**, 105–812 (2020)
76. N.B. Younes, T. Hamrouni and S.B. Yahia, Bridging conjunctive and disjunctive search spaces for mining a new concise and exact representation of correlated patterns. Proceedings of the 13th International Conference on Discovery Science (Springer, Berlin, 2010), pp. 189–204

77. U. Yun, D. Kim, E. Yoon, H. Fujita, Damped window based high average utility pattern mining over data streams. Knowl.-Based Syst. **144**, 188–205 (2018)
78. U. Yun, H. Ryang, G. Lee, H. Fujita, An efficient algorithm for mining high utility patterns from incremental databases with one database scan. Knowl.-Based Syst. **124**, 188–206 (2017)
79. M.J. Zaki, Scalable algorithms for association mining. IEEE Trans. Knowl. Data Eng. **12**(3), 372–390 (2000). https://doi.org/10.1109/69.846291
80. D. Zhang, K. Lee, I. Lee, Hierarchical trajectory clustering for spatio-temporal periodic pattern mining. Expert Syst. Appl. **92**, 1–11 (2018)
81. S. Zida, P. Fournier-Viger, J.C.W. Lin, C.W. Wu, V.S. Tseng, Efim: a fast and memory efficient algorithm for high-utility itemset mining. Knowl. Inf. Syst. **51**(2), 595–625 (2017)

Mining Periodic High-Utility Sequential Patterns with Negative Unit Profits

Ut Huynh, Bac Le, Duy-Tai Dinh, and Van-Nam Huynh

Abstract This chapter focuses on mining periodic high-utility sequential patterns where external utility values may be positive or negative (PHUSPN). This type of patterns not only yields a high-utility (e.g., high profit) but also appears regularly in a sequence database. Finding PHUSPN is useful for several applications such as market basket analysis, where it can reveal recurring items sold with a negative profit in a package with other items at a higher positive return. Several efficient algorithms have been proposed for the task of mining periodic high-utility sequential patterns or mining high-utility sequential patterns with negative item values. But no work considers the combination of two tasks in the literature although such items occur in many real-life sequence databases. We propose an algorithm name PHUSN to discover such kinds of patterns efficiently. An experimental evaluation was performed on real-life datasets to compare the performance of PHUSN with state-of-the-art algorithms in terms of execution time, memory usage, and the number of generated patterns. Experimental results show that the PHUSN can efficiently discover the complete set of PHUSPN and faster than compared algorithms since it can prune many redundant patterns.

1 Introduction

High-utility sequential pattern mining (HUSPM) is the task of finding sequential patterns that have a high-utility in sequence databases where items may appear zero, once or multiple times in each itemset [1]. In addition, items are associated with weights indicating their unit profit (external utility) or relative importance. HUSPM

U. Huynh · B. Le (✉)
Faculty of Information Technology, University of Science,
Ho Chi Minh City, Vietnam
e-mail: lhbac@fit.hcmus.edu.vn

Vietnam National University, Ho Chi Minh City, Vietnam

D.-T. Dinh · V.-N. Huynh
Japan Advanced Institute of Science and Technology, Nomi, Japan

© The Author(s), under exclusive license to Springer Nature Singapore Pte Ltd. 2021
R. Uday Kiran et al. (eds.), *Periodic Pattern Mining*,
https://doi.org/10.1007/978-981-16-3964-7_9

has been commonly studied for various tasks such as market basket analysis, website clickstream analysis, customer behavior analysis, and stock market analysis. However, a problem of traditional HUSPM algorithms is that they often generate a large amount of patterns, such that a majority of them may be considered uninteresting or redundant depending on the applications and user requirements. For this challenge, many extensions of HUSPM have been proposed by using various constraints in HUSP [2–4]. Such algorithms not only reduce the number of patterns found but also discover more interesting patterns.

A periodic high-utility sequential pattern (PHUSP) is a high-utility pattern that appears regularly in a sequence database. PHUSP mining (PHUSPM) considers the periodic appearance of patterns as a criterion to select interesting HUSP. In customer behavior analysis, when analyzing customer transactions, a retail store manager may be interested in finding the high-utility patterns that appear regularly and have a high sale volume. Detecting these purchase patterns is useful for understanding the behavior of customers and thus adopting effective sales and marketing strategies. In market basket analysis, marketers can use PHUSPM algorithms to detect some sets of products that are sold on approximately a daily or weekly basis. From that, they can better understand the behavior of customers and thus adapt efficient marketing strategies. In website clickstream analysis, the number of clicks or time spent on each web page or user interface element can be viewed as the quantities of item in sequences. Then, administrators can discover the web pages or user interface elements where users spend most of their time and utilize periodically. Based on that, administrators can improve the functions and user interface of websites to better suit these important periodic behaviors.

Although the above algorithm can discover concise and interesting HUSP, they have not designed for the task of finding HUSP where external utility values may be positive or negative [2]. Such patterns also appear commonly in real life. For example, in cross-selling, a product may be sold at negative profit when it is packed with another one with much higher positive return. But again, such algorithms may discover large numbers of patterns that may redundant in some cases. Thus, this chapter considers both periodicity and negative unit profits in HUSPM. It proposes an efficient algorithm named PHUSN to determine whether a HUSP occurs periodically, irregularly, or mostly in specific time intervals in a sequence database that contains item with negative unit profits. To the best of our knowledge, this is the first work that takes account of periodic and negative properties in discovering high-utility patterns. The main objective is to provide a theoretical framework for the community of the same field, as well as a tool for practical uses.

The rest of this chapter is organized as follows. Section 2 reviews related work; Sect. 3 introduces the preliminaries and problem statement; Sect. 4 describes the proposed algorithm; Sect. 5 shows comparative experiment; Sect. 6 draws a conclusion and outlines the direction for the future work.

2 Related Work

2.1 High-Utility Sequential Pattern Mining

The goal of HUSPM is to find all sequential patterns that have a utility greater than or equal to a minimum utility threshold (minUtil) in a sequence database. HUSPM is quite challenging as the utility measure is neither monotone nor anti-monotone unlike the support measure traditionally used in SPM. Numerous algorithms have been proposed for HUSPM and its extension [3–16]. Yin et al. [13] proposed an algorithm named USpan for HUSPM. This algorithm builds a lexicographic q-sequence tree (LQS-Tree) to maintain all generated sequences during the mining process. In addition, it uses two concatenation mechanisms: I-Concatenation and S-Concatenation, in combination with two pruning strategies: width and depth pruning. Wang et al. [11] proposed an algorithm named HUS-Span. The algorithm uses a utility-chain structure to represent the search space of HUSPM. It also introduces two tight utility upper bounds: prefix extension utility (PEU) and reduced sequence utility (RSU), as well as two companion pruning strategies to identify HUSPs. The experimental evaluation showed that HUS-Span outperforms USpan in terms of execution time. The reason is that using PEU and RSU, HUS-Span can generate less candidates than USpan.

Recently, Le et al. [10] proposed two algorithms named AHUS and AHUS-P. The algorithms use a pure array structure (PAS) to represent sequences. This data structure is very compact and contains sufficient information of sequences, thus it can reduce memory usage and effectively support to the mining process. Moreover, the two algorithms use two upper bounds to prune the search space. AHUS-P uses a parallel mining strategy to concurrently discover patterns by sharing the search space to multiple processors. Each processor independently performs its mining task and does not wait for other tasks. AHUS-P is more efficient than the serial AHUS algorithm for large-scale dataset. More recently, Gan et al. [6] proposed an algorithm named ProUM that uses a projection-based utility mining on sequence data. It uses a data structure named utility-array to keep the position, time order, and utility of sequences from the projected sequence dataset. It also utilizes an upper bound named sequence extension utility (SEU) in combination with two pruning strategies called PUO and PUK to reduce the search space. The results show that the ProUM outperforms USpan and HUS-Span algorithms.

For the task of mining HUSP with negative unit profits, Xu et al. [2] proposed an algorithm named HUSP-NIV for mining HUSPN. This algorithm extends the USpan algorithm for the mining process. It redefines the concepts of sequence utility and remaining utility of utility matrix to maintain the utility of a sequence that contains items with negative external utility values.

2.2 Periodic High-Utility Sequential Pattern Mining

For the task of frequent-based mining, Kiran et al. [17] proposed the PFP-growth and the PFP-growth++ algorithms for mining periodic-frequent patterns (PFP). PFP-growth compresses the database into a PF-tree structure and recursively mines the PF-tree to discover all PFP. PFP-growth++ is similar to PFP-growth but it employs an improved PF-tree++ structure instead of the PF-tree. In 2017, Kiran et al. [18] proposed a new interestingness measure called periodic-ratio and a pattern-growth algorithm named GPF-growth to discover partial periodic-frequent patterns. More-over, Kirin et al. [19] proposed a model to find partial periodic itemsets in temporal databases. A measure called periodic-frequency was used to determine the periodic interestingness of itemsets by taking into account their number of cyclic repetitions in the entire data. Recently, Kiran et al. [20] proposed an algorithm to find fuzzy periodic-frequent patterns in a quantitative temporal database. The algorithm uses a pruning technique called improved maximum scalar cardinality to reduce the search space and the computational cost of the mining process. In general, the above algorithms only use the support (frequency) of patterns for PFPM. Thus, they are unable to discover patterns that yield a high profit.

For the task of utility-based mining, Fournier-Viger et al. [21] proposed an algorithm named PHM for mining periodic high-utility itemsets (PHUI) on transaction database. This algorithm combines the concept of periodic itemsets with the concept of high-utility itemsets. It also introduces two novel measures named minimum periodicity and average periodicity to more precisely assess the periodic behavior of patterns. On sequence database, Dinh et al. [3] proposed an algorithm named PHUSPM for mining periodic HUSP. It relies on the USpan algorithm to discover HUSP and defines several properties to obtain PHUSP. However, it is time consuming since it does not use any pruning strategy to reduce the search space of PHUSPM. Moreover, Dinh et al. [4] proposed an algorithm named PHUSN that uses a novel structure called the PUSP structure to facilitate the mining process. It uses a periodic pruning strategy named MPP to prune non-periodic patterns and thus speed up the discovery of PHUSP. Experimental results indicate that PHUSN outperforms four other algorithms in terms of execution time and memory usage.

3 Preliminaries

Given a set of m distinct items $I = \{i_1, i_2, \ldots, i_\Omega\}$. A quantitative (q-) item is a pair of the form (i, q) where $i \in I$ and q is a positive number indicating how many units of this item were purchased, so-called internal utility. Each item $i_k \in I$ ($1 \leq k \leq \Omega$) is associated with a weight denoted as $p(i_k)$ representing the unit profit or relative importance of i_k, so-called external utility. For the problem of HUSPM with negative unit profits, $p(i_k)$ is either a positive or a negative value. A q-itemset $X = [(i_1, q_1)(i_2, q_2) \ldots (i_m, q_m)]$ is a set of one or more q-items where (i_k, q_k) is a

Table 1 External utility values

Item	Quality
a	2
b	5
c	-3
d	4
e	6
f	-1
g	7

Table 2 A q-sequence database

sid	tid	Transactions	tu	su
1	1	$(a, 5)(c, 2)(g, 5)$	45	91
	2	$(a, 3)(b, 1)(c, 3)(f, 2)$	11	
	3	$(b, 3)(d, 2)(e, 2)$	35	
2	1	$(c, 2)(e, 1)$	6	96
	2	$(a, 2)(b, 2)(f, 5)$	14	
	3	$(b, 2)(c, 1)(e, 4)(g, 6)$	76	
3	1	$(a, 1)(b, 1)(e, 3)$	25	82
	2	$(c, 3)(d, 2)(g, 3)$	29	
	3	$(b, 2)(e, 1)$	16	
	4	$(d, 3)$	12	
4	1	$(b, 1)c(1)(e, 2)(g, 5)$	52	114
	2	$(a, 3)(b, 2)(e, 4)(f, 2)$	40	
	3	$(b, 2)(c, 1)(e, 2)$	22	
5	1	$(a, 4)(d, 2)(f, 2)(g, 10)$	86	86

q-item $(1 \leq k \leq m)$. Without loss of generality, assume that q-items in a q-itemset are sorted according to a total order $\prec$ (e.g., the lexicographical order). In addition, the quantity of a q-item i in a q-sequence s is denoted as $q(i, s)$. A q-sequence s is an ordered list of q-itemsets $s = \langle X_1 X_2 \ldots X_n \rangle$ where $X_j (1 \leq j \leq n)$ is a q-itemset. A q-sequence database $SDB = \{s_1, s_2, \ldots, s_N\}$ is a set of N q-sequences where each sequence $s_{id} \in SDB$ $(1 \leq id \leq N)$ is a subset of I.

Example 1 Table 1 shows the items with their external utilities, in which items c and f have negative unit profits. Table 2 represents a q-sequence dataset with five sequences, denoted from s_1 to s_5. Each q-sequence consists of one or several q-itemsets (transactions). Each transaction consists of one or several q-items. For example, the q-sequence s_1 contains three q-itemsets $[(a, 5) (c, 2) (g, 5)]$, $[(a, 3) (b, 1) (c, 3) (f, 2)]$ and $[(b, 3) (d, 2) (e, 2)]$, where the internal utility of q-item c in the first q-itemset and the second q-itemset are 2 and 3, respectively. In the following,

Table 3 The utility matrix of s_1

Item	tid_1	tid_2	tid_3
a	10	6	0
b	0	5	15
c	−6	−9	0
d	0	0	8
e	0	0	12
f	0	−2	0
g	35	0	0

Table 4 The remaining utility matrix of s_1

Item	tid_1	tid_2	tid_3
a	81	40	35
b	81	35	20
c	81	35	20
d	81	35	12
e	81	35	0
f	81	35	0
g	46	35	0

the notation i_{tid} will be used to refer to the occurrence of the item i in the tid-th q-itemset of a q-sequence. For example, in the q-sequence s_1, the notation c_1 means that the q-item c appears in the first q-itemset of s_1, that is $(c, 2)$. Similarly, c_2 represents $(c, 3)$ in the second q-itemset of s_1. Without lost of generality, the lexicographical order $\prec$ is used to sort q-items in each q-itemset of a sequence. For example, $a_1 \prec c_1$, $a_1 \prec a_2$ in q-sequence s_1.

Definition 1 (*Q-sequence utility*) The utility of a q-item (i, q) in a q-sequence s is calculated as $u(i, q) = p(i) \times q(i)$. The utility of a q-itemset X in s is calculated as $u(X) = \sum\limits_{i_k \in X \wedge p(i_k) > 0}^{m} u(i_k, q_k)$. The utility of a q-sequence s is calculated as $u(s) = \sum\limits_{j=1}^{n} u(X_j)$.

Example 2 The utility of a_1 in s_1 is $u(a, 5) = 2 \times 5 = 10$. The utility of $[(a, 5) (c, 2) (g, 5)]$ in s_1 is $u([(a, 5) (c, 2) (g, 5)]) = u(a, 5) + u(g, 5) = 2 \times 5 + 7 \times 5 = 45$. The utility of s_1 is $u(s_1) = u([(a, 5) (c, 2) (g, 5)]) + u([(a, 3) (b, 1) (c, 3) (f, 2)]) + u([(b, 3) (d, 2) (e, 2)]) = 45 + 11 + 35 = 91$.

Definition 2 (*Remaining utility*) Given a q-sequence $s = \langle X_1 X_2 \ldots X_n \rangle$ where $X_k = [(i_{k_1},q_{k_1}) (i_{k_2},q_{k_2}) \ldots (i_{k_m},q_{k_m})]$ is a q-itemset of s. The remaining utility of q-item i_{k_m} in s is denoted and defined as $ru(i_{k_m}, s) = \sum\limits_{i' \in s \wedge i_{k_m} \prec i' \wedge p(i_k) > 0} u(i')$.

Example 3 The values $ru(a_1, s_3)$, $ru(b_1, s_3)$ and $ru(b_3, s_3)$ are respectively equal to 89, 84 and 18.

Definition 3 (*Utility and Remaining utility matrices*) A utility matrix/remaining utility matrix of a q-sequence $s = \langle X_1 X_2 \dots X_n \rangle$ is a $m \times n$ matrix, where m and n are respectively the number of q-items and q-itemsets in s. The element at the position (k, j) $(0 \leq k < m, 0 \leq j < n)$ of the utility matrix stores the utility $u(i_k, q)$ of the q-item (i_k, q) in the q-itemset j. The element at the position (k, j) $(0 \leq k < m, 0 \leq j < n)$ of the remaining utility matrix stores the $ru(i_k, s)$ of q-item (i_k, q) in q-itemset j.

Example 4 The utility and remaining utility matrices of q-sequence s_1 are shown in Tables 3 and 4, respectively.

Definition 4 (*Q-subsequence*) Given two q-itemsets $X = [(i_1, q_1)(i_2, q_2) \dots (i_m, q_m)]$ and $X' = [(i'_1, q'_1)(i'_2, q'_2) \dots (i'_{m'}, q'_{m'})]$, where i_k and $i'_{k'} \in I$. If there exist positive integers $1 \leq j_1 \leq j_2 \leq \dots \leq j_m \leq m'$, such that $i_1 = i'_{j_1} \wedge q_1 = q'_{j_1}, i_2 = i'_{j_2} \wedge q_2 = q'_{j_2}, \dots, i_m = i'_{j_m} \wedge q_m = q'_{j_m}$ then X' is said to contain X, which is denoted as $X \subseteq X'$.

Given two q-sequences $s = \langle X_1 X_2 \dots X_n \rangle$ and $s' = \langle X'_1 X'_2 \dots X'_{n'} \rangle$ $(n \leq n')$, where $X_k, X'_{k'}$ are q-itemsets $(1 \leq k \leq n, 1 \leq k' \leq n')$. If there exists positive integers $1 \leq j_1 \leq j_2 \leq \dots \leq j_n \leq n'$, such that $X_1 \subseteq X'_{j_1}, X_2 \subseteq X'_{j_2}, \dots, X_n \subseteq X'_{j_n}$, then s is a q-subsequence of s' and s' is a q-supersequence of s, denoted as $s \subseteq s'$.

Example 5 The q-sequences $\langle [(a, 5)(c, 2)(g, 5)] \rangle$ and $\langle [(b, 3)(d, 2)(e, 2)] \rangle$ are two q-subsequences of s_1.

Definition 5 (*Matching*) Given a q-sequence $s = \langle (i_1, q_1)(i_2, q_2) \dots (i_\alpha, q_\alpha) \rangle$ and a sequence $t = \langle t_1 t_2 \dots t_\beta \rangle$, s is said to match t iff $\alpha = \beta$ and $i_k = t_k$ for $1 \leq k \leq \alpha$, denoted as $t \sim s$.

Example 6 Sequence $\langle (ce)(abf)(bceg) \rangle$ matches the q-sequence s_2.

Definition 6 (*Ending q-item maximum utility*) Given a q-sequence $s = \langle X_1 X_2 \dots X_n \rangle$ where $X_j (1 \leq j \leq n)$ is a q-itemset and a sequence $t = \langle t_1 t_2 \dots t_m \rangle$. If any q-subsequence $s_a = \langle X_{a_1} X_{a_2} \dots X_{a_m} \rangle$ $(s_a \subseteq s$ and $s_a \sim t)$ where $X_{a_m} = [(i_{a_1}, q_{a_1})(i_{a_2}, q_{a_2}) \dots (i_{a_m}, q_{a_m})]$, then (i_{a_m}, q_{a_m}) is called the ending q-item of sequence t in q-sequence s. The ending q-item maximum utility of a sequence t in a q-sequence s is denoted and defined as $u(t, i, s) = \max\{u(s')|s' \sim t \wedge s' \subseteq s \wedge i \in s'\}$.

Example 7 The ending q-items of $t = \langle (ac) \rangle$ in q-sequence s_1 are c_1, c_2 and their ending q-item maximum utility are respectively $u(\langle ac \rangle, c_1, s_1) = \max(4) = 4$, $u(\langle ac \rangle, c_2, s_1) = \max(-3) = -3$.

Definition 7 (*Sequence utility*) The sequence utility of a sequence $t = \langle t_1, t_2, \dots, t_m \rangle$ in a q-sequence $s = \langle X_1 X_2 \dots X_n \rangle$ is denoted and defined as $v(t, s) = \bigcup_{s' \sim t \wedge s' \subseteq s} u(s')$.

The utility of t in SDB is denoted as $v(t)$ and defined as a utility set: $v(t) = \bigcup_{s \in SDB} v(t, s)$.

Example 8 The utility of the sequence $t = \langle (ac) \rangle$ in the q-sequence s_1 is $v(t, s_1) = \{u(\langle ((a, 5)(c, 2)) \rangle), u(\langle ((a, 3)(c, 3)) \rangle)\} = \{4, -3\}$. Since $\langle (ac) \rangle$ only appears in s_1, the utility of t in the dataset is also $\{4, -3\}$.

Definition 8 (*Sequence maximum utility*) The maximum utility of a sequence t in a q-sequence s is denoted and defined as $u_{\max}(t, s) = \max\{u(t, i, s) : \forall i \in s' \wedge s' \sim t \wedge s' \subseteq s\}$. The maximum utility of a sequence t in SDB is denoted and defined as $u_{\max}(t) = \sum u_{\max}(t, s) : \forall s \in SDB\}$.

Example 9 The maximum utility of $t = \langle (ac) \rangle$ in the sequence dataset is $u_{\max}(t) = u_{\max}(\langle (ac) \rangle, s_1) = 4$.

Definition 9 (*High-utility sequential pattern*) A sequence t is said to be a high-utility sequential pattern if $u_{\max}(t) \geq minUtil$, where $minUtil$ is a given user-specified minimum utility threshold.

Example 10 Given $minUtil = 30, 50, 80$, and 100, we found 2; 854; 1; 892; 488, and 97 HUSPs from the dataset shown in Table 2, respectively.

Definition 10 (*Period set of sequence*) Given a q-sequence database $SDB = \{s_1, s_2, \ldots, s_N\}$ and a sequence t. The set of q-sequences containing t is denoted as $S(t) = \{s_{\alpha_1}, s_{\alpha_2}, \ldots, s_{\alpha_k}\}$, where $1 \leq \alpha_1 < \alpha_2 < \cdots < \alpha_k \leq N$.

Definition 11 (*Consecutive q-sequences*) Given two q-sequences s_α, s_β and a sequence t such that $t \sim s' \wedge s' \subseteq s_\alpha \wedge s_\alpha \in S(t)$ and $t \sim s'' \wedge s'' \subseteq s_\beta \wedge s_\beta \in S(t)$. s_α and s_β are said to be consecutive with respect to t iff there is not a q-sequence $s_\gamma \in S(t)$ such that $\alpha < \gamma < \beta$.

The period of two consecutive q-sequence s_α and s_β is denoted and defined as $pe(s_\alpha, s_\beta) = \beta - \alpha$. In other words, $pe(s_\alpha, s_\beta)$ is the number of q-sequences between s_α and s_β.

Example 11 The sequence $\langle (ag) \rangle$ appears in two q-sequences s_1 and s_5. Hence, $pe(s_1, s_5) = 5 - 1 = 4$.

Definition 12 (*Periods of a sequence*) Given a sequence t and $S(t) = \{s_{\alpha_1}, s_{\alpha_2}, \ldots, s_{\alpha_k}\}$ $(1 \leq \alpha_1 < \alpha_2 < \cdots < \alpha_k \leq n)$. The periods of a sequence t is a list of periods denoted and defined as $pes(t) = \bigcup_{1 \leq z \leq k+1} pe(s_{\alpha_{z-1}}, s_{\alpha_z})$, where α_0 and α_{k+1} are constants defined as $\alpha_0 = 0$ and $\alpha_{k+1} = n$.

Example 12 The sequence $\langle ac \rangle$ has $pes(\langle ac \rangle) = \{1, 1, 1, 1, 1\}$. The sequence $\langle (ag) \rangle$ has $pes(\langle (ag) \rangle) = \{1, 4, 0\}$.

Definition 13 (*Maximum, minimum and average periodicity measures*) The maximum periodicity, minimum periodicity and average periodicity of a sequence t are denoted and defined respectively as $maxPer(t) = max(pes(t))$, $minPer(t) = min(pes(t))$ and $avgPer(t) = \sum x \in pes(t)/|pes(t)|$.

Example 13 The periods of $\langle (af)b \rangle$ are $pes(\langle (af)b \rangle) = \{1, 1, 2, 1\}$. Thus, $maxPer(\langle (af)b \rangle) = 2, minPer(\langle (af)b \rangle) = 1$ and $avgPer(\langle (af)b \rangle) = 5/4 = 1.25$.

Property 1 (Relationship between average periodicity and support [4]) Given a sequence t and a q-sequence database SDB. An alternative and equivalent method of calculating the average periodicity of t is $avgPer(t) = |SDB| / (|S(t)| + 1)$.

The rationale for combining three measures is to avoid discovering patterns that occur with too short or too long periods [4]. Specifically, the average periodicity measure is used to avoid finding patterns that infrequently appear in a dataset. The minimum and maximum periodicity thresholds allow to specify that the periods of patterns must not be too short and long, respectively. We give multiple options to the users of the proposed algorithm, users can choose to use or not use the minimum and average periodicity measures. If the minimum periodicity measure is not used, the minimum periodicity threshold is set to 0. If the average periodicity measure is not used, the minimum and maximum average periodicity threshold can be set to 0 and $+\infty$, respectively.

Definition 14 (*Periodic high-utility sequential patterns with negative unit profits*) Given five positive user-specified thresholds: $minUtil, minAvg, maxAvg, minPer$ and $maxPer$. A sequence t is a periodic high-utility sequential pattern iff t is a HUSP and $minAvg \leq avgPer(t) \leq maxAvg, minPer(t) \geq minPer$ and $maxPer(t) \leq maxPer$.

The goal of PHUSPN mining is to discover the set of PHUSPNs that satisfies definition 14.

Example 14 Let $minUtil = 140, minPer = 1.0, maxPer = 3.0, minAvg = 1.0$ and $maxAvg = 2.0$. The complete set of PHUSPN is $\langle (ab)(be) \rangle$: 147, $\langle (cg)(be) \rangle$: 150, $\langle (g)(abf)(be) \rangle$: 142, $\langle (g)(ab)(be) \rangle$: 146, $\langle (g)(be) \rangle$: 168.

Definition 15 (*ULPN: utility list structure*) Assume that a sequence t has k $(k > 0)$ ending q-items i in a q-sequence s where $i_1 < i_2 < \ldots < i_k$. The ULPN of t in s is a list of k elements, where the $\alpha^{th} (1 \leq \alpha \leq k)$ element in the ULPN contains

$$\begin{cases} tid : \text{ is the itemset ID of } i_\alpha \text{ of } t \text{ in } s \\ acu : \text{ is the maximum utility of } i_\alpha \text{ in } t \\ link : \text{ is a pointer pointing to either the } (\alpha + 1)^{th} \text{ element or } null \end{cases}$$

Definition 16 (*UCPN: utility-chain structure*) Given a sequence t and a q-sequence s. The $UCPN$ of t in s is denoted and defined as

$$UCPN(t, s) = \begin{cases} peuts : \text{ is the prefix extension utility of } t \text{ in } s \\ ULPN : \text{ is the ULPN of sequence } t \text{ in } s \end{cases}$$

Definition 17 (*PHUSN: node structure*) Given a sequence t, the $PHUSN$ of t in SDB is denoted and defined as

$$PHUSN(t) = \begin{cases} sidSet : \text{ is the set of sequence IDs of all q-sequences containing } t \text{ in } SDB \\ ucpSet : \text{ is the set of } UCPN(t, s) \text{ in } SDB, \text{ which is defined as } ucpSet = \bigcup_{s \in SDB} (UCPN(t, s)) \end{cases}$$

Definition 18 (*Concatenation*) Given a sequence t, there are two types of concatenation of t:

$$\begin{cases} I - Extension : & \text{is the action of inserting an item into the last itemset of } t \\ S - Extension : & \text{is the action of adding a new itemset having one item at the end of } t \end{cases}$$

Example 15 For example, $\langle (acg) \rangle$ and $\langle (ac)(a) \rangle$ are generated by performing an I-Extension and a S-Extension of sequence $\langle (ac) \rangle$, respectively.

Definition 19 (*SWU: sequence weighted utilization*) The SWU of a sequence t in SDB is defined as

$$SWU(t) = \sum_{s' \sim t \wedge s' \subseteq s \wedge s \subseteq SDB} u(s).$$

For example, $SWU(\langle a(be) \rangle) = u(s_1) + u(s_2) + u(s_3) + u(s_4) = 91 + 96 + 82 + 114 = 383$.

Theorem 1 (Sequence weighted downward closure property) *Given t_1 and t_2, if t_2 contains t_1, then $SWU(t_2) \leq SWU(t_1)$.*

Theorem 1 can be used to evaluate whether an item is promising [3, 4, 13]. In PHUSN algorithm, this theorem is also used to prune all items that have a SWU $<$ *minUtil*.

Definition 20 (*PEU: prefix extension utility*) Given a sequence t and a q-sequence s. The PEU of t in s is denoted and defined as

$$PEU(t, s) = \max\{PEU(t, i_k, s) : \forall i_k \text{ that is an ending q-item of } t \text{ in } s\}$$

$$PEU(t, i_k, s) = \begin{cases} u(t, i_k, s) + ru(i_k, s), & \text{if } ru(i_k, s) > 0, \\ 0, & \text{otherwise.} \end{cases}$$

The PEU of t in SDB is denoted and defined as

$$PEU(t) = \sum_{s' \sim t \wedge s' \subseteq s \wedge s \subseteq SDB} PEU(t, s).$$

Given t_1 and t_2, if t_2 contains t_1 then $u(t_2) \leq PEU(t_1)$.

Definition 21 (*RSU: reduced sequence utility*) Given a sequence t and a q-sequence s. The RSU of t in s is denoted and defined as

$$RSU(t, s) = \begin{cases} PEU(t')|t' \subseteq t \wedge s_1 \sim t \wedge s_1 \subseteq s \wedge s_2 \sim t \wedge s_2 \subseteq s, \\ 0, & \text{otherwise.} \end{cases}$$

The RSU of sequence t in SDB is denoted and defined as

$$RSU(t) = \sum_{s' \sim t \wedge s' \subseteq s \wedge s \subseteq SDB} RSU(t, s).$$

Given t_1 and t_2, if t_2 contains t_1 then $u(t_2) \leq RSU(t_1)$.

Theorem 2 (Pruning strategy by PEU and RSU [4, 11]) *Given a pattern t, $PEU(t)$ and $RSU(t)$ are considered as upper bounds on the utility of t and its descendants. If $PEU(t) < minUtil$ or $RSU(t) < minUtil$, then t and its descendants can be pruned from the search space without affecting the result of the mining process.*

Theorem 3 (MPP: maximum periodicity pruning [4]) *Given a sequence t, if $maxPer(t) > maxPer$, then the sequence t and its descendants are not PHUSPN.*

4 The PHUSN Algorithm

The pseudo code of the PHUSN algorithm is shown in Algorithm 1. The input is a sequence t, a q-sequence database SDB and five predefined thresholds: $minUtil$, $minPer$, $maxPer$, $minAvg$ and $maxAvg$. Initially, PHUSN scans the SDB to calculate the total number of q-sequences ($|SDB|$) and calculates the SWU of all items in SDB (line 1). It then selects all items having a SWU that is no less than $minUtil$ and builds the PHUSN structure as well as the lexicographic tree required by the mining process (line 2). The topmost node in that tree is the root node, where its children are q-sequences that contain a single item. Each node other than the root stores a sequence t, the $PHUSN$ structure of t, utility matrices and remaining utility matrices of q-items in q-sequences of SDB. If $PEU(t)$ is less than $minUtil$, then the algorithm will consider t as a leaf and will not expand the lexicographic tree using node t, i.e., all its descendants will be pruned (lines 3–4). In the next step, PHUSN scans the projected database that includes the PHUSN of t in SDB to collect all items that can be combined with t to form a new sequence by I-Extension or S-Extension (line 5). Each item having a RSU value that is lower than $minUtil$ is discarded from the mining process (line 6). Then, PHUSN performs a loop over all items in the iExts and sExts. For each item i in the iExts, the algorithm performs an I-Extension with this item to form a new sequence t' by inserting i in the last itemset of t. In addition, the $PHUSN$ structure and maximum utility of t' are constructed and calculated by extending the $PHUSN$ of t. In the next step, PHUSN applies the MPP strategy (Theorem 3) to discard non-periodic patterns. If yes, PHUSN stops considering these patterns and backtracks to the previous step. Otherwise, it checks whether t' is a HUSPN (line 11) and calls the $check\_Periodic$ procedure to check the periodicity of t' (line 12). The input of this procedure is the set $S(t'), minPer, maxPer, minAvg$ and $maxAvg$. The procedure first calculates the minimum periodicity, maximum periodicity and average periodicity of t' (Property 1) by scanning $S(t')$ that is the $sidSet$ in the PHUSN structure (lines 1 to 3). If t' is a periodic pattern (line 4), the procedure returns true (line 5). Otherwise, it returns false (line 7). Next, PHUSN recursively calls itself to expand t' (line 14). A similar process is performed for all items in sExts. Note that, PHUSN passes a sequence and its projected database to each recursive call as input parameters. The sequence database SDB and lines 1, 2 are used only for initializing the algorithm and are not performed during recursive calls. For each item in sExts a new pattern is generated by performing an S-Extension (lines 15 to 22). When the algorithm terminates, it has an output of all PHUSPN.

Algorithm 1: The PHUSN algorithm

---

input : SDB: a q-sequence database, $minUtil$: the minimum utility threshold, t: a sequence with its PHUSN, $minPer, maxPer, minAvg, maxAvg$
output: The set of PHUSPN

1 Scan SDB to calculate $|SDB|$ and SWU for all items
2 Remove all items that have $SWU < minUtil$
3 **if** $(PEU(t) < minUtil)$ **then**
4 | return

5 Scan the projected database to:
   a. put I-Extension items into iExts,
   b. put S-Extension items into sExts

6 Remove low RSU items from iExts and sExts
7 **foreach** *item* $i \in iExts$ **do**
8 | $(t', v(t')) \leftarrow$ I-Extension(t, i)
9 | Construct the $PHUSN$ structure of t'
10 | **if** $(maxPer(t') \leq maxPer)$ **then**
11 | **if** $(u_{\max}(t') \geq minUtil)$ **then**
12 | **if** $(check\_Periodic(S(t'), minPer, maxPer, minAvg, maxAvg))$ **then**
13 | output t'

14 | $PHUSN(t', minUtil, minPer, maxPer, minAvg, maxAvg)$

15 **foreach** *item* $i \in sExts$ **do**
16 | $(t', v(t')) \leftarrow$ S-Extension(t, i)
17 | Construct the $PHUSN$ structure of t'
18 | **if** $(maxPer(t') \leq maxPer)$ **then**
19 | **if** $(u_{\max}(t') \geq minUtil)$ **then**
20 | **if** $(check\_Periodic(S(t'), minPer, maxPer, minAvg, maxAvg))$ **then**
21 | output t'

22 | $PHUSN(t', minUtil, minPer, maxPer, minAvg, maxAvg)$

23 **return**;

Algorithm 2: *check_Periodic* procedure

---

input : $S(t)$ a set of $sid, minPer, maxPer, minAvg, maxAvg$
output: return *true* if t is a periodic sequential pattern. Otherwise, return *false*

1 Calculate $minPer(t) = min(pes(t))$ based on $S(t)$
2 Calculate $maxPer(t) = max(pes(t))$ based on $S(t)$
3 Calculate $avgPer(t) = |S|/(|S(t)| + 1)$
4 **if** $(minPer(t) \geq minPer \wedge maxPer(t) \leq maxPer \wedge minAvg \leq avgPer(t) \leq maxAvg)$
   then
5 | **return** *true*;

6 **else**
7 | **return** *false*;

---

5 Comparative Experiment

Experiments were performed to evaluate the performance of PHUSN on a computer with a 64 bit Intel(R) Xeon(R) Silver 4116 CPU @ 2.10GHz, 8 GB of RAM, running Windows 10 Enterprise LTSC. The source code can be found at https://github.com/uthuyn/PHUSN. All the algorithms were implemented in C#. The proposed algorithm was compared with two algorithms. The first algorithm is an extension of the HUS-Span algorithm [11] for mining HUSPNs, namely HUS-Span-NIV. The second algorithm is HUSP-NIV [2] that is designed for mining HUSPN. The performance of the three algorithms has been compared on both synthetic and real datasets, which were previously used in [4, 10]. For each dataset, we randomly selected 1/20 the number of items to change their external utility to negative values. The characteristics of these datasets are shown in Table 5. They are eight real-life datasets and a synthetic dataset named Sd1000k. They have varied characteristics such as sparse and dense datasets; short and long sequences. For each dataset, the $minUtil$ was decreased until a clear winner was observed or algorithms became too long to execute. In some cases, a constraint on the maximum length of PHUSPN ($maxLength$) was used to speed up the experiments. For each other positive user-specified thresholds including $minAvg$, $maxAvg$, $minPer$, and $maxPer$, a suitable empirical value was chosen for each dataset to ensure that the algorithms discovered a certain number of PHUSPN.

First, the execution time of PHUSN is compared with HUSP-NIV and HUSP-Span-NIV. Figure 1 show that PHUSN outperforms the compared algorithms on all datasets. In each subfigure, the vertical axis and horizontal axis represent the execution time (in millisecond) and minimum utility threshold values, respectively. In general, for all datasets, when the minimum utility threshold is decreased or when datasets contain more sequences or longer sequences, the running time of the algorithms increase. In that case, PHUSN can be much more efficient than the two algorithms, especially on Sign, Bible, Foodmart, Online retail, and Tafeng datasets.

Table 5 Characteristics of the datasets

Dataset	Size	#Sequence	#Item	Avg. seq length	Type
Sign	375 KB	800	310	51.99	Realistic
Kosarak10k	0.98 MB	10, 000	10, 094	8.14	Realistic
Fifa	7.21 MB	20, 450	2, 990	34.74	Realistic
Bible	8.56 MB	36, 369	13, 905	21.64	Realistic
BMSwebview2	5.46 MB	77, 512	3, 340	4.62	Realistic
Tafeng	8.00 MB	32, 266	23, 811	3.71	Realistic
Foodmart	3.00 MB	8, 842	1559	6.59	Realistic
Online retail	4.60 MB	4, 335	3, 928	4.24	Realistic
Sd1000k	156 MB	920, 467	10, 000	5.0	Synthetic

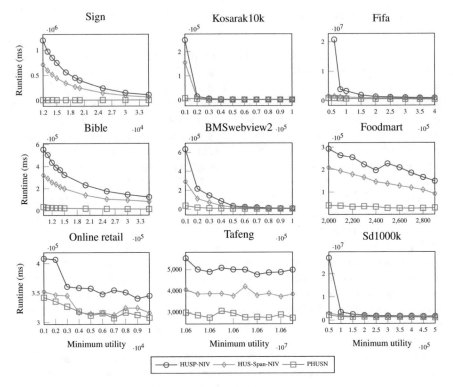

Fig. 1 Runtimes for various minimum utility threshold values

It can be observed that the PHUSN runs faster than the HUS-Span-NIV algorithm. It means that the MPP pruning strategy of PHUSN is effective and can prune many non-periodic patterns. In addition, the results show that the PHUSN runs faster than the HUSP-NIV algorithm. It means that the modified q-sequence utility, utility matrix, and remaining utility matrix are suitable for the framework of mining PHUSPN. Especially, the SWU of items in PHUSN is less than ones in HUSP-NIV. Thus, PHUSN can prune more non-candidates than the HUSP-Span-NIV. Generally, the PHUSN structure is more compact and efficient than the one used in HUSP-NIV.

Second, the compared algorithms have been also compared in terms of memory performance for the nine datasets for the same $minUtil$, $minAvg$, $maxAvg$, $minPer$, $maxPer$ and $maxLength$ values as in the runtime experiment. Results are shown in Fig. 2 in terms of memory usage (vertical axes) for various minimum utility values (horizontal axes). In general, PHUSN consumes less memory than HUSP-NIV and HUS-Span-NIV in most cases, except for the Foodmart dataset. Generally, for each dataset, the memory usage increases when the minimum utility threshold is decreased, and it is also greater for larger datasets. It is observed that PHUSN consumes less memory than HUSP-NIV in most cases. It means that the PHUSN structure is more effective than the structure used by the HUSP-NIV

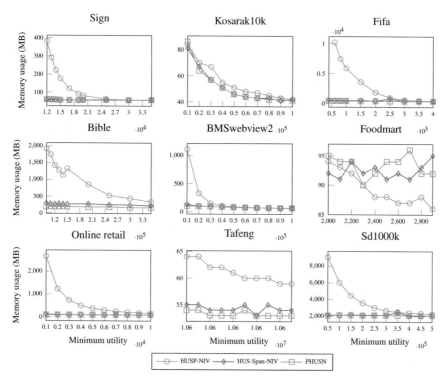

Fig. 2 Memory usage for various minimum utility threshold values

algorithm. In addition, the PHUSN consumes less memory than HUS-Span-NIV, although they are very close in some cases. Thus, the MPP pruning strategy used by PHUSN helps to reduce the memory usage of this algorithm.

Finally, the number of patterns was measured for various maximum periodicity threshold values on each dataset. In Fig. 3, vertical axes denote the number of patterns and horizontal axes indicate the corresponding maximum periodic threshold values. In general, HUSP-NIV and HUS-Span-NIV generate the same number of HUSPN for each dataset. For all datasets, the values $minPer = 1$ and $minAvg = 1$ were used. On Sign, the $maxPer$, $maxAvg$ thresholds were set to 20 and 5, respectively. For $minUtil$ from 35,000 to 12,000, PHUSN found 2, 7, 30, 90, 122, 179, 225, 264, 310 and 361 PHUSPN, respectively. On Kosarak10k, the $maxPer$, $maxAvg$ thresholds were set to 100 and 20, respectively. In addition, the $maxLength = 3$ is used for $minUtil = 10,000$ and 20,000. For $minUtil$ from 10,000 to 100,000, PHUSN found 2, 2, 3, 3, 3, 4, 6, 10, 20 and 20 PHUSPN, respectively. On Fifa, the $maxPer$, $maxAvg$ thresholds were set to 100 and 5, respectively. In addition, the $maxLength = 3$ is used for all cases. For $minUtil$ from 400,000 to 40,000, PHUSN found 11, 36, 86, 185, 325, 530, 666, 707, 735 and 752 PHUSPN, respectively. On Bible, the $maxPer$, $maxAvg$ thresholds were set to 100 and 5, respectively. In addi-

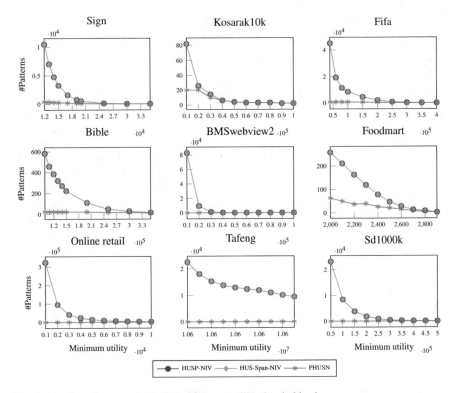

Fig. 3 Number of patterns for various minimum utility threshold values

tion, the $maxLength$=5 is used for all cases. For $minUtil$ from 350,000 to 100,000, PHUSN found 10, 16, 17, 22, 23, 23, 23, 23, 24 and 24 PHUSPN, respectively. On BMSwebview2, the $maxPer$, $maxAvg$ thresholds were set to 1000 and 70, respectively. For $minUtil$ from 100,000 to 10,000, PHUSN found 1, 2, 5, 7, 10, 17, 23, 25, 26 and 25 PHUSPN, respectively. On Foodmart, the $maxPer$, $maxAvg$ thresholds were set to 300 and 50, respectively. In addition, the $maxLength$=3 is used for all cases. For $minUtil$ from 2,900 to 2,000, PHUSN found 3, 5, 9, 15, 21, 26, 39, 37, 50 and 63 PHUSPN, respectively. On Online retail, the $maxPer$, $maxAvg$ thresholds were set to 1,000 and 10, respectively. In addition, the $maxLength$=2 is used for all cases. For $minUtil$ from 10,000 to 1,000, PHUSN found 27, 28, 28, 29, 29, 29, 29, 29, 29 and 30 PHUSPN, respectively. On Tafeng, the $maxPer$, $maxAvg$ thresholds were set to 18,000 and 100,000 respectively. In addition, the $maxLength$=4 is used for all cases. For $minUtil$ from 10,623,900 to 10,623,000, PHUSN found 24 PHUSPN for all cases. On Sd1000k, the $maxPer$, $maxAvg$ thresholds were set to 1,000 and 50, respectively. In addition, the $maxLength$=3 is used for all cases. For $minUtil$ from 500,000 to 50,000, PHUSN found 11, 11, 12, 12, 12, 12, 13, 15 16, 16 PHUSPN, respectively. From these results, it can be observed that the maximum periodicity pruning strategy (Theorem 3) eliminate many non-candidate patterns from the search space and thus reduce the runtime and memory usage of the PHUSN algorithm.

6 Conclusion

This paper has proposed an algorithm named PHUSN for mining periodic high-utility sequential pattern with negative unit profits. The proposed algorithm extends the PUSP structure [4] to the PHUSN structure for efficiently mining PHUSPN. Experimental results indicate that PHUSN outperforms HUSP-NIV and HUS-Span-NIV algorithms in terms of execution time and memory usage. The number of patterns generated by the three algorithms was also measured for various minimum utility threshold values. The results show that all the pruning strategies used in PHUSN can eliminate many non-PHUSPN and thus speed up the mining process. In future work, we will design a parallel framework that can enhance the computational cost of PHUSN, as well as extend the pattern mining framework for other tasks [22, 23].

References

1. P. Fournier-Viger, J.C.-W. Lin, R.U. Kiran, Y.S. Koh, R. Thomas, A survey of sequential pattern mining. Data Sci. Pattern Recognit. **1**(1), 54–77 (2017)
2. X. Tiantian, X. Dong, X. Jianliang, X. Dong, Mining high utility sequential patterns with negative item values. Int. J. Pattern Recognit Artif Intell. **31**(10), 1750035 (2017)
3. T. Dinh, V.-N. Huynh, B. Le, Mining periodic high utility sequential patterns, in *Asian Conference on Intelligent Information and Database Systems*, pp. 545–555 (2017)
4. D.-T. Dinh, B. Le, P. Fournier-Viger, V.-N. Huynh, An efficient algorithm for mining periodic high-utility sequential patterns. Appl. Intell. **48**(12), 4694–4714 (2018)
5. T. Dinh, M.N. Quang, B. Le, A novel approach for hiding high utility sequential patterns, in *Proceedings of the 6th International Symposium on Information and Communication Technology*, pp. 121–128 (2015)
6. W. Gan, J.C.-W. Lin, J. Zhang, H.-C. Chao, H. Fujita, S.Y Philip, ProUM: projection-based utility mining on sequence data. Inf. Sci. **513**, 222–240 (2020)
7. M.N. Quang, T. Dinh, U. Huynh, B. Le, MHHUSP: an integrated algorithm for mining and hiding high utility sequential patterns, in *Proceedings of the 8th International Conference on Knowledge and Systems Engineering*, pp. 13–18 (2016)
8. M.N. Quang, U. Huynh, T. Dinh, N.H. Le, B. Le, An approach to decrease execution time and difference for hiding high utility sequential patterns, in *Proceedings of the 5th International Symposium on Integrated Uncertainty in Knowledge Modelling and Decision Making*, pp. 435–446 (2016)
9. B. Le, D.-T. Dinh, V.-N. Huynh, Q.-M. Nguyen, P. Fournier-Viger, An efficient algorithm for hiding high utility sequential patterns. Int. J. Approx. Reason. **95**, 77–92 (2018)
10. B. Le, U. Huynh, D.-T. Dinh, A pure array structure and parallel strategy for high-utility sequential pattern mining. Expert Syst. Appl. **104**, 107–120 (2018)
11. J.-Z. Wang, J.-L. Huang, Y.-C. Chen, On efficiently mining high utility sequential patterns. Knowl. Inf. Syst. **49**(2), 597–627 (2016)
12. D.-T. Dinh, V.-N. Huynh, k-PbC: an improved cluster center initialization for categorical data clustering. Appl. Intell., 1–23 (2020)

13. J. Yin, Z. Zheng, L. Cao, Uspan: an efficient algorithm for mining high utility sequential patterns, in *Proceedings of the 18th ACM SIGKDD International Conference on Knowledge Discovery and Data Mining*, pp. 660–668 (2012)
14. D.-T. Dinh, V.-N. Huynh, B. Le, P. Fournier-Viger, U. Huynh, Q.-M. Nguyen, A survey of privacy preserving utility mining, in *High-Utility Pattern Mining* (Springer, Berlin, 2019), pp. 207–232
15. P. Fournier-Viger, J.C.-W. Lin, T. Dinh, H.B. Le, Mining correlated high-utility itemsets using the bond measure, in *International Conference on Hybrid Artificial Intelligence Systems*, pp. 53–65 (2016)
16. P. Fournier-Viger, Y. Zhang, J.C.-W. Lin, D.-T. Dinh, H.B. Le, Mining correlated high-utility itemsets using various measures. Logic J. IGPL **28**(1), 19–32 (2020)
17. R.U. Kiran, M. Kitsuregawa, P.K. Reddy, Efficient discovery of periodic-frequent patterns in very large databases. J. Syst. Softw. **112**, 110–121 (2016)
18. R.U. Kiran, J.N. Venkatesh, M. Toyoda, M. Kitsuregawa, P.K. Reddy, Discovering partial periodic-frequent patterns in a transactional database. J. Syst. Softw. **125**, 170–182 (2017)
19. R.U. Kiran, H. Shang, M. Toyoda, M. Kitsuregawa, Discovering partial periodic itemsets in temporal databases, in *Proceedings of the 29th International Conference on Scientific and Statistical Database Management*, pp. 1–6 (2017)
20. R.U. Kiran, C. Saideep, P. Ravikumar, K. Zettsu, M. Toyoda, M. Kitsuregawa, P.K. Reddy, Discovering fuzzy periodic-frequent patterns in quantitative temporal databases, in *2020 IEEE International Conference on Fuzzy Systems (FUZZ-IEEE)*, pp. 1–8 (2020)
21. P. Fournier-Viger, J.C.-W. Lin, Q.-H. Duong, T.-L. Dam, Phm: mining periodic high-utility itemsets, in *Industrial Conference on Data Mining*, pp. 64–79 (2016)
22. D.-T. Dinh, T. Fujinami, V.-N. Huynh, Estimating the optimal number of clusters in categorical data clustering by silhouette coefficient, in *International Symposium on Knowledge and Systems Sciences*, pp. 1–17 (2019)
23. L. Nhat-Vinh, T.-N. Vuong, D.-T. Dinh, Combining correlation-based feature and machine learning for sensory evaluation of saigon beer. Int. J. Knowl. Syst. Sci. (IJKSS) **11**(2), 71–85 (2020)

Hiding Periodic High-Utility Sequential Patterns

Ut Huynh, Bac Le, and Duy-Tai Dinh

Abstract Periodic high-utility sequential pattern (PHUSP) is a subset of HUSP that appears commonly in various applications such as market basket analysis, healthcare, and gen analytics. The sensitive PHUSP can be leaked by adversaries if datasets are released without using any privacy-preserving methods. The traditional high-utility sequential patterns hiding algorithms can hide the completed set of HUSP. However, such methods are maybe not appropriate to protect PHUSP since the number of found PHUSP is much less than HUSP. Thus, they take high computational cost, produce high missing cost with low quality sanitized datasets. This chapter addresses their limitations by introducing a novel algorithm named PHUS-Hiding for hiding PHUSP. The experiment was conducted on several real-life datasets to compare the performance of PHUS-Hiding with state-of-the-art algorithms in terms of execution time, memory usage, and missing cost. The experimental results show that PHUS-Hiding efficiently hides all PHUSP and faster than the compared algorithms.

1 Introduction

Privacy-preserving and information security are major challenges in the I nternet era. A company has a high risk of information leakage when the datasets that contain sensitive or valuable information are taken illegally. For such reasons, privacy-preserving algorithms have been designed to protect sensitive information from hackers or competitors. These algorithms can be classified into two groups: data hiding and knowledge hiding. Data hiding techniques handle directly on the raw data by using manipulation methods such as encrypt, anonymization, randomization to transform data into sanitized copies. Such algorithms are also called as Privacy-Preserving Data Pub-

U. Huynh · B. Le (✉)
Faculty of Information Technology, University of Science, Ho Chi Minh City, Vietnam
e-mail: lhbac@fit.hcmus.edu.vn

Vietnam National University, Ho Chi Minh City, Vietnam

D.-T. Dinh
Japan Advanced Institute of Science and Technology, Nomi, Japan

© The Author(s), under exclusive license to Springer Nature Singapore Pte Ltd. 2021
R. Uday Kiran et al. (eds.), *Periodic Pattern Mining*,
https://doi.org/10.1007/978-981-16-3964-7_10

lishing (PPDP) [1]. However, PPDP may reduce the usability and utility of data. In other words, it may lead to inaccurate or non-retrievable knowledge. Hiding knowledge aims to protect the knowledge from data mining algorithms. It is also called as Privacy-Preserving Data Mining (PPDM). PPDM methods design a strategy to completely hide the knowledge without loosing the usability and utility of data. The knowledge inside the data can be association rules, useful patterns, classifications, high-utility itemset, high-utility patterns, among others.

The utility-based mining (UBM) has been extensively studied by researchers in recent years. The UBM algorithms can be classified into two main branches: high-utility itemset mining (HUIM) [2–8] and high-utility sequential pattern mining (HUSPM) [9–14]. Periodic high-utility sequential patterns mining (PHUSPM) is a topic in UBM. This kind of patterns not only comes with a high-utility (e.g., high profit) value but also encounters frequently in a sequence database. PHUSPM can be applied into many applications such as customer behavior, market basket and biomedical analysis. In the traditional retail or the new e-commerce, understanding the customer behaviors is a key success factor for retailers. Detecting these periodic purchased patterns from the market basket helps sale managers knows the frequently or seasonally sale product chains. With this support, the managers can make suitable marketing strategies. The frequently purchased products with high profit may need short storage cycle and a routinely marketing campaign. As a supply chain managers, they are always looking for an effective way to predict the customer demands in sort and long term periods. Based on the customer needs, they can prepare a propitiate plan to deliver goods and optimize the progress of products produce. Nowadays, most of the consumers do online shopping. The operations of users on the website such as the number of clicks or the time that customers spent on a page can be tracked easily. By website click-stream analysis, the administrators can understand the behaviors of customers and thus can improve their functions to better suit customer needs. For example, they can arrange or suggest periodic products in ways that are more convenient for users based on personal behaviors. In biological sciences, genetics is a field that attempts to analysis and predicts how mutations, individual genes, and genetic interactions can affect organisms. Disease resistance is the ability to prevent or reduce the presence of diseases. It can arise from genetic or environmental factors. Genetic analysis may be done to identify genetic/inherited disorders and the relationships between the important genes or select relevant sequences according to a specific disease. Biologists can discover the important genes by using HUSPM algorithms. Furthermore, the importance and periodic instance of gene patterns in DNA sequences can be detected by PHUSPM algorithms.

The data mining algorithms can extract the useful knowledge from the data [15–18]. However, in terms of privacy preserving, the sensitive information can be also disclosed by such algorithms. Privacy-Preserving Utility Mining (PPUM) aims to protect the sensitive information from UBM algorithms. Several algorithms have been proposed for such tasks [19–22]. These algorithms were designed for hiding HUSPs in sequence datasets. Thus, they may not suitable for hiding PHUSPs although this type of patterns appears commonly in real-life applications. The experimental results in [13, 23] have shown that the cardinality of PHUSPs is much less than that

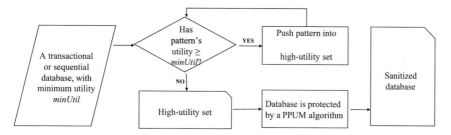

Fig. 1 The general model of PPUM on a utility database

of HUSPs. Thus, the data structure was used in previous works is not optimal for the problem of hiding PHUSPs. In addition, hiding inappropriate patterns may lead to high computation cost and also high missing cost (MC). Obviously, the utility of the sanitized data is also decreased. To address these limitations, this chapter proposes a novel algorithm named periodic high-utility sequential patterns hiding (PHUS-Hiding) for hiding PHUSPs. An extensive experiment was conducted on real-life datasets to compare the performance of the proposed algorithm with the state-of-the-art HUS-Hiding [24]. The experimental results revealed that PHUS-Hiding can efficiently hide PHUSPs and outperforms the HUS-Hiding algorithm in terms of computational cost and missing cost.

The rest of this chapter is organized as follows: Section 2 reviews the related work. Section 3 introduces the preliminaries. Section 4 proposes a hiding algorithm. Section 5 discusses the experimental results. Finally, Sect. 6 summaries the findings of this chapter.

2 Related Work

The general process for PPUM is described in Fig. 1 [20–22]. First, a mining algorithm is used on a quantitative transaction database or a sequence database to discover all HUIs or HUSPs, respectively. Next, the sensitive patterns are selected from the full set of high-utility patterns based on the hiding goals or business requirements. This is an optional step and patterns can be arranged according to the descending order of utility [22]. Finally, a PPUM algorithm is used to sanitize the original dataset. Each sensitive pattern is modified by decreasing its utility until less than the minimum utility threshold *minUtil*. Consequently, the sanitized database contains no sensitive patterns. Thus, it can be shared with the partners without worrying that the sensitive information can be disclosed by HUPM algorithms.

The previous surveys [25, 26] give an overview of PPUM algorithms. Figure 2 shows a taxonomy of PPUM algorithms. In general, these algorithms can be grouped into two groups. In the first group, algorithms are classified based on the approaches they use to hide sensitive patterns: transaction protection and item protection. The

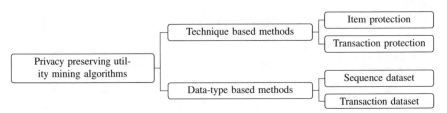

Fig. 2 A taxonomy of PPUM methodologies

transaction protection uses heuristic or genetic algorithms to modify the original transactions by inserting or deleting appropriate new items or transactions. The item protection modifies one or more appropriate properties of item in HUIs or HUSPs. In the second group, PPUM algorithms are designed for transaction or sequence dataset.

Several PPUM algorithms have been proposed for hiding high-utility sequential patterns [20–22, 24]. Regarding the workflow, they can be divided into two types: separate and integrated models. In the separate model, the mining and hiding steps perform individually. First, all HUSPs are discovered by using a HUSPM algorithm. The HUSPs are then selected based on business requirements as an optional step. Next, a HUSPH algorithm is applied to modify the selected patterns. HHUSP and MSPCF algorithms [20] first find HUSPs by using an extension of the USpan algorithm [10]. HUSPs are then modified by two strategies. HHUSP algorithm modifies the item that has the maximal utility in each HUSP, whereas MSPCF modifies the item that appears most frequently in each HUSP. The results have shown that HHUSP outperforms MSPCF in terms of computation cost. HHUSP-A and HHUSP-D [22] arrange HUSPs based on their utilities before the modification. Specifically, HHUSP-A sorts HUSPs by the ascending order, whereas HHUSP-D sorts them by the descending order of utilities. The results have shown that HHUSP-D outperforms HHUSP-A and HHUSP algorithm. For the group of the integrated model, the algorithms first discover a HUSP and then modify that HUSP such that its utility is lower than $minUtil$. The process is performed iteratively until all patterns are found and sanitized. This model has the advantage of decreasing the computation complexity on large-scale datasets. MHHUSP algorithm [21] compresses the search space into a LQS-Tree and uses a utility matrix structure to maintain the operation state. For each found HUSP, the procedure named modifyProcedure is used to reduce the utility of the pattern to lower than the $minUtil$.

Recently, HUS-Hiding algorithm [24] uses a new structure called utility chain for hiding (UCH) to maintain the search space and supports for the hiding process. First, HUS-Hiding scans the database and builds the UCH for items that have the sequence weighted utility (SWU) greater than or equal to $minUtil$. For each pattern p on LQS-Tree, the algorithm stops expanding the search space for patterns that take p as the prefix if the prefix extension utility (PEU) of p is lower than $minUtil$. From the 1-sequences UCH, HUS-Hiding expands p to p' by concatenating p with the I-Extension and S-Extension items through scanning p's projected dataset. An utility upper bound called reduced sequence utility (RSU) is used to prune unpromising items from I-Extension and S-Extension items from the mining process. If p' is an HUSP, then the algorithm modifies its utility. The algorithm calls itself recursively to discover and modify all HUSPs. HUS-Hiding is proven to be superior compared to other algorithms in terms of computational cost and scalability.

Discovering periodic patterns is a common task in pattern mining. On transaction dataset, PFP-growth and PFP-growth++ [27] algorithms were proposed for mining periodic frequent patterns (PFPs). Recently, Kiran et al. [28] proposed a pattern-growth algorithm named GPF-growth to find partial PFPs in a database. Fournier-Viger et al. [29] designed an algorithm named PHM to discover periodic high-utility itemsets (PHUIs) in a transaction database. They proposed several measures named minimum, maximum and average periodicity to find PHUIs. On sequence dataset, Dinh et al. [13] proposed an algorithm named PHUSPM. This algorithm uses the Uspan algorithm for mining HUSPs. They also define several measures to filter PHUSPs from the full set of HUSPs. Recently, Dinh et al. [23] proposed an algorithm name PUSOM that utilizes a data structure named PUSP to maintain the search space and a pruning strategy named MPP to cut down non-periodic patterns. The results have shown that the PUSOM algorithm can efficiently discover the complete set of PHUSPs and reduce the computational cost.

3 Preliminaries and Problem Statement

Let $I = \{i_1, i_2, \ldots, i_n\}$ is a set of all distinct items or products where $i_k \in I$ ($1 \leq k \leq n$) is associated with a weight or also called external utility and denoted as $w(i_k)$. $w(i_k)$ value represents for the unit profit, the unit cost or the importance ratio of i_k. Let $SDB = \{s_1, s_2, \ldots, s_t\}$ be a quantitative sequence database or q-sequence database for short, where s_{sid} ($1 \leq sid \leq t$) such that s is a q-sequence and sid is its unique identifier. A q-sequence s is a list of quantitative transactions or q-itemsets for short, $s = \langle t_1 t_2 \ldots t_m \rangle$ where t_k ($1 \leq k \leq m$) is a q-itemset. Basically, q-itemsets of q-sequence is ordered by purchased time. A q-itemset $t = [(i_1, q_1)(i_2, q_2) \ldots (i_n, q_n)]$ is a set of one or more q-items where (i_k, q_k) is a q-item ($1 \leq k \leq n$) such that i is a product where $i \in I$ and q is purchased quantity or also called internal utility. The total of quantity of a q-item i in a q-sequence s is denoted as $q(i, s)$. In the rest of this chapter, the brackets are omitted if a q-itemset contains only one q-item for the sake of brevity. If there are two or more q-items in a q-itemset, they are sorted according to a definite order $\succ$. The common order in use $\succ$ is the alphabetical order. In the examples of remaining work, we use Tables 1 and 2 for the cases. Table 2 consists five q-sequences. For example, the q-sequence s_1 shows 3 q-itemsets, such that $b(3)$, $[a(5)b(2)]$, and $[c(1)e(6)]$. In this q-sequence, the internal utility of q-item b in the first and the second q-itemset are, respectively, 3 and 2. Table 1 presents the items and it's external utility or also called weight. There are five items includes a, b, c, d, and e appear with weight values are 2, 10, 4, 3, 1, respectively.

Table 1 External utility values

Item i	Weight $w(i)$
a	2
b	10
c	4
d	3
e	1

Table 2 A quantitative sequence database

SID	Q-sequence
1	$b(3)[a(5)b(2)][c(1)e(6)]$
2	$[c(5)e(4)][a(3)b(1)]d(6)[a(2)d(8)]$
3	$[c(2)d(1)][c(5)d(6)e(1)]e(1)a(2)$
4	$[a(1)d(5)][a(3)b(2)][b(1)d(4)][a(2)b(1)][d(6)e(1)]$
5	$[a(1)c(2)e(3)][a(40)c(1)e(1)][a(5)c(1)e(1)]d(1)$

The utility of a q-item (i, q) denoted as $u(i, q)$, the utility of a q-itemset t denoted as $u(t)$, and the utility of a q-sequence s denoted as $u(s)$ are defined as follow:

$$u(i, q) = w(i) \times q(i) \tag{1}$$

$$u(t) = \sum_{k=1}^{n} u(i_k, q_k) \tag{2}$$

$$u(s) = \sum_{k=1}^{n} u(t_k) \tag{3}$$

Definition 1 (*q-itemsets subset*) Given two q-itemsets $t_a = [(i_{a_1}, q_{a_1}) (i_{a_2}, q_{a_2})\ldots (i_{a_n}, q_{a_n})]$ and $t_b = [(i_{b_1}, q_{b_1}) (i_{b_2}, q_{b_2})\ldots(i_{b_m}, q_{b_m})]$. Iff for each integer $k \in [1, n]$, there exists an integer $l \in [1, m]$ such that $i_{a_k} = i_{b_l}$ and $q_{a_k} = q_{b_l}$ then t_a is called a subset of t_b and denoted as $t_a \subseteq t_b$.

Definition 2 (*q-sequence sub-sequence*) Given a q-sequence $s = \langle t_1, t_2,\ldots, t_n \rangle$ and a q-sequence $s' = \langle t'_1, t'_2,\ldots,t'_m \rangle$ where $1 \leq m \leq n$. Iff there exist a set of integers $1 \leq j_1 < j_2 < \cdots < j_k \leq m$ such that $t'_k \subseteq t_{j_k}$ then q-sequence s is called super-sequence of s' or s' is called sub-sequence of s and denoted as $s' \subseteq s$.

Definition 3 (*Matching*) Given a sequence $p = \langle t_1 t_2 \ldots t_n \rangle$ and a q-sequence $s = \langle (s_1, q_1)(s_2, q_2) \ldots (s_n, q_n) \rangle$, s is said to match p iff $s_k = t_k$ for $1 \leq k \leq n$, denoted as $p \sim s$.

Definition 4 (*Utility of a sequence*) The utility of a sequence can be defined in several methodologies, based on business requirements to conduct utility functions

[9–11]. In this study, the utility of a sequence will be calculated by the maximum utility function which proposed in [11]. Given a sequence $p = \langle t_1, t_2, \ldots, t_m \rangle$, the utility of p in a q-sequence s is defined as $u(p, s) = max\{u(s')|s' \sim p \wedge s' \subseteq s\}$. Sequence p occurs multiple times in SDB. Therefore, the utility of p in SDB is defined and denoted as $u(p) = \sum_{s \in SDB} u(p, s)$.

Definition 5 (*Upper bound utility of a sequence—SWU*) Sequence weighted utility (SWU) is used as upper bound utility of a sequence. SWU of a sequence p in SDB is maximum utility value that p can reach that is $u(p) \leq SWU(p)$. SWU of p defined as $SWU(p) = \sum_{s' \sim p \wedge s' \subseteq s \wedge s \subseteq SDB} u(s)$.

Definition 6 (*High-utility sequential pattern—HUSP*) A sequential pattern p is a HUSP if $u(p) \geq minUtil$, where $minUtil$ is a predefined minimum utility threshold.

Definition 7 (*Periods of sequence*) Given a sequential pattern p in a q-sequence database $SDB = \{s_1, s_2, \ldots, s_n\}$. The q-sequences contains p is selected into a set which is denoted as $T(p) = \{s_{t_1}, s_{t_2}, \ldots, s_{t_k}\}$ where $1 \leq t_1 < t_2 < \cdots < t_k \leq n$. There are two q-sequences s_l and s_m where $s_l \in T(p) \wedge s_m \in T(p) \wedge l < m$ are sustained sequences of p if there is without q-sequence $s_r \in T(p)$ such that $l < r < m$. The number of q-sequences between two sustained sequences s_l and s_m is called a periodic of p which is denoted and defined as $pn(s_l, s_m) = (m - l)$. If p appears in multiple q-sequences, the periods of p are denoted as $ps(p) = \{t_1 - t_0, t_2 - t_1, t_3 - t_2, \ldots, t_k - t_{k-1}\}$ where $t_0 = 0$ and $t_k = n$ are constant. Mathematically, $ps(p) = \bigcup_{1 \leq z \leq k} (t_z - t_{z-1})$.

To assess the periodicity of p, we also used three measures as in [23]. The maximum periodicity is denoted and defined as $maxPer(p) = max(ps(p))$. The minimum periodicity is denoted and defined as $minPer(p) = min(ps(p))$. The average periodicity is denoted and defined as $avgPer(p) = \sum_{t \in ps(p)} t/|ps(p)|$. The average periodicity can be easily calculated as proposed in [23] by property $avgPer(p) = |SDB|/(|T(p)| + 1)$.

For example, considering the sequence $\langle (ce) \rangle$ occurs in s_1, s_2, s_3, and s_5 then $T(\langle (ce) \rangle) = \{s_1, s_2, s_3, s_5\}$ and $ps(\langle (ce) \rangle) = \{1, 1, 1, 2, 0\}$. So the $maxPer(\langle (ce) \rangle) = 2$, the $minPer(\langle (ce) \rangle) = 0$ and the $avgPer(\langle (ce) \rangle) = 1$.

The rationale for combining three measures is to avoid discovering patterns that occur with too short or too long periods [23]. Specifically, the average periodicity measure is used to avoid finding patterns that infrequently appear in a dataset. The minimum and maximum periodicity thresholds allow to specify that the periods of patterns must not be too short and long, respectively. We give multiple options to the users of the proposed algorithm, users can choose to use or not use the minimum and average periodicity measures. If the minimum periodicity measure is not used, the minimum periodicity threshold is set to 0. If the average periodicity measure is not used, the minimum and maximum average periodicity threshold can be set to 0 and $+\infty$, respectively.

Table 3 Six PHUSPs of the example

Pattern p	$u(p)$	$minPer$	$maxPer$	$avgPer$
$\langle(a)(d)\rangle$	137	1	2	1.25
$\langle(ce)(a)\rangle$	146	1	2	1.25
$\langle(ce)(a)(a)\rangle$	135	3	3	5/3
$\langle(ce)(a)(d)\rangle$	148	3	3	5/3
$\langle(c)(a)\rangle$	138	1	2	1.25
$\langle(c)(a)(d)\rangle$	141	3	3	5/3

Definition 8 (*Periodic high-utility sequential patterns—PHUSP*) Given a pre-defined minimum utility threshold $minUtil$ and periodical thresholds $minAvg$, $maxAvg$, $minPer$, $maxPer$. A HUSP p is also a PHUSP iff $minAvg \leq avgPer$ $(p) \leq maxAvg$, $minPer(p) \geq minPer$, $maxPer(p) \leq maxPer$.

For example, if predefined utility thresholds $minUtil = 135$, $minPer = 1$, $maxPer = 3$, $minAvg = 1$, $maxAvg = 2$, the complete set of six PHUSPs of the above example show in Table 3.

4 The Proposed Hiding Periodic High-Utility Sequential Pattern

Definition 9 (*Remaining utility*) Given a q-itemset $t_m=[(i_{m_1},q_{m_1})\ (i_{m_2},q_{m_2})\ \cdots\ (i_{m_k},q_{m_k})]$ of a q-sequence $s = \langle t_1 t_2 \ldots t_n \rangle$ where $1 \leq m \leq n$, then the remaining utility of q-item i_{m_l} where $1 \leq l \leq k$ is defined as $ru(i_{m_l}, s) = \sum\limits_{i' \in s \wedge i_{m_l} \prec i'} u(i', s)$.

For example, $ru(b_1, s_1)$, $ru(a_2, s_1)$ and $ru(b_2, s_1)$ values are 40, 30 and 10, respectively. A remaining utility chain of q-sequence s_1 is depicted in Table 4.

Table 4 Remaining utility chain of q-sequence s_1

Item indexes	1	2	3	4	5
Sequence	b(3)	[a(5)	b(2)]	[c(1)	e(6)]
Remaining utility	40	30	10	6	0

4.1 Index-Chain Support Hiding Structure

Utility chain for Hiding (UCH) was introduced by Le et al. [24]. UCH structure supports effectively for hiding task but field $tidset$ of it still is not direct points to items in a q-sequence. In this study, we developed a new structure called index-chain support hiding (ICH) that is defined as follow:

Definition 10 (*Index-set* ($idxSet$) *of a sub-sequence p in a q-sequence s*) Given a q-sequence $s = \langle t_1, t_2, \ldots, t_n \rangle$ and a sub-sequence $p = \langle t'_1, t'_2, \ldots, t'_m \rangle$ where $1 \leq m \leq n \wedge p \subseteq s$. If all items of s included the duplication are collected into a ordering preserved string and denoted as $str(s) = \{i_1, i_2, \ldots, i_l\}$ then $idxSet(s) = \{1, 2, \ldots, l\}$ is set of all index values. Sub-sequence p occurs multiple times in s. A set of index values where corresponding to the instance k of p in s is denoted as $idxSet(p_k, s)$ where $idxSet(p_k, s) \subseteq idxSet(s)$. Then, all index-sets of sub-sequence p in sequence s is denoted and defined as $idxSet(p, s) = \bigcup_{p_k \subseteq s} idxSet(p_k, s)$

Definition 11 (*Index-chain support hiding—ICH*) A sequence p occurs multiple times in q-sequences of SDB. A instance k of p in a q-sequence is presented by a tuple in ICH. Each tuple in ICH contains four dimensions:

- **sid** is the identity of q-sequence s.
- **idxSet** is the index-set of p in s that is corresponding to the instance p_k.
- **u** is the utility of p in s that is corresponding to the instance p_k.
- **ru** is the remaining utility of p in s that is corresponding to the instance p_k.

Table 5 shows the ICH of sequence $\langle b \rangle$. Sequence $\langle b \rangle$ appears two times in q-sequence s_1 at $index = 1$ and $index = 3$, respectively. The utility and the remaining utility of $\langle b \rangle$ in q-sequence s_1 (Table 4) at $index = 1$ are $u = 30$, $ru = 40$ and at $index = 3$ are $u = 20$, $ru = 10$, respectively. Next, sequence $\langle (b)(e) \rangle$ is formed by an external concatenation $\langle b \rangle$ with $\langle e \rangle$. Furthermore, in q-sequence s_1, sequence $\langle (b)(e) \rangle$ can be formed two times by composing sequence $\langle b \rangle$ at $index = 1$ with $\langle e \rangle$ at $index = 5$ and sequence $\langle b \rangle$ at $index = 3$ with $\langle e \rangle$ at $index = 5$, respectively. Therefore, $idxSet(\langle (b)(e) \rangle, s_1) = [1, 5]$ and $[3, 5]$ that are corresponding to the utility values are $u = 36$ and $u = 26$. Moreover, the item e is placed at $index = 5$ where is the end of s_1 then $ru = 0$ for the two instances (Table 6).

Table 5 The ICH of sequence $\langle b \rangle$

sid	idxSet	u	ru
1	[1]	30	40
	[3]	20	10
2	[4]	10	46
4	[4]	20	55
	[5]	10	45
	[8]	10	19

Table 6 The ICH of sequence $\langle (b)(e) \rangle$

sid	idxSet	u	ru
1	[1, 5]	36	0
	[3, 5]	26	0
4	[4, 10]	21	0
	[5, 10]	11	0
	[8, 10]	11	0

4.2 PHUS-Hiding Algorithm

The pseudo-code of PHUS-Hiding algorithm is shown in Algorithm 1. In the first phase, the algorithm scans the database and calculates the SWU for each item $i_k \in I$. An item i is pruned if $SWU(i) < minUtil$. At this phase, the algorithm also collects the ICH of each item $i_k \in I$. For each item, PHUS-Hiding gets an input including SDB, $minUtil$ and a sequence p that contains one item i with its ICH. First, the algorithm collects all items that can be concatenated with p into two lists named iExts and sExts (line 1). For each item in iExts list, new sequence p' is generated by using the I-Extension mechanism (line 3), which also constructs the ICH of p' (line 4). If p' is a HUSP, $fnIsPeriodic$ function is called to determine that p' is a PHUSP (line 5). The pseudo-code of $fnIsPeriodic$ function is shown in Algorithm 2. This function gets inputs including a sequence p' with its ICH and the user-specified minPer, maxPer, minAvg, maxAvg thresholds. The minimum periodicity, maximum periodicity are calculated by scanning the ICH of p' to create the $ps(p')$. The minimum, maximum, and average periodicity values of p' are then determined from $ps(p')$. If p' satisfies the user-specified conditions, then p' is a PHUSP (line 5). The function $fnHidingOne$ is then used to modify p' (line 6). The algorithm invokes itself recursively to process all sequences prefixed by p' (line 7). All items in the sExts list are processed by the S-Extension mechanism in the same manner as the items in the iExts list (lines 8–13). Finally, the algorithm can hide the complete set of PHUSP in the input dataset and returns the sanitized dataset (line 14).

The pseudo-code of $fnHidingOne$ function is shown in Algorithm 3. This function modifies utility of a PHUSP p in the same manner as algorithms in [20] and [22]. The input includes the ICH of p and utility value of p. This function first calculates the total reduction utility value that is denoted by $diff = u(p) - minUtil$ (line 1). A loop is performed to modify p until $diff \leq 0$ (line 2). The function selects the item i_m that has the largest utility value among items in p (line 3). This step is performed by using the $idxSet$ in the ICH of p which points directly to the indexes in q-sequence. It is worth noting that an item may occur multiple times in a q-sequence with different indexes and quantities. Thus, each occurrence is considered as a separate item. A PHUSP also appears multiple times in q-sequences. In the next step, the function calculates the reduction rate denoted by α (line 4). This rate is used to

---

Algorithm 1: The PHUS-Hiding Algorithm

---

input : A quantitative sequence database of SDB or p-projected database for the next phase, the minimum utility threshold $minUtil$, sequence p with its ICH

output: A sanitized database SDB'

1 Scan the projected database once to:
   a. Put I-Extension items into iExts, or
   b. Put S-Extension items into sExts

2 **foreach** *item* $i \in iExts$ **do**
3     $p' \leftarrow$ I-Extension(p, i)
4   Construct the ICH of p'
5   **if** $u(p') \geq minUtil \wedge fnIsPeriodic(p')$ **then**
6        $fnHidingOne(p', u(p'))$
7   PHUS-Hiding(p')

8 **foreach** *item* $i \in sExts$ **do**
9     $p' \leftarrow$ S-Extension(p, i)
10   Construct the ICH of p'
11   **if** $u(p') \geq minUtil \wedge fnIsPeriodic(p')$ **then**
12        $fnHidingOne(p', u(p'))$
13   PHUS-Hiding(p')

14 **return** SDB';

Algorithm 2: fnIsPeriodic Function

---

input : ICH of $p, minPer, maxPer, minAvg, maxAvg$

output: $True$ if p is a periodic sequential patterns, otherwise $False$

1 **if** $min(ps(p)) < minPer$ **then**
2   **return** $False$

3 **if** $max(ps(p)) > maxPer$ **then**
4   **return** $False$

5 $avgPer(p) = |SDB|/(|T(p)| + 1)$
6 **if** $minAvg \leq avgPer(p) \leq maxAvg$ **then**
7   **return** $True$

8 **return** $False$

---

determine the quantity to be decreased of an item i_m on each q-sequence where p occurs. For each q-sequence s in the ICH of p, the quantity of i_m is then modified based on α and the reduction quantity (dq) (lines 5–8). As a special case, the quantity of i_m is assigned as one if $\alpha \geq 1$ to assure that the function does not remove i_m from its q-sequence. When $diff \leq 0$, p is no longer a PHUSP.

Algorithm 3: fnHidingOne Function

---

input : A PHUSP p with its ICH and $u(p)$

1   $diff = u(p) - minUtil$

2   **while** $diff > 0$ **do**

3   Select $i_m \in p$ where $\sum u(i_m) = max\{ \sum\limits_{s \in UCH(p)} u(i_p, s)\}$

4   Calculate the reduction rate of i_m: $\alpha = diff / \sum u(i_m)$

5   **foreach** $s|s \in ICH(p)$ **do**

6   Declare $dq = \begin{cases} q(i_m, s) \times \alpha, if\ \alpha < 1 \\ q(i_m, s) - 1, if\ \alpha \geq 1 \end{cases}$

7   Reduce internal utility of i_m: $q(i_m, s) = q(i_m, s) - dq$

8   Set $diff = diff - dq \times w(i)$

4.3   Missing Cost (MC)

PPUM algorithms transform the original database into a sanitized version. As a result, the transformation produces side effects on the sanitized database. In this work, missing cost (MC) is used to measure the amount of side effects that the PPUM algorithm produces after the hiding process. In other words, MC represents the similarity between the original database and the sanitized database. The PPUM algorithm that produces a lower MC value is a better one. Let SDB and SDB' be the original quantitative sequence database and a sanitized version, respectively. The MC is defined and measured as following:

$$MC = su\,(SDB) - su\left(SDB'\right) \qquad (4)$$

According to the Eq. 4, the MC is calculated by subtracting the sanitized database utility from the original database utility.

5   Experiment

Experiments were performed to evaluate the performance of PHUS-Hiding algorithm and HUS-Hiding on a computer with a Intel Core i7-6770HQ CPU @ 2.60GHz, 16 GB of RAM, running Windows 10 Pro 64bit. The source code is provided at https://github.com/uthuyn/PHUS-Hiding. All algorithms were implemented in C#. The proposed algorithm was compared with state-of-the-art HUS-Hiding algorithm [24]. The performance of the algorithms has been compared on nine realistic datasets that were obtained from the SPMF data mining library website [30] and in [14, 23]. These datasets have varied characteristics such as short and long sequences; sparse and dense datasets. The internal utility of items in each sequence was randomly generated from 1 to 9. The external utility of each item was generated using a log-normal

Table 7 Characteristics of the experimental datasets

Dataset	Size	#Sequence	#Item	Avg. seq length	Type
Sign	375 KB	800	310	51.99	Realistic
Kosarak10K	0.98 MB	10, 000	10, 094	8.14	Realistic
Fifa	7.21 MB	20, 450	2, 990	34.74	Realistic
Bible	8.56 MB	36, 369	13, 905	21.64	Realistic
BMS-WebView-1	2.8 MB	59, 601	497	2.51	Realistic
BMS-WebView-2	5.46 MB	77, 512	3, 340	4.62	Realistic
Tafeng	8.00 MB	32, 266	23, 811	3.71	Realistic
Foodmart	3.00 MB	8, 842	1559	6.59	Realistic
Online retail	4.60 MB	4, 335	3, 928	4.24	Realistic

distribution as described in [23]. The characteristics of these datasets are shown in Table 7. For each dataset, the minimum utility threshold $minUtil$ was decreased until a clear winner was observed or algorithms became too long to execute. In some cases, a constraint on the maximum length threshold of PHUSP ($maxLength$) was used to speed up the experiments. For each other positive predefined thresholds including $minAvg$, $maxAvg$, $minPer$, and $maxPer$, appropriate empirical values were selected for each dataset to ensure that the algorithms discovered a certain number of PHUSPs.

5.1 Runtime and Memory Usage

The execution time of PHUS-Hiding is compared with the HUS-Hiding algorithm. The results are presented in Fig. 3. In each of these charts, the vertical axis and horizontal axis represents the execution time and minimum utility threshold values respectively. It can be observed that PHUS-Hiding outperforms the state-of-the-art HUS-Hiding algorithm for all datasets. On BMS-Webview-2, HUS-Hiding needs 12,469, 22,016, and 45,203 ms for hiding patterns for the $minUtil$ from 30,000, 20,000, and 10,000, respectively. PHUS-Hiding needs 4,640, 7,719, and 11,781 ms for hiding all patterns for the same $minUtil$ and with the $minPer$, $maxPer$, $minAvg$, $maxAvg$ were set to 1, 1000, 1, 70, respectively. On Tafeng with $maxLength = 4$, HUS-Hiding needs 37,937, 93,781 and 1,805,281 ms for hiding patterns for the $minUtil$ from 400,000, 300,000, and 200,000, respectively. PHUS-Hiding needs 48,765, 109,890 and 343,391 ms for hiding all patterns for the same $minUtil$ and with the $minPer$, $maxPer$, $minAvg$, $maxAvg$ were set to 1,

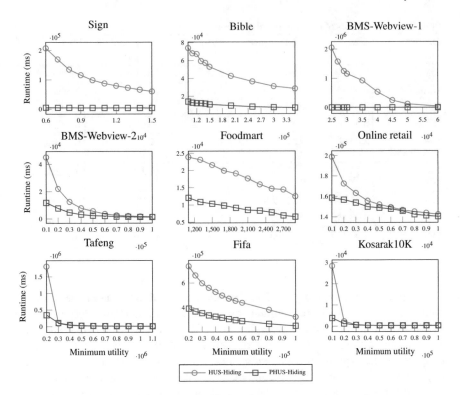

Fig. 3 Runtimes for various minimum utility threshold values

9000, 1 and 3000, respectively. Similar observations can be applied for the other datasets.

The memory usage of PHUS-Hiding is compared with HUS-Hiding algorithm. The results are presented in Fig. 4. In each of these charts, the vertical axis and horizontal axis represent the memory usages in megabytes and minimum utility threshold values, respectively. Generally, for each dataset, memory usage increases when the minimum utility threshold is low or for larger datasets. In general, PHUS-Hiding consumes less memory and more stable than HUS-Hiding in most cases. On Sign, HUS-Hiding requires 45, 46, 45, 46, 45, 47, 50, 47, 47, and 48 MB to hide all patterns for the $minUtil$ from 15,000 to 6,000, respectively. PHUS-Hiding requires 45 MB to hide all patterns at the same $minUtil$ and with the $minPer$, $maxPer$, $minAvg$, $maxAvg$ were set to 1, 20, 1, and 5, respectively. On Foodmart, the two algorithms require similar memory usage. Specifically, they need 75 MB for $minUtil$ from 2,900 to 1,100. A special case happens on the Bible dataset where HUS-Hiding requires less memory usage than PHUS-Hiding. For the $minUtil$ from 350,000 to 100,000 with $maxLength = 5$, HUS-Hiding needs 207, 210, 231, 239, 242, 320, 236, 263, 273, and 236 MB, respectively, whereas PHUS-Hiding needs 243, 259, 260, 268, 273, 279, 276, 281, 285, and 296 MB with the $minPer$, $maxPer$,

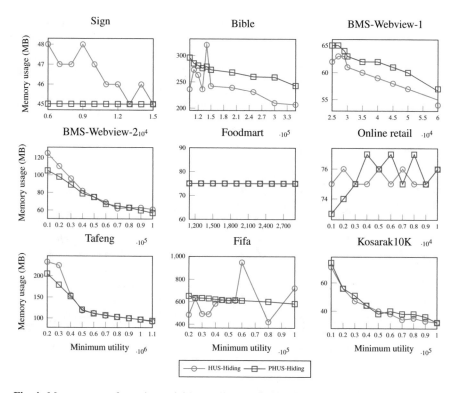

Fig. 4 Memory usage for various minimum utility threshold values

$minAvg$, $maxAvg$ were set to 1, 100, 1 and 5, respectively. Similar observations can be applied for the other datasets.

5.2 Number of Patterns

The experimental results in Fig. 5 show that the cardinality of PHUSP is much less than that of HUSP. In each of these charts, the vertical axis represents the total number of HUSPs for HUS-Hiding and PHUSPs for PHUS-Hiding, respectively. The horizontal axis represents minimum utility threshold values. For each dataset, the number of HUSPs found were depended on the minimum utility threshold value, whereas the number of PHUSPs found were also depended on the $minPer, maxPer$, $minAvg, maxAvg$ values. On BMS-Webview-2, HUS-Hiding found 1, 2, 5, 8, 15, 24, 51, 106, 318, and 2765 HUSPs for the $minUtil$ from 100,000 to 10,000, respectively. PHUS-Hiding found 1, 2, 5, 7, 8, 14, 21, 22, 26, and 26 PHUSPs for the same $minUtil$ and with the $minPer, maxPer, minAvg, maxAvg$ were set to 1, 1000, 1, 70, respectively. On Bible with $maxLength = 5$, HUS-Hiding found 10, 15, 19,

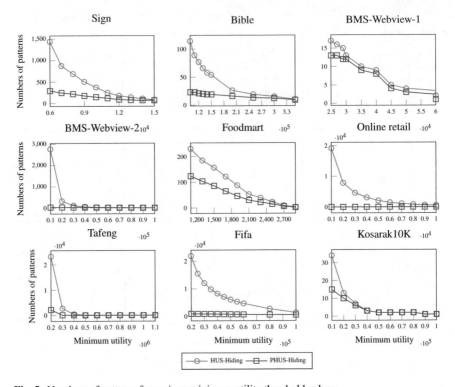

Fig. 5 Numbers of patterns for various minimum utility threshold values

26, 54, 58, 66, 77, 89, and 114 HUSPs for the $minUtil$ from 350,000 to 100,000, respectively. PHUS-Hiding found 9, 12, 13, 16, 19, 19, 20, 21, 23, and 23 PHUSPs for the same $minUtil$ and with the $minPer$, $maxPer$, $minAvg$, $maxAvg$ were set to 1, 100, 1 and 5, respectively. On Online retail with $maxLength = 2$, HUS-Hiding found 509, 654, 840, 1,096, 1,469, 2,084, 3,078, 4,586, 7,849 and 19,118 HUSPs for the $minUtil$ from 10,000 to 1,000, respectively. PHUS-Hiding found 28, 29, 29, 30, 30, 30, 30, 30, 30, and 31 PHUSPs for the same $minUtil$ and with the $minPer$, $maxPer$, $minAvg$, $maxAvg$ were set to 1, 1000, 1 and 10, respectively. The results shown that the cardinality of hiding PHUSPs is much less than that of hiding HUSPs in all cases. Similar observation can be applied for other datasets.

5.3 Missing Cost (MC)

In this section, the missing costs of PHUS-Hiding and HUS-Hiding algorithms were evaluated as shown in Fig. 6. PHUS-Hiding modifies less patterns than HUS-Hiding. Thus, PHUS-Hiding outperforms HUS-Hiding in most cases. It means that PHUS-

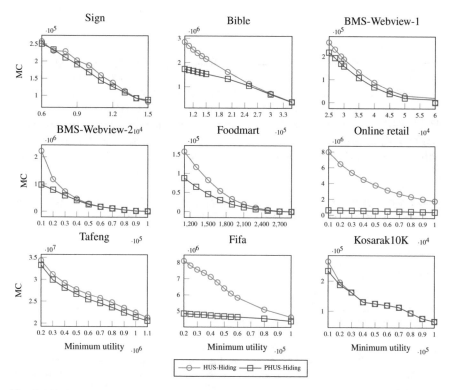

Fig. 6 MC for various minimum utility threshold values

Hiding produces less side effects than HUS-Hiding. On BMS-Webview-2, HUS-Hiding produces the MC of 728,503, 1,173,778, and 2,216,717 for the $minUtil$ of 30,000, 20,000, and 10,000, respectively. PHUS-Hiding produces the MC of 581,288, 786,427, and 974,993 for the same setting and with the $minPer$, $maxPer$, $minAvg$, $maxAvg$ were set to 1, 1000, 1, 70, respectively. On Bible with $maxLength = 5$, HUS-Hiding produces the MC of 2,533,699, 2,686,156, and 2,846,887 for the $minUtil$ of 120,000, 110,000, and 100,000, respectively. PHUS-Hiding produces the MC of 1,653,442, 1,691,325, and 1,734,577 for the same setting and with the $minPer$, $maxPer$, $minAvg$, $maxAvg$ were set to 1, 100, 1, and 5, respectively. Online retail with $maxLength = 2$, HUS-Hiding produces the MC of 5,357,576, 6,445,502, and 7,982,640 for the $minUtil$ of 3,000, 2,000 and 1,000, respectively. PHUS-Hiding produces the MC of 578,798, 600,065, and 617,733 for the same setting and with the $minPer$, $maxPer$, $minAvg$, $maxAvg$ were set to 1, 1000, 1, and 10, respectively. Similar observation can be applied for the other datasets. In some cases, if the number of HUSPs is equal to the number of PHUSPs, the missing costs produced by two algorithms are comparative.

6 Conclusion

This chapter has proposed an algorithm named PHUS-Hiding to efficiently hide PHUSPs from quantitative sequence datasets. A new data structure called ICH was designed to speed up the hiding step. An extensive experiments were conducted on nine real-life datasets to evaluate the performance of the PHUS-Hiding algorithm. The results have shown that PHUS-Hiding outperforms the state-of-the-art HUS-Hiding [24] in terms of runtime, memory usage, and side effect. In future work, we will improve the PHUS-Hiding to achieve better privacy with lower computation costs.

Acknowledgements This research is funded by Vietnam National Foundation for Science and Technology Development (NAFOSTED) under grant number 102.05-2018.307.

References

1. B.C.M. Fung, K. Wang, R. Chen, P.S. Yu, Privacy-preserving data publishing: a survey of recent developments. ACM Comput. Surv. **42**(4) (2010)
2. P. Fournier-Viger, J.C.-W. Lin, B. Vo, T.T. Chi, J. Zhang, H.B. Le, A survey of itemset mining. Wiley Interdiscip. Rev.: Data Min. Knowl. Discov. **7**(4), e1207 (2017)
3. V.S. Tseng, B.-E. Shie, C.-W. Wu, S.Y. Philip, Efficient algorithms for mining high utility itemsets from transactional databases. IEEE Trans. Knowl. Data Eng. **25**(8), 1772–1786 (2013)
4. B. Shen, Z. Wen, Y. Zhao, D. Zhou, W. Zheng, Ocean: fast discovery of high utility occupancy itemsets, in *Pacific-Asia Conference on Knowledge Discovery and Data Mining*, pp. 354–365 (2016)
5. W. Song, Y. Liu, J. Li, Bahui: fast and memory efficient mining of high utility itemsets based on bitmap. Int. J. Data Warehous. Min. (IJDWM) **10**(1), 1–15 (2014)
6. U. Yun, H. Ryang, K.H. Ryu, High utility itemset mining with techniques for reducing over-estimated utilities and pruning candidates. Expert. Syst. Appl., **41**(8), 3861–3878 (2014)
7. P. Fournier-Viger, C.-W. Wu, S. Zida, V.S. Tseng, Fhm: faster high-utility itemset mining using estimated utility co-occurrence pruning, in *International Symposium on Methodologies for Intelligent Systems*, pp. 83–92 (2014)
8. P. Fournier-Viger, Y. Zhang, J.C.-W. Lin, D.-T. Dinh, H.B. Le, Mining correlated high-utility itemsets using various measures. Logic J. IGPL **28**(1), 19–32 (2020)
9. C.F. Ahmed, S.K. Tanbeer, B.-S. Jeong, A novel approach for mining high-utility sequential patterns in sequence databases. ETRI J. **32**(5), 676–686 (2010)
10. J. Yin, Z. Zheng, L. Cao, USpan: an efficient algorithm for mining high utility sequential patterns, in *Proceedings of the 18th ACM SIGKDD International Conference on Knowledge Discovery and Data Mining* (ACM, 2012), pp. 660–668
11. G.-C. Lan, T.-P. Hong, V.S. Tseng, S.-L. Wang, Applying the maximum utility measure in high utility sequential pattern mining. Expert. Syst. Appl. **41**(11), 5071–5081 (2014)
12. J.-Z. Wang, J.-L. Huang, Y.-C. Chen, On efficiently mining high utility sequential patterns. Knowl. Inf. Syst. **49**(2), 597–627 (2016)
13. T. Dinh, V.-N. Huynh, B. Le, Mining periodic high utility sequential patterns, in *Asian Conference on Intelligent Information and Database Systems*, pp. 545–555 (2017)
14. B. Le, U. Huynh, D.-T. Dinh, A pure array structure and parallel strategy for high-utility sequential pattern mining. Expert Syst. Appl. **104**, 107–120 (2018)

15. D.-T. Dinh, T. Fujinami, V.-N. Huynh, Estimating the optimal number of clusters in categorical data clustering by silhouette coefficient, in *International Symposium on Knowledge and Systems Sciences*, pp. 1–17 (2019)
16. D.-T. Dinh, V.-N. Huynh, k-PbC: an improved cluster center initialization for categorical data clustering. Appl. Intell., 1–23 (2020)
17. L. Nhat-Vinh, T.-N. Vuong, D.-T. Dinh, Combining correlation-based feature and machine learning for sensory evaluation of saigon beer. Int. J. Knowl. Syst. Sci. (IJKSS) **11**(2), 71–85 (2020)
18. D.-T. Dinh, V.-N. Huynh, S. Sriboonchitta, Clustering mixed numerical and categorical data with missing values. Inf. Sci. **571**, 418–442 (2021)
19. C. Saravanabhavan, R.M.S. Parvathi, Privacy preserving sensitive utility patterns mining. J. Theor. Appl. Inf. Technol. **49**(2), 496–506 (2013)
20. T. Dinh, M.N. Quang, B. Le, A novel approach for hiding high utility sequential patterns, in *Proceedings of the 6th International Symposium on Information and Communication Technology*, pp. 121–128 (2015)
21. M.N. Quang, T. Dinh, U. Huynh, B. Le, MHHUSP: an integrated algorithm for mining and hiding high utility sequential patterns, in *Proceedings of the 8th International Conference on Knowledge and Systems Engineering*, pp. 13–18 (2016)
22. M.N. Quang, U. Huynh, T. Dinh, N.H. Le, B. Le, An approach to decrease execution time and difference for hiding high utility sequential patterns, in *Proceedings of the 5th International Symposium on Integrated Uncertainty in Knowledge Modelling and Decision Making*, pp. 435–446 (2016)
23. T. Dinh, B. Le, P. Fournier Viger, V.-N. Huynh, An efficient algorithm for mining periodic high-utility sequential patterns. Appl. Intell. **48**, 06 (2018)
24. B. Le, D.-T. Dinh, V.-N. Huynh, Q.-M. Nguyen, P. Fournier-Viger, An efficient algorithm for hiding high utility sequential patterns. Int. J. Approx. Reason. **95**, 77–92 (2018)
25. W. Gan, J. Chun-Wei, H.-C. Chao, S.-L. Wang, S.Y. Philip, Privacy preserving utility mining: a survey, in *2018 IEEE International Conference on Big Data (Big Data)*, pp. 2617–2626 (2018)
26. D.-T. Dinh, V.-N. Huynh, B. Le, P. Fournier-Viger, U. Huynh, Q.-M. Nguyen, A survey of privacy preserving utility mining, in *High-Utility Pattern Mining* (Springer, Berlin, 2019), pp. 207–234
27. R.U. Kiran, M. Kitsuregawa, P.K. Reddy, Efficient discovery of periodic-frequent patterns in very large databases. J. Syst. Softw. **112**, 110–121 (2016)
28. R.U. Kiran, J.N. Venkatesh, M. Toyoda, M. Kitsuregawa, P.K. Reddy, Discovering partial periodic-frequent patterns in a transactional database. J. Syst. Softw. **125**, 170–182 (2017)
29. P. Fournier-Viger, J.C.-W. Lin, Q.-H. Duong, T.-L. Dam, Phm: mining periodic high-utility itemsets, in *Industrial Conference on Data Mining*, pp. 64–79 (2016)
30. P. Fournier-Viger, A. Gomariz, T. Gueniche, A. Soltani, C.-W. Wu, V.S Tseng, Others. SPMF: a Java open-source pattern mining library. J. Mach. Learn. Res. **15**(1), 3389–3393 (2014)

NetHAPP: High Average Utility Periodic Gapped Sequential Pattern Mining

Youxi Wu, Meng Geng, Yan Li, Lei Guo, and Philippe Fournier-Viger

Abstract Sequential pattern mining is a key data mining task, where the aim is to find subsequences appearing frequently in sequences of items (symbols). To provide more flexibility and reveal more valuable patterns, sequential pattern mining with a periodic gap has emerged as an important extension. Algorithms for this task identify repetitive gapped subsequences (patterns) in a sequence. Although this has many applications, patterns are only selected based on their occurrence frequency and the external utility (relative importance) of each symbol is ignored. Consequently, these methods can find many unimportant frequent patterns and neglect some low frequency but extremely important patterns. To address this problem, this chapter presents a novel task of High Average Utility Periodic Gapped Sequential Pattern (HAPP) mining and proposes an efficient algorithm called Nettree for HAPP (NetHAPP), which involves two key steps: support calculation and candidate pattern generation. To calculate the support of patterns, a backtracking strategy is adopted that effectively reduces the time complexity of algorithm. To reduce the number of candidate patterns, an average utility upper bound method is combined with a pat-

Y. Wu
School of Artificial Intelligence, Hebei University of Technology, Tianjin 300401, China
e-mail: wuc567@163.com

Hebei Key Laboratory of Big Data Computing, Tianjin 300401, China

M. Geng
School of Artificial Intelligence, Hebei University of Technology, Tianjin 300401, China

Y. Li
School of Economics and Management, Hebei University of Technology,
Tianjin 300401, China

L. Guo
State Key Laboratory of Reliability and Intelligence of Electrical Equipment,
Hebei University of Technology, Tianjin 300401, China

P. Fournier-Viger (✉)
Shenzhen University, Shenzhen, China
e-mail: philfv@szu.edu.cn

© The Author(s), under exclusive license to Springer Nature Singapore Pte Ltd. 2021   191
R. Uday Kiran et al. (eds.), *Periodic Pattern Mining*,
https://doi.org/10.1007/978-981-16-3964-7_11

tern join strategy. A wide range of experimental results show that NetHAPP is not only more efficient than competitive algorithms but can also discover more valuable patterns.

1 Introduction

Sequences of items (symbols) are found in many domains. To analyse such data, a key task is sequential pattern mining (SPM) [13, 32, 56] which consists of identifying frequently occurring subsequences (also called patterns) in sequences. SPM has been widely applied in fields such as biological sequence analysis [45, 50], behavioural analysis of customer purchases [29, 60], time series analysis [16, 46, 51], web detection [6] and inspection report analysis [20, 21]. To meet the requirements of different applications, algorithms have been developed for different variations of the task of SPM such as high utility pattern mining [14, 15, 62], maximal frequent pattern mining [25, 27], tri-partition pattern mining [30], negative sequential pattern mining [9, 18], tree pattern mining [48], and closed pattern mining [7, 24]. In traditional SPM, a sequence is described in the form of multiple set items that are sequentially ordered. For instance, the sequence <c(abd)(bd)(cd)> indicates that item c was followed by items a, b and d, then followed by b and d, and then by c and d. But a limitation of traditional SPM algorithms is that they only consider whether or not a pattern such as bd appears in a sequence, and ignore the fact that a pattern may appear multiple times in the same sequence. A solution to this problem is to consider a sequence representation where simultaneous items are forbidden, such as <cabdbdcd>, and to count all occurrences of each pattern.

Since there are no constraints on the gaps between items in SPM, a phenomenon arises in which the gap between two consecutive items of a pattern can be relatively large, thus failing to meet the users needs. Hence, several algorithms have been developed for SPM with gap constraints [17, 47] in recent years. Although SPM becomes more difficult when gap constraints are added, it can reveal patterns that describe well the behaviour of sequences, are tailored to the user's needs, and can be applied in many real-world applications. This study focuses on mining patterns with periodic gaps [55], that is, where all the gap constraints are the same. For example, a pattern $C[0, 2]T[0, 2]T$ indicates that an item C is followed by an item T zero to two items after, and then followed by another item T, up to two items after. Generally, the above pattern can be described as having the form $\mathbf{p} = p_1[min_1, max_1]p_2[min_2, max_2]p_3$ where the notation $[min_i, max_i]$ indicates the minimum and maximum gap after the i-th item p_i. Since the size of the gaps in this pattern is always the same ($min_1 = min_2$ and $max_1 = max_2$), the gap constraint is said to be periodic.

Table 1 compares the characteristics of four pattern mining tasks: association rule mining [3], SPM [13, 32, 56], gapped SPM [17, 47] and periodic gapped SPM [55].

The SPM methods mentioned above only take into account the occurrence frequencies of patterns to find frequent patterns. They do not consider other factors that can help to evaluate the importance of patterns such as purchase quantities, unit

Table 1 Comparison of different types of mining

Types	Characteristics
Association rule mining	Considers the internal relations between items, but not their ordering
SPM	Focuses on the order between items, but does not consider repetitive patterns or gaps between items
Gapped SPM	Pays attention not only to the order between items and repetitions, but also considers constraints on the gaps between items
Periodic gapped SPM	Requires the same gap constraints any two consecutive items.

profits of items [39], or the interest and weight of an item. As a result, the information extracted by traditional SPM algorithms is insufficient for many applications. For example, in biological sequences, the frequency may not allow for the discovery of a gene sequence related to a certain disease, since even though a gene may not appear frequently, its high expression may cause the gene to be very significant. Conversely, an inhibitory gene may occur frequently but may have less effect. Researchers have therefore proposed a pattern mining task called high utility SPM [43], which extends traditional SPM by adding external utility values to items to not only consider the occurrence frequencies of patterns but also the importance of each item. However, this method has a serious flaw, as it does not take into account the lengths of patterns, meaning that it is easy to mine long but meaningless patterns. Inspired by the high average utility method [26], we propose a task called High Average Utility Periodic Gapped Sequential Pattern (HAPP) mining. The main contributions of this work are as follows:

1. This chapter describes HAPP mining and proposes an efficient mining algorithm named NetHAPP which performs two key steps: support calculation and candidate pattern reduction.
2. To calculate the support effectively, this chapter proposes the NetBTM algorithm, which adopts a backtracking matching strategy and employs a Nettree data structure [37, 53].
3. We impose an upper bound on the average utility and combine this with a pattern join strategy to generate candidate patterns. These strategies can effectively reduce the number of candidate patterns.
4. Experimental results on the DNA and VIRUS datasets show that not only NetHAPP outperforms competitive algorithms but also that HAPPs have significance.

The structure of this chapter is as follows: Section 2 introduces related work. Section 3 defines the problem. Section 4 proposes the NetHAPP algorithm, and provides a complexity analysis. Section 5 reports results of comparative experiments on the DNA and VIRUS datasets, and analyses the experimental results. Section 6 presents the conclusion to this chapter.

Table 2 Comparison of patterns found by different mining methods

Mining method	Results	Count
Frequent SPM	A, G, T, AA, AG, GA, GT, TG, AAA, AGT, GAA, GAG, GTG, AGTG, GAGT, GAGTG	16
High utility SPM	C, T, AA, AG, GA, GT, TG,AAA, ACG, ACT, AGA, AGG, AGT, ATA, ATC, ATG, ATT, CGG, CGT, CTG, ..., GAGTAAGAATCGTG	357,913,712
High average utility SPM	G, T, GT, TG, GTG	5

2   Related Work

SPM [2] has been used in numerous fields, and various mining methods have been proposed. For example, web log mining [22] and transaction flow mining [12] have been proposed for different types of databases, and rare SPM [33], maximum SPM [34] and closed SPM [10, 58] have been designed for different pattern features. Tri-partition alphabet SPM [30], negative SPM [8] and high utility SPM [1] have been designed for different mining tasks. The last of these has been widely used in many essential fields, such as customer purchase behaviour analysis [40], disease diagnosis [35], sentiment classification [5] and event log discovery [11]. Ryang and Yun [36] proposed an efficient algorithm for mining high utility patterns based on a novel indexed list data structure, and experimental results showed that the proposed algorithm effectively mined high utility patterns. Nam et al. [31] proposed the DHUP algorithm, which applied the concept of a damped window model for increased memory usage and scalability testing. Table 2 compares patterns found by different mining methods, namely, frequent SPM, high utility SPM and high average utility SPM in the sequence $s = $ GAGTAAGAATCGTG under the nonoverlapping condition (occurrences of a pattern are not allowed to overlap for the purpose of counting their occurrences). Note that it is assumed that some utility values (positive numbers) are assigned to each symbol (A, G, T and C) of that sequence to indicate their relative importance (not shown in Table 2).

As observed in Table 2, frequent SPM finds patterns that appear frequently. In this example, 16 frequent patterns were discovered such as AAA and AGT. High utility SPM mines patterns whose utilities are greater than a minimum threshold, but ignores the influence of the pattern length, and several long and useless patterns are discovered. HAPP mining reveals only five patterns that have a high average utility (defined as the utility of a pattern divided by its length). In this example, A, AA and AG are frequent patterns but not HAPPs, since the utility of item A is low. GAGTAAGAATCGTG is a high utility pattern, but not a HAPP, due to its long pattern length. Generally, high average utility SPM has more applications than SPM

```
         1   2   3   4   5
  s =     C   T   C   T   T
         C   T       T         <1, 2, 4>   The first occurrence
         C   T           T     <1, 2, 5>   The second occurrence
         C           T   T     <1, 4, 5>   The third occurrence
                 C   T   T     <3, 4, 5>   The fourth occurrence
```

Fig. 1 All occurrences of pattern **p** in sequence **s**

and high utility SPM [38, 41], and so has become a topic of intense research interest since it comprehensively considers the effects of pattern length and utility.

The SPM algorithms with gap constraints can be categorized into three types: using no condition [55], the one-off condition [49, 57] and the nonoverlapping condition [7]. The following examples illustrate the differences between these three constraints.

Example 1 Consider the sequence $s = s_1 s_2 s_3 s_4 s_5 = CTCTT$, containing five items. The pattern $p = p_1[0, 3] p_2[0, 3] p_3 = C[0, 2]T[0, 2]T$ has four occurrences in that sequence, which are presented in Fig. 1. In that figure, the numbers 1, 2, 3, 4 and 5 represent positions in the sequence **s**. For instance, the first occurrence $< 1, 2, 4 >$ of pattern **p** appears at position 1, 2 and 4 of **s**.

As shown in Fig. 1, no condition means that there are no constraints on the occurrences of a pattern. Thus, all occurrences of **p** are considered, that is, $<1,2,4>$, $<1,2,5>$, $<1,4,5>$ and $<3,4,5>$. However, the support (occurrence frequency) does not satisfy the Apriori property [23]. This means that the number of occurrences of a pattern may be greater than that of its sub-patterns. But it is necessary to use some form of Apriori property to reduce the search space and efficiently discover sequential patterns [61]. The one-off condition means that any character in the sequence can be used at most once. In this example, there is only one occurrence under the one-off condition, i.e. $<1,2,4>$. Although this constraint satisfies the Apriori property, the support calculation is an NP-hard problem [4], and mining algorithms for this scenario are thus approximate. The nonoverlapping condition means that although any character (item) in the sequence can be matched multiple times, it must not be in the same position. There are two occurrences of **p** in **s** under the nonoverlapping condition, i.e. $<1,2,4>$ and $<3,4,5>$. This example shows that the nonoverlapping condition is stricter than no condition and looser than the one-off condition. Nonoverlapping SPM not only satisfies the Apriori property but also can find all frequent patterns, and therefore outperforms the other two competitive methods. Table 3 gives a comparison of studies in this area.

It can be seen from Table 3 that the work in [54] is the closest to the scheme presented here. The differences between the approaches in [54] and in this chapter can be summarized as follows:

Table 3 Comparison of related studies

Literature	Pattern type	Gap constraint	Type of condition	Pruning strategy	Mining type
Wu et al. [55]	Frequent	Periodic gap	No condition	Apriori-like	Exact
Wu et al. [38]	Frequent	Periodic gap	One-off condition	Apriori	Approximate
Ding et al. [25]	Frequent	No	Nonoverlapping condition	Apriori	Approximate
Yao et al. [59]	High utility	No	–	Other	–
Tseng et al. [42]	High utility	No	–	Other	–
Hong et al. [19]	High average utility	No	–	Other	–
Lu et al. [28]	High average utility	No	–	Other	–
Wu et al. [54]	Frequent	Periodic gap	Nonoverlapping condition	Apriori	Exact
This chapter	High average utility	Periodic gap	Nonoverlapping condition	Apriori	Exact

1. The work in [54] focuses on mining frequent patterns, while the method in this chapter mines HAPPs. We find that HAPP mining has more actual significance than frequent SPM.
2. The work in [54] employs the NETGAP algorithm to calculate the support. It first creates a whole Nettree and then prunes the invalid nodes after a nonoverlapping occurrence is found. In this chapter, we propose the NetBTM algorithm, which employs a backtracking strategy to calculate the pattern support without pruning invalid nodes. Therefore, NetBTM outperforms NETGAP.
3. The pattern support in [54] satisfies the Apriori property, and uses it to generate candidate patterns. The average utility in this chapter does not satisfy the Apriori property, and an upper bound on the average utility is defined to improve the mining efficiency and avoid incomplete mining results.

3 Problem Definition

A sequence of length n is denoted as $\mathbf{s} = s_1 s_2 \ldots s_n$, where $s_i (0 \leq i \leq n) \in \Sigma$, Σ represents the set of items in sequence $\mathbf{s}$, and the size of Σ can be expressed as $|\Sigma|$. For example, in a DNA sequence, Σ is $\{A,T,C,G\}$ and $|\Sigma| = 4$.

Definition 1 (*Periodic Gap Pattern*) A periodic gap pattern $\mathbf{p}$ of length m is denoted as $\mathbf{p} = p_1[a, b]p_2 \ldots [a, b]p_m$ (or abbreviated as $\mathbf{p} = p_1 p_2 \ldots p_m$ with $gap = [a, b]$), where a and b $(0 \leq a \leq b)$ are integers indicating the minimum and maximum gaps between any two consecutive items $p_{(j-1)}$ and p_j, respectively.

Definition 2 (*Occurrence and Nonoverlapping Occurrence*) Suppose we have a sequence $\mathbf{s} = s_1 s_2 \ldots s_n$ and a pattern $\mathbf{p} = p_1[a, b]p_2 \ldots [a, b]p_m$. Then, $L =< l_1, l_2, \ldots, l_m >$ is an occurrence of pattern $\mathbf{p}$ in sequence $\mathbf{s}$ if and only if $p_1 = s_{l_1}$, $p_2 = s_{l_2}, \ldots p_m = s_{l_m}$ $(0 < l_1 < l_2 < \cdots < l_m \leq n)$ and $a \leq l_j - l_{j-1} - 1 \leq b$. Suppose we have another occurrence $L' =< l'_1, l'_2, \ldots l'_m >$. Then, L and L' are two nonoverlapping occurrences if and only if $\forall\, 1 \leq j \leq m$ and $l_j \neq l'_j$.

Definition 3 (*Support*) The support of pattern $\mathbf{p}$ in sequence $\mathbf{s}$ is the number of nonoverlapping occurrences, which is denoted as $sup(\mathbf{p}, \mathbf{s})$.

Example 2 The occurrences of pattern $\mathbf{p} = C[0, 2]T[0, 2]T$ in sequence $\mathbf{s} = $ CTCTTG are $<1,2,4>$, $<1,2,5>$, $<1,4,5>$ and $<3,4,5>$. The nonoverlapping occurrences are $<1,2,4>$ and $<3,4,5>$. The fourth item of the sequence (s_4) exists in both occurrences but matches the positions p_3 and p_2 of $\mathbf{p}$, respectively. Hence, $sup(\mathbf{p}, \mathbf{s}) = 2$.

Definition 4 (*Utility and Average Utility*) Let $U(p_j)$ be a positive number that is assigned to each item p_j, and represents its relative importance. The utility of a pattern $\mathbf{p}$ of length m in a sequence $\mathbf{s}$ is the sum of the utilities of its item multiplied by its support, that is:

$$PU(\mathbf{p}, \mathbf{s}) = \sum_{j=1}^{m} U(p_j) \times sup(\mathbf{p}, \mathbf{s}) . \tag{1}$$

The average utility is the ratio of the pattern's utility to its length, which is denoted as

$$PAU(\mathbf{p}, \mathbf{s}) = \frac{PU(\mathbf{p}, \mathbf{s})}{m} = \frac{\sum_{j=1}^{m} U(p_j) \times sup(\mathbf{p}, \mathbf{s})}{m} . \tag{2}$$

Example 3 In Example 1, the nonoverlapping occurrences of $\mathbf{p}$ in $\mathbf{s}$ are $<1,2,4>$ and $<3,4,5>$, and the utilities of C and T are 8 and 3, respectively. We therefore know that $PU(\mathbf{p}, \mathbf{s}) = \sum_{j=1}^{m} U(p_j) \times sup(\mathbf{p}, \mathbf{s}) = (8 + 3 + 3) \times 2 = 28$. Hence, $PAU(\mathbf{p}, \mathbf{s}) = \frac{PU(\mathbf{p}, \mathbf{s})}{m} = \frac{28}{3} = 9.3$.

Definition 5 (*HAPP*) If the average utility of the periodic gap pattern $\mathbf{p}$ is not less than the minimum threshold *minpau*, then pattern $\mathbf{p}$ is a HAPP; otherwise, it is a non-HAPP.

Our aim is to mine all HAPPs in a sequence.

Example 4 Suppose we have a sequence $\mathbf{s} = $ CTCTTG, *gap* = [0, 2], the utilities of T, C and G are 3, 8 and 8, respectively, and *minpau*=8. Then, all HAPPs are {C, T, G, C[0, 2]C, C[0, 2]T, C[0, 2]G, C[0, 2]C[0, 2]G, C[0, 2]T[0, 2]T} .

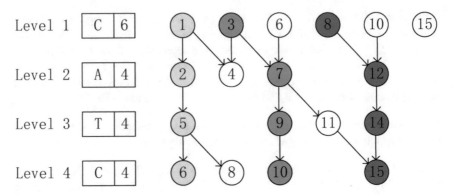

Fig. 2 Nettree of pattern **p** in sequence **s**. There are four levels, since the pattern length is four. Nodes n_1^1, n_1^3, n_1^6, n_1^8, n_1^{10} and n_4^{15} are roots, and nodes n_4^6, n_4^8, n_4^{10} and n_4^{15} are leaves. The same node ID can appear at different levels, for example, n_1^6 and n_4^6 both have ID 6, and are created in the first and fourth levels, respectively. Some nodes may have more than one parent, for example, node n_2^4 has two parents, n_1^1 and n_1^3

4 Proposed Algorithm

To discover all HAPPs in a sequence, the main factors affecting the performance are the calculation of the average utility and the generation of candidate patterns. The core difficulty in calculating the utility is associated with calculating the support [44]. Section 4.1 describes the design of the NetBTM algorithm, which employs a backtracking strategy to calculate the support. Then, Sect. 4.2 presents the pattern join strategy that relies on an upper bound on the average utility to efficiently generate candidate patterns. Finally, Sect. 4.3 describes the NetHAPP algorithm, and analyses its time and space complexities.

4.1 Calculation of the Average Utility

In calculating the average utility of a pattern, the key issue is the calculation of the support of the pattern. Wu et al. [52] proposed the NETGAP algorithm, which was based on Nettree, thus giving a complete method for calculating the support. However, the weakness of NETGAP is its low efficiency. We therefore design a more effective algorithm called NetBTM to calculate a pattern's support which employs a backtracking strategy. Examples 5 and 6 will illustrate the principles of NETGAP and NetBTM, respectively.

Example 5 Suppose we have sequence $\mathbf{s} = s_1 s_2 s_3 s_4 s_5 s_6 s_7 s_8 s_9 s_{10} s_{11} s_{12} s_{13} s_{14} s_{15} =$ CACATCAC
TGTAGTC, a pattern $\mathbf{p} = p_1 p_2 p_3 p_4 = $ CATC, and $gap = [0, 3]$. The principle of NETGAP is as follows:

1. First, we create a Nettree based on the sequence **s** and pattern **p**, as shown in Fig. 2.
2. The next step is to prune the invalid nodes, i.e. non-leaf nodes without children. In this example, the invalid nodes are n_2^4 and n_1^{15}.
3. We then start from the leftmost root n_1^1, and find the leftmost child n_2^2, until the leaf is found. It is easy to obtain the first path $< n_1^1, n_2^2, n_3^5, n_4^6 >$ (marked in yellow), whose corresponding occurrence is $< 1, 2, 5, 6 >$.
4. Nodes n_1^1, n_2^2, n_3^5 and n_4^6 are then pruned. Following this, NETGAP prunes invalid nodes in the new Nettree. In this example, the invalid node is n_4^8.
5. The above steps are iterated, and NETGAP obtains the second occurrence $< 3, 7, 9, 10 >$, (marked in green). Nodes n_1^3, n_2^7, n_3^9 and n_4^{10} are then pruned, and the invalid nodes n_1^6 and n_3^{11} are also pruned. Finally, the third occurrence $< 8, 12, 14, 15 >$ (marked in blue) is found. Hence, NETGAP finds three nonoverlapping occurrences.

From the above example, it can be seen that the NETGAP algorithm is inefficient, since invalid nodes must be pruned after an occurrence is found. The NetBTM algorithm proposed in this chapter employs the backtracking strategy to avoid this pruning operation.

Example 6 Consider again the sequence and pattern of Example 5. The principle of NetBTM is as follows:

1. As in Example 5, we first create a Nettree based on the sequence **s** and pattern **p**, as shown in Fig. 2.
2. NetBTM does not need to prune invalid modes, such as n_2^4 and n_1^{15}, and finds the first occurrence $< 1, 2, 5, 6 >$.
3. The next step is to find the next occurrence from the second root n_1^3. Node n_2^4 is the first child of n_1^3, and has no child. In this case, the algorithm backtracks to the parent node n_1^3 and finds the second child node, n_2^7. Thus, NetBTM obtains the second occurrence $< 3, 7, 9, 10 >$.
4. Similarly, NetBTM finds the third occurrence $< 8, 12, 14, 15 >$. There are no further occurrences. Hence, NetBTM also finds three nonoverlapping occurrences.

From Examples 5 and 6, we know that both NETGAP and NetBTM find the same nonoverlapping occurrences. However, NetBTM employs the backtracking strategy, and does not need to prune invalid nodes. Pseudocode for NetBTM is given in Algorithm 1.

Theorem 1 *The time and space complexities of NetBTM are both $O(m \times n \times w)$ in the worst case, and the average time and space complexities are $O(m \times n \times w / r / r)$, where m, n, w and r represent the pattern length, sequence length, gap width ($w = b - a + 1$, gap $= [a, b]$) and item number $|\Sigma|$, respectively.*

Proof A Nettree has m levels. Each level has no more than n nodes, and each node has no more than w children. Each node can be visited no more than once. Hence, the time and space complexities of NetBTM are both $O(m \times n \times w)$ in the worst

---

Algorithm 1 NetBTM

Input: Sequence **s** with length n, pattern **p** with length m, $gap = [a, b]$
Output: $sup(\mathbf{p}, \mathbf{s})$
1: Create a Nettree according to **p** and **s**;
2: **for** *each root* **do**
3:    $occ \leftarrow$ Get an occurrence by iteratively selecting the leftmost child with the backtracking strategy;
4:    $sup(\mathbf{p}, \mathbf{s})$++;
5:    Delete occ;
6: **end for**
7: **return** $sup(\mathbf{p}, \mathbf{s})$

---

case. In the average case, each level has n/r nodes, and each node has w/r children. The time and space complexities of NetBTM are therefore $O(m \times n \times w \, / \, r \, / \, r)$ in the average case.

4.2 Candidate Pattern Generation

In this subsection, we introduce our method for candidate pattern generation based on an average utility upper bound. Example 7 illustrates that the average utility does not satisfy the Apriori property.

Example 7 Suppose we have sequence **s** = GAGTAAGAATCGTG, pattern $\mathbf{p}_1$ = GA, $\mathbf{p}_2$ = GAG, $gap = [0, 3]$ and $minpau = 25$. The utilities of each item for this example are shown in Table 4.

The supports of patterns $\mathbf{p}_1$ and $\mathbf{p}_2$ are $sup(\mathbf{p}_1, \mathbf{s}) = 3$ and $sup(\mathbf{p}_2, \mathbf{s}) = 3$, respectively. We calculate the average utility of the two patterns using Definition 4: $PAU(\mathbf{p}_1, \mathbf{s}) = \frac{PU(\mathbf{p}_1, \mathbf{s})}{2} = \frac{15 \times 3}{2} = 22.5$, $PAU(\mathbf{p}_2, \mathbf{s}) = \frac{PU(\mathbf{p}_2, \mathbf{s}) \times 3}{3} = \frac{75}{3} = 25$. $\mathbf{p}_1$ is not a HAPP, since $PAU(\mathbf{p}_1, \mathbf{s}) < minpau$. Although $\mathbf{p}_2$ is a super-pattern of $\mathbf{p}_1$, $\mathbf{p}_2$ is a HAPP, since $PAU(\mathbf{p}_2, \mathbf{s}) \geq minpau$. It can be seen from this example that the average utility does not satisfy the Apriori property. But an enumeration tree strategy can be used to generate candidate patterns. In Example 7, $\mathbf{p}_1$ =GA can generate four candidate patterns: GAA, GAT, GAC and GAG. However, using the enumeration tree strategy may yield an exponential number of candidate patterns. To tackle this issue, we propose using a pattern join strategy based on an average utility upper bound. We first give a definition of the upper bound on the average utility.

Table 4 Utility of each item

Item	A	T	C	G
Utility value	5	3	8	10

Definition 6 (*Upper Bound on the Average Utility and High Average Utility Upper Bound Pattern (HABP)*) The upper bound on the average utility of pattern **p** is the product of its support and the maximum utility, as follows:

$$SPU(\mathbf{p}, \mathbf{s}) = sup(\mathbf{p}, \mathbf{s}) \times U_{max} , \tag{3}$$

where U_{max} represents the maximum utility among items. If $SPU(\mathbf{p,s})$ is not less than *minpau*, then pattern **p** is a HABP.

Example 8 The upper bound on the average utility for the patterns $\mathbf{p}_1$ and $\mathbf{p}_2$ of Example 7 is calculated as follows. According to Table 4, $U_{max} = 10$. From Eq. 3, $SPU(\mathbf{p}_1, \mathbf{s}) = 3 \times 10 = 30$ and $SPU(\mathbf{p}_2, \mathbf{s}) = 3 \times 10 = 30$.

Theorem 2 *If pattern p is a non-HABP, then it is a non-HAPP.*

Proof We know that $\sum_{i=1}^{m} U(p_i) \leq m \times U_{max}$. Hence, $PAU(\mathbf{p}, \mathbf{s}) = \frac{\sum_{i=1}^{m} U(p_i) \times sup(\mathbf{p},\mathbf{s})}{m} \leq \frac{m \times U_{max} \times sup(\mathbf{p},\mathbf{s})}{m} = sup(\mathbf{p}, \mathbf{s}) \times U_{max} = SPU(\mathbf{p}, \mathbf{s})$. $SPU(\mathbf{p}, \mathbf{s}) < minpau$ since pattern **p** is a non-HABP. Thus, $PAU(\mathbf{p,s})$ is less than *minpau*, and pattern **p** is a non-HAPP.

Theorem 3 *The upper bound on the average utility satisfies the Apriori property.*

Proof According to Eq. 3, the upper bound on the average utility of a pattern is its support multiplied by a fixed value. It is known that the pattern support satisfies the Apriori property. Hence, the upper bound on the average utility also satisfies the Apriori property.

According to Theorems 2 and 3, candidate patterns can be generated by employing a pattern join strategy.

Definition 7 (*Maximum Prefix Pattern, Maximum Suffix Pattern and PatternJoin*) Suppose we have a pattern $\mathbf{p} = \mathbf{p}_1[M, N]\mathbf{p}_2 \ldots [M, N]\mathbf{p}_m$, and items r and l. If $\mathbf{q} = \mathbf{p}r = \mathbf{p}_1[M, N]\mathbf{p}_2 \ldots [M, N]\mathbf{p}_m[M, N]r$, then **p** is called the maximum prefix pattern of **q**, and is denoted as $prefix(\mathbf{q}) = \mathbf{p}$. Similarly, if $\mathbf{d} = l\mathbf{p} = l[M, N]\mathbf{p}_1[M, N]\mathbf{p}_2 \ldots [M, N]\mathbf{p}_m$, then **p** is called the maximum suffix pattern of **d**, and is denoted as $suffix(\mathbf{d}) = \mathbf{p}$. We obtain a new pattern **t** of length $m+2$, denoted as $\mathbf{t} = \mathbf{d} \oplus \mathbf{q} = l\mathbf{p}r$, and this process is called pattern join.

Example 9 Suppose we have two patterns $\mathbf{p}_1 = ACC$ and $\mathbf{p}_2 = CCA$, $gap = [0, 2]$, $Prefix(\mathbf{p}_2) = CC$, and $suffix(\mathbf{p}_1) = CC$. Then, since $Prefix(\mathbf{p}_2) = Suffix(\mathbf{p}_1)$, pattern $\mathbf{p} = \mathbf{p}_1 \oplus \mathbf{p}_2 = ACCA$ can be generated by pattern join.

Theorem 4 *Suppose HU_m is the set of HABP patterns. The candidate pattern set C_{m+1} can be generated from HU_m using the pattern join strategy. We can safely say that $HU_{m+1} \subseteq C_{m+1}$.*

Table 5 Number of candidate patterns with different lengths

Pattern length	Length = 1	Length = 2	Length = 3	Length = 4	Length = 5	Length = 6	Total
Number of candidate patterns from enumeration tree	4	12	28	28	12	4	88
Number of candidate patterns from pattern join	4	9	23	5	1	0	42
Number of HAPPs	2	1	1	0	0	0	4

Proof The proof is by contradiction. Suppose $\mathbf{p}_{m_{ij}}$ is a HABP, but $\mathbf{p}_{m_{ij}}$ is not in the set C_{m+1}. The maximum prefix and suffix patterns of $\mathbf{p}_{m_{ij}}$ are $\mathbf{p}_{m_i}$ and $\mathbf{p}_{m_j}$, respectively, i.e. $Prefix(\mathbf{p}_{m_{ij}}) = \mathbf{p}_{m_i}$ and $Suffix(\mathbf{p}_{m_{ij}}) = \mathbf{p}_{m_j}$. According to Theorem 3, $\mathbf{p}_{m_i}$ and $\mathbf{p}_{m_i}$ are two HABPs. According to Definition 7, $\mathbf{p}_{m_{ij}} = p_{m_i} \oplus p_{m_j}$. Thus, $\mathbf{p}_{m_{ij}}$ is in the set C_{m+1}, which contradicts the hypothesis.

According to Theorem 4, we know that all candidate patterns can be generated by the pattern join strategy. The following example shows that the pattern join strategy outperforms the enumeration tree strategy.

Example 10 Suppose we have a sequence s $=$GAGTAAGAATCGTG, $gap = [0, 3]$ and $minpau$ =25. The utilities are shown in Table 4. The set of length 3 HABPs is $HU_3 = \{$AAA, AGT, ATG, GAA, GAG, GAT, GTG$\}$. We will generate $7 \times 4 = 28$ candidate patterns using the enumeration tree strategy, since each HABP will generate four candidate patterns. However, according to the pattern join strategy, there are five candidate patterns: $C_4 = \{$AAAA, AGTG, GAAA, GAGT, GATG$\}$. From Table 5, we can see that the number of candidate patterns generated by the pattern join strategy is much smaller than using the enumeration tree strategy. Hence, the pattern join strategy is more effective than the enumeration tree strategy.

4.3 NetHAPP Algorithm

In this subsection, we propose the NetHAPP algorithm and analyse its time and space complexities. The steps of the NetHAPP algorithm are as follows:

Step 1: Scan the sequence database (SDB), calculate the average utility of each item, and store the HAPPs of length 1 into HAU and the HABPs of length 1 into HU_1.

Step 2: Generate the candidate pattern set C_{m+1} from HU_m using the pattern join strategy.

Step 3: For each pattern $\mathbf{p}$ in set C_{m+1}, calculate its average utility and the upper bound on the average utility. If pattern $\mathbf{p}$ is a HAPP, then store it in the sets HAU and HU_{m+1}. If pattern $\mathbf{p}$ is an HABP, then store it in the HABP set HU_{m+1}.

Step 4: Repeat Steps 2 and 3 until C_{m+2} or HU_{m+1} is empty. The HAPPs are now in set HAU. Pseudocode for the NetHAPP algorithm is shown in Algorithm 2.

Algorithm 2 NetHAPP

Input: Sequence database **SDB**, *minpau*, $gap = [a, b]$ and the utilities
Output: The set of HAPPs HAU
1: Scan the SDB, calculate the average utility of each character, store the HAPPs with length 1 into HAU_1, and store the HABPs with length 1 into HU_1;
2: $m \leftarrow 1$;
3: **while** $C_{m+1} \neq$ null **do**
4:    **for** each pattern **p** in C_{m+1} **do**
5:       $sup(\mathbf{p}, \mathbf{s}) \leftarrow$ NetBTM(SDB, **p**, *gap*);
6:       Calculate $PAU(\mathbf{p}, \mathbf{s})$ and $SPU(\mathbf{p}, \mathbf{s})$ according to Equations 2 and 3, respectively;
7:       **if** $PAU(\mathbf{p}, \mathbf{s}) \geq minpau$ **then**
8:          $HAU_{m+1} \leftarrow HAU_{m+1} \bigcup \mathbf{p}$;
9:          $HU_{m+1} \leftarrow HU_{m+1} \bigcup \mathbf{p}$;
10:      **else if** $SPU(\mathbf{p}, \mathbf{s}) \geq minpau$ **then**
11:         $HU_{m+1} \leftarrow HU_{m+1} \bigcup \mathbf{p}$;
12:      **end if**
13:   **end for**
14:   $m \leftarrow m + 1$;
15:   $C_{m+1} \leftarrow$ Patternjoin (HU_m);
16: **end while**
17: **return** HAU;

Theorem 5 *The time complexity of the NetHAPP algorithm is $O(m \times n \times w \times L)$ in the worst case, and $O(m \times n \times w \times L/r/r)$ in the average case, where m, n, w, L and r are the pattern length, sequence length, gap width ($w = b - a + 1$, $gap = [a, b]$), number of candidate patterns and item number $|\Sigma|$, respectively.*

Proof According to Theorem 1, the time complexity of the NetBTM algorithm in the worst case is $O(m \times n \times w)$, and the average time complexity is $O(m \times n \times w \times L/r/r)$. Since each candidate pattern runs once, the time complexity of NetHAPP is $O(m \times n \times w \times L)$ and $O(m \times n \times w \times L/r/r)$ in the average case.

Theorem 6 *The space complexity of the NetHAPP algorithm is $O(m \times (n \times w + L))$ in the worst case, and $O(m \times (n \times w/r/r + L))$ in the average case.*

Proof The space used by NetHAPP consists of the space for candidate patterns and the space required by NetBTM. Obviously, the space complexity of candidate patterns is $O(m \times L)$. According to Theorem 1, the space complexity of the NetBTM algorithm in the worst case is $O(m \times n \times w)$, and the average time complexity is $O(m \times n \times w/r/r)$. Hence, the space complexity of NetHAPP is $O(m \times (n \times w + L))$ in the worst case and $O(m \times (n \times w/r/r + L))$ in the average case.

5   Experiments and Analysis

This section presents the experimental evaluation. Section 5.1 describes the experimental data, and Sect. 5.2 introduces the competitive algorithms. Section 5.3 compares and analyses the mining ability of the proposed algorithm and competitive algorithms. Section 5.4 analyses the running performance for different strategies for generating candidate patterns and calculating the support. Section 5.5 compares the mining results of NetHAPP and NOSEP, and analyses the advantages of our method.

All experiments were carried out on a computer with a Intel(R)Core(TM)i5-4210U processor, 8 GB memory, Windows 8.1 operating system, and VC++6.0 as the experimental environment.

5.1   Database

Table 6 describes the experimental database used in this chapter. It contains sequences having various lengths.

5.2   Baseline Method

To evaluate the effectiveness of NetHAPP, three experiments were designed. In the first, we applied NetHAPP to different lengths of DNA and virus sequences to explore the mining ability of this algorithm. The second experiment was designed to evaluate the running performance by reporting the running time and the number of candidate patterns. The third experiment was designed to analyse the superiority of adding a high average utility to traditional SPM. Brief descriptions of these algorithms are given below:

Table 6 Database

Dataset	Type	From	Length
DNA1[a]	DNA	Homo sapiens AL158070	6,000
DNA2	DNA	Homo sapiens AL158070	8,100
DNA3	DNA	Homo sapiens AL158070	10,000
VIRUS1[b]	Virus	Severe acute respiratory syndrome corona virus 2	7,000
VIRUS2	Virus	Severe acute respiratory syndrome corona virus 2	9,000
VIRUS3	Virus	Severe acute respiratory syndrome corona virus 2	11,000

[a] DNA13 were used in [44], and can be downloaded from http://www.ncbi.nlm.nih.gov/nuccore/AL158070.11

[b] VIRUS13 are the gene sequences of the virus causing COVID-19, and can be downloaded from https://www.ncbi.nlm.nih.gov/nuccore/MN908947

1. NetHAPP_nogap: To analyse the impact of the introduction of periodic gap constraint to NetHAPP on the mining results, the NetHAPP_nogap algorithm was used to mine the high average utility patterns with no gap constraints.
2. NetHAPP_nc and NetHAPP_oo: To explore the mining performance when the nonoverlapping condition is introduced to NetHAPP, the NetHAPP_nc and NetHAPP_oo algorithms were used to mine the high average utility patterns with gap constraints with no condition and the one-off condition, respectively.
3. NOSEP [44]: To analyse the difference between NetHAPP and the classical repetitive SPM algorithm, the NOSEP algorithm was run as a competitor.
4. NetHAPP_bf and NetHAPP_df: To determine the efficiency of the pattern join strategy in NetHAPP to generate candidate patterns, we applied the NetHAPP_bf and NetHAPP_df algorithms, which used breadth-first and depth-first strategies to generate candidate patterns, respectively.
5. NetHAPP_ng: To demonstrate the superiority of NetHAPP due to the use of the backtracking strategy to calculate the support, we used the NetHAPP_ng algorithm, which adopts the NETGAP algorithm in the support calculation step.

5.3 Mining Ability

Three competitive algorithms (NetHAPP_nogap, NetHAPP_nc and NetHAPP_oo) were run to compare the mining ability of the NetHAPP algorithm for different lengths of sequences. Six datasets were selected: DNA1, DNA2, DNA3, VIRUS1, VIRUS2 and VIRUS3. The parameters were $gap = [0, 3]$, $minpau = 210$, $U(A) = 0.2$, $U(G) = 0.3$, $U(C) = 0.2$ and $U(T) = 0.3$. The running times and numbers of patterns are shown in Figs. 3 and 4, respectively.

The following experimental results were found. The introduction of periodic gap constraints can effectively improve the mining ability of the algorithm. For example, for the DNA3 data in Fig. 4, we can see that the NetHAPP_nogap and NetHAPP algorithms mined 7 and 68 HAPPs, respectively. All of the other experiments gave similar results. The reason for this is that NetHAPP introduces periodic gap constraints, which makes the mining results more flexible and targeted.

The mining ability of the proposed algorithm under nonoverlapping condition is better than with no condition and the one-off condition. For example, from Fig. 4, we can see that the numbers of HAPPs mined by NetHAPP_nc, NetHAPP_oo and NetHAPP were 3E+02, 12 and 25, respectively, and the running times of these three algorithms were 2E+03, 168.6 and 3.5 s for the VIRUS1 database, respectively. The reason for this result is that the nonoverlapping condition is stricter than with no condition and looser than the one-off condition. Therefore, this method will neither generate too many valueless patterns nor ignore meaningful patterns. NetHAPP employs the backtracking strategy to calculate pattern support, which effectively improves the efficiency of the algorithm in terms of running time. In this way, NetHAPP can mine a moderate number of HAPPs and is faster than both NetHAPP_nc and NetHAPP_oo.

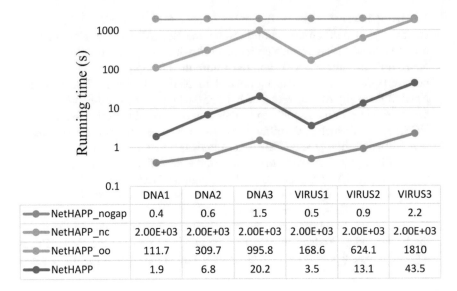

Fig. 3 Comparison of running times

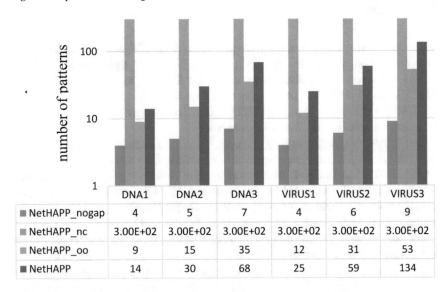

Fig. 4 Comparison of numbers of patterns

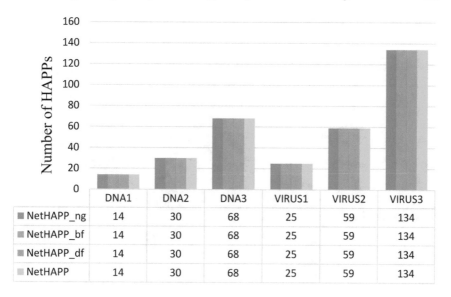

	DNA1	DNA2	DNA3	VIRUS1	VIRUS2	VIRUS3
■ NetHAPP_ng	14	30	68	25	59	134
■ NetHAPP_bf	14	30	68	25	59	134
■ NetHAPP_df	14	30	68	25	59	134
■ NetHAPP	14	30	68	25	59	134

Fig. 5 Comparison of numbers of HAPPs

5.4 Running Performance

In this subsection, we will compare the running performance of the NetHAPP algorithm using different strategies for candidate pattern generation and support calculation. Six datasets were used: DNA1, DNA2, DNA3, VIRUS1, VIRUS2 and VIRUS3. The parameters were $gap = [0, 3]$, $minpau = 210$, $U(A) = 0.2$, $U(G) = 0.3$, $U(C) = 0.2$ and $U(T) = 0.3$. The numbers of HAPPs, running times and numbers of candidate patterns are shown in Figs. 5, 6 and 7, respectively.

The following observations were made:

The pattern join strategy outperforms the breadth-first and depth-first strategies. For example, from the VIRUS3 data in Fig. 5, we know that all of the algorithms discovered 134 HAPPs. However, NetHAPP was faster than NetHAPP_bf and NetHAPP_df, and the running times for NetHAPP_bf, NetHAPP_df as well as NetHAPP are 104.5, 105 and 43.5 s, respectively, as shown in Fig. 6. From Fig. 7, it can be seen that the numbers of candidate patterns for the three algorithms were 996, 996 and 681, respectively. NetHAPP_bf and NetHAPP_df generated more candidate patterns than NetHAPP. All of the other experiments showed similar results. Thus, our experiment verified that the pattern join strategy can effectively prune candidate patterns, and therefore gives better performance than breadth-first and depth-first strategies.

The backtracking strategy can effectively improve the mining performance of the algorithm. From Figs. 5 and 7, we can see that the numbers of HAPPs and candidate patterns for NetHAPP_ng and NetHAPP were the same. However, NetHAPP was faster than NetHAPP_ng, as shown in Fig. 6. For example, from the DNA2 data in

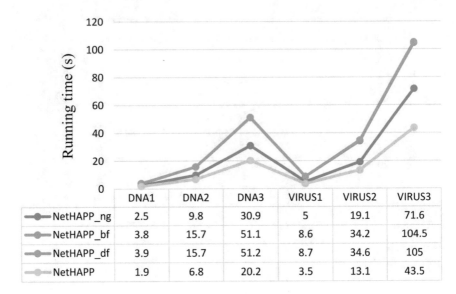

	DNA1	DNA2	DNA3	VIRUS1	VIRUS2	VIRUS3
NetHAPP_ng	2.5	9.8	30.9	5	19.1	71.6
NetHAPP_bf	3.8	15.7	51.1	8.6	34.2	104.5
NetHAPP_df	3.9	15.7	51.2	8.7	34.6	105
NetHAPP	1.9	6.8	20.2	3.5	13.1	43.5

Fig. 6 Comparison of running times

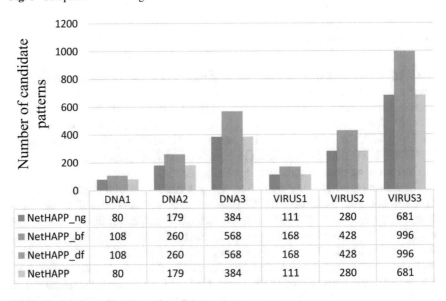

	DNA1	DNA2	DNA3	VIRUS1	VIRUS2	VIRUS3
NetHAPP_ng	80	179	384	111	280	681
NetHAPP_bf	108	260	568	168	428	996
NetHAPP_df	108	260	568	168	428	996
NetHAPP	80	179	384	111	280	681

Fig. 7 Comparison of numbers of candidate patterns

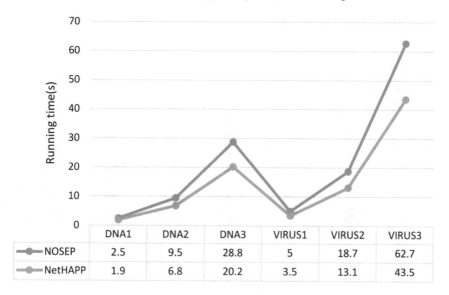

	DNA1	DNA2	DNA3	VIRUS1	VIRUS2	VIRUS3
NOSEP	2.5	9.5	28.8	5	18.7	62.7
NetHAPP	1.9	6.8	20.2	3.5	13.1	43.5

Fig. 8 Comparison of running times

Fig. 6, the running times of NetHAPP_ng and NetHAPP were 9.8 and 6.8 s, respectively. The reason for this is that NetHAPP employs the backtracking strategy, which does not need to prune invalid nodes.

5.5 High Average Utility Pattern Mining and Frequent Pattern Mining

In this subsection, we compare the mining results from NetHAPP and NOSEP, and analyse the benefits of adding high average utility to traditional SPM. To ensure a fair comparison, we set $minpau = 210$ in NetHAPP and $minsup = 210/0.3 = 700$ in NOSEP to ensure that the numbers of candidate patterns generated by the two algorithms are the same. Six datasets were selected: DNA1, DNA2, DNA3, VIRUS1, VIRUS2 and VIRUS3. The parameters were $gap = [0, 3]$, $minpau = 210$, $U(A) = 0.2$, $U(G) = 0.3$, $U(C) = 0.2$ and $U(T) = 0.3$. The running times and numbers of patterns are shown in Figs. 8 and 9.

The following experimental results were found.

For each of the different databases, the mining ability of NetHAPP was superior to that of NOSEP. For example, it can be seen from Fig. 8 that NetHAPP is faster than NOSEP. This is because NetHAPP uses the NetBTM algorithm to calculate the pattern support, which effectively reduces the calculation time. From the VIRUS2 data in Fig. 9, we can see that the numbers of patterns mined by NOSEP and NetHAPP were 107 and 59, respectively. The reason for this is because NOSEP only uses the

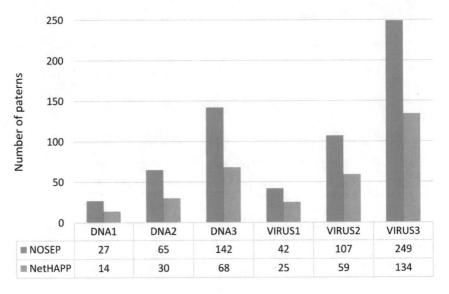

	DNA1	DNA2	DNA3	VIRUS1	VIRUS2	VIRUS3
■ NOSEP	27	65	142	42	107	249
■ NetHAPP	14	30	68	25	59	134

Fig. 9 Comparison of numbers of patterns

support to evaluate the importance, while NetHAPP employs both the support and the utilities to evaluate the importance. The latter can thus better meet the needs of users.

6   Conclusion

To solve the high average utility periodic gapped SPM problem, we addressed the issue of HAPP mining with the following three characteristics: (i) the pattern has periodic gap constraints, i.e. each gap constraint is the same; (ii) the nonoverlapping constraint is used in the support calculation, i.e. the characters in the sequence can be used repeatedly in different positions and (iii) the mining method mines patterns with high average utility. The concept of high average utility is interesting as it takes into account not only the frequency of each pattern and the utility value of each item, but also the pattern length. To efficiently find all HAPPs, we proposed the NetHAPP algorithm. When calculating the average utility of a pattern, the difficulty lies in calculating the pattern's support. To overcome this problem, a backtracking strategy was adopted that effectively reduces the time complexity. When generating candidate patterns, the difficulty lies in reducing the number of candidate patterns as the average utility does not satisfy the Apriori property. To address this problem, an upper bound on the average utility and a pattern join strategy were proposed, which effectively reduces the numbers of candidate patterns. Extensive experiments were conducted on DNA and virus sequences. The experimental results showed that in terms of mining ability, the NetHAPP algorithm was superior to the traditional SPM algorithm, as it not only improved the running time significantly but could also mine infrequent but important patterns. In terms of mining performance, both the backtracking and the pattern join strategies significantly reduced the time and space complexities, thus greatly improving the efficiency of the algorithm.

There are several possibilities for future work such as applying the proposed algorithm in various applications, designing faster and more memory-efficient algorithms, and to study algorithms for mining sequential patterns respecting other types of constraints.

Acknowledgements This work was partly supported by National Natural Science Foundation of China (61976240, 52077056), and Graduate Student Innovation Program of Hebei Province (CXZZBS2020024).

References

1. Z. Abdullah, O. Adam, T. Herawan, M.M. Deris, A review on sequential pattern mining algorithms based on apriori and patterns growth, in *Proceedings of the International Conference on Data Engineering 2015 (DaEng-2015)*, eds. by J. Abawajy, M. Othman, R. Ghazali, M. Deris, H. Mahdin, T. Herawan. Lecture Notes in Electrical Engineering (Springer, Singapore, 2019), pp. 111–118
2. R. Agrawal, R. Srikant, Mining sequential patterns, in *Proceedings of the Eleventh International Conference on Data Engineering* (IEEE, Taipei, Taiwan, 1995), pp. 3–14
3. R. Agarwal, R. Srikant, Fast algorithms for mining association rules, in *Proceedings of the 20th VLDB Conference*, pp. 487–499
4. X. Chai, X. Jia, Y. Wu, H. Jiang, X. Wu, Strict pattern matching with general gaps and One-Off condition. J. Softw. **26**(5), 1096–1112 (2015)
5. X. Chen, Y. Rao, H. Xie, F. Wang, Y. Zhao, J. Yin, Sentiment classification using negative and intensive sentiment supplement information. Data Sci. Eng. **4**(2), 109–118 (2019)
6. M. D'Andreagiovanni, F. Baiardi, J. Lipilini, S. Ruggieri, F. Tonelli, Sequential pattern mining for ICT risk assessment and management. J. Log. Algebr. Methods Program **102**, 1–16 (2019)
7. B. Ding, D. Lo, J. Han, S. Khoo, Efficient mining of closed repetitive gapped subsequences from a sequence database, in *2009 IEEE 25th International Conference on Data Engineering* (IEEE, 2009), pp. 1024–1035
8. X. Dong, Y. Gong, L. Cao, e-RNSP: an efficient method for mining repetition negative sequential patterns. IEEE T. Cybern. **50**(5), 2084–2096 (2020)
9. X. Dong, Q. Qiu, J. Lu, L. Cao, T. Xu, Mining top-k useful negative sequential patterns via learning. IEEE Trans. Neural Netw. Learn. Syst. **30**(9), 2764–2778 (2019)
10. F. Fumarola, P.F. Lanotte, M. Ceci, D. Malerba, CloFAST: closed sequential pattern mining using sparse and vertical ID-lists. Knowl. Inf. Syst. **48**(2), 429–463 (2016)
11. P. Fournier-Viger, J. Li, J.C.W. Lin, T.T. Chi, R.U. Kiran, Mining cost-effective patterns in event logs. Knowl.-Based Syst. **191**, 105241 (2020)
12. P. Fournier-Viger, P. Yang, J.C.W. Lin, P.U. Kiran, Discovering stable periodic-frequent patterns in transactional data, in *Advances and Trends in Artificial Intelligence*, eds. by F. Wotawa, G. Friedrich, I. Pill, R. Koitz-Hristov, M. Ali. From Theory to Practice. IEA/AIE. Lecture Notes in Computer Science (Springer, Berlin, 2019), pp. 230–244
13. P. Fournier-Viger, J.C.W. Lin, R.U. Kiran, Y.-S. Koh, A survey of sequential pattern mining. Data Sci. Pattern Recognit. **1**(1), 54–77 (2017)

14. W. Gan, J.C.W. Lin, P. Fournier-Viger, H.C. Chao, P.S. Yu, HUOPM: high-utility occupancy pattern mining. IEEE T. Cybern. (2019). https://doi.org/10.1109/TCYB.2019.2896267
15. W. Gan, J.C.W. Lin, J. Zhang, H. Chao, H. Fujita, P.S. Yu, ProUM: Projection-based utility mining on sequence data. Inf. Sci. **513**, 222–240 (2020)
16. J. Ge, Y. Xia, J. Wang, C.H. Nadungodage, S. Prabhakar, Sequential pattern mining in databases with temporal uncertainty. Knowl. Inf. Syst. **51**(3), 821–850 (2017)
17. D. Guo, X. Hu, F. Xie, X. Wu, Pattern matching with wildcards and gap-length constraints based on a centrality-degree graph. Appl. Intell. **29**, 57–74 (2013)
18. T. Guyet, R. Quiniou, NegPSpan: efficient extraction of negative sequential patterns with embedding constraints. Data Min. Knowl. Disc. **34**, 563–609 (2020)
19. T.P. Hong, C.H. Lee, S.L. Wang, Mining high average-utility itemsets, in *Proceedings of the IEEE International Conference on Systems* (IEEE, San Antonio, 2009), pp. 2526–2530
20. H. Jiang, X. Chen, T. He, Z. Chen, X. Li, Fuzzy clustering of crowdsourced test reports for apps. ACM Trans. Internet Technol. **18**(2), 1–28 (2018)
21. H. Jiang, X. Li, Z. Ren, J. Xuan, Z. Jin, Toward better summarizing bug reports with crowdsourcing EliciteWd attribute. IEEE Trans. Reliab. **68**(1), 2–22 (2019)
22. B.C. Kachhadiya, B. Patel, A survey on sequential pattern mining algorithm for web log pattern data, in *2018 2nd International Conference on Trends in Electronics and Informatics (ICOEI)* (IEEE, Tirunelveli, 2018), pp. 1269–1273
23. H.T. Lam, F. Moerchen, D. Fradkin, T. Calders, Mining compressing sequential patterns. Statal Anal. Data Min. **71**(1), 34–52 (2014)
24. B. Le, H. Duong, T. Truong, P. Fournier-Viger, FGenSM: two efficient algorithms for mining frequent closed and generator sequences using the local pruning strategy. Knowl. Inf. Syst. **52**, 71–107 (2017)
25. G. Lee, U. Yu, Performance and characteristic analysis of maximal frequent pattern mining methods using additional factors. Soft. Comput. **22**, 4267–4273 (2018)
26. J.C.W. Lin, J.M. Wu, P. Fournier-viger, T. Hong, T. Li, Efficient mining of high average-utility sequential patterns from uncertain databases, in *2019 IEEE International Conference on Systems, Man and Cybernetics (SMC)* (IEEE, Bari, Italy, 2019) pp. 1989–1994
27. S. Lin, Y. Chen, D. Yang, J. Wu, Discovering long maximal frequent pattern, in *2016 Eighth International Conference on Advanced Computational Intelligence (ICACI)* (IEEE, Chiang Mai, Thailand, 2016), pp. 136–142
28. T. Lu, B. Vo, H.T. Nguyen, T.Z. Hong, A new method for mining high average utility itemsets, in *Computer Information Systems and Industrial Management*, eds. by K. Saeed, V. Snel. CISIM 2015. Lecture Notes in Computer Science (Springer, Heidelberg, 2014), pp. 33–42
29. A.R. Maske, B. Joglekar, An algorithmic approach for mining customer behavior prediction in market basket analysis, in *Innovations in Computer Science and Engineering*, eds. by H. Saini, R. Sayal, A. Govardhan, R. Buyya. Lecture Notes in Networks and Systems (Springer, Singapore, 2019), pp. 31–38
30. F. Min, Z. Zhang, W. Zhai, R. Shen, Frequent pattern discovery with tri-partition alphabets. Inf. Sci. **507**, 715–732 (2020)
31. H. Nam, U. Yun, E. Yoon, J.C.W. Lin, Efficient approach of recent high utility stream pattern mining with indexed list structure and pruning strategy considering arrival times of transactions. Inf. Sci. **529**, 1–27 (2020)
32. J. Pei, J. Wang, W. Wang, Constraint-based sequential pattern mining: the pattern-growth methods. J. Intell. Inf. Syst. **28**, 133–160 (2007)
33. A. Rahman, Y. Xu, K. Radke, E. Foo, Finding anomalies in SCADA logs using rare sequential pattern mining, in *Network and System Security*, eds. by J. Chen, V. Piuri, C. Su, M. Yung. NSS 2016. Lecture Notes in Computer Science (Springer, Cham, 2016), pp. 499–506
34. J. Ren, Y. Sun, S. Guo, Maximal sequential pattern mining based on simultaneous monotone and anti-monotone constraints, in *Third International Conference on Intelligent Information Hiding and Multimedia Signal Processing (IIH-MSP 2007)* (IEEE, Kaohsiung, 2007), pp. 143–146

35. C.B. Rjeily, G. Badr, A.H.E. Hassani, E. Andres, Medical data mining for heart diseases and the future of sequential mining in medical field, in *Machine Learning Paradigms*, eds. by G. Tsihrintzis, D. Sotiropoulos, L. Jain. Intelligent Systems Reference Library (Springer, Cham, 2019), pp. 71–99
36. H. Ryang, U. Yun, Indexed list-based high utility pattern mining with utility upper-bound reduction and pattern combination techniques. Knowl. Inf. Syst. **51**(2), 627–659 (2017)
37. Q. Shi, J. Shan, W. Yan, Y. Wu, X. Wu, NetNPG: nonoverlapping pattern matching with general gap constraints. Appl. Intell. **50**(6), 1832–1845 (2020)
38. A. Soltani, M. Soltani, A new algorithm for high average-utility itemset mining. J. AI Data Min. **7**(4), 537–550 (2019)
39. W. Song, Y. Liu, J. Li, Mining high utility itemsets by dynamically pruning the tree structure. Appl. Intell. **40**, 29–43 (2014)
40. W. Song, B. Jiang, Y. Qiao, Mining multi-relational high utility itemsets from star schemas. Intell. Data Anal. **22**(1), 143–165 (2018)
41. T. Truong, H. Duong, B. Le, P. Fournier-Viger, U. Yun, Efficient high average-utility itemset mining using novel vertical weak upper-bounds. Knowledge-Based Syst. **183**, 104847 (2019)
42. V.S. Tseng, B.E. Shie, C.W. Wu, P.S. Yu, Efficient algorithms for mining high utility itemsets from transactional databases. IEEE Trans. Knowl. Data Eng. **25**(8), 1772–1786 (2013)
43. J. Wang, J. Huang, Y. Chen, On efficiently mining high utility sequential patterns. Knowl. Inf. Syst. **49**, 597–627 (2016)
44. X. Wang, L. Chai, Q. Xu, Y. Yang, J. Li, J. Wang, Y. Chai, Efficient subgraph matching on large RDF graphs using mapreduce. Data Sci. Eng. **4**(1), 24–43 (2019)
45. Y. Wang, W. Hou, F. Wang, Mining co-occurrence and sequence patterns from cancer diagnoses in New York State. PLoS ONE (2018). https://doi.org/10.1371/journal.pone.0194407
46. Q. Xu, D. He, N. Zhang, C. Kang, J. Bai, J. Huang, A short-term wind power forecasting approach with adjustment of numerical weather prediction input by data mining. IEEE Trans. Sustain. Energy **6**(4), 1283–1291 (2015)
47. X. Wu, J. Qiang, F. Xie, Pattern matching with flexible wildcards. J. Comput. Sci. Technol. **29**(5), 740–750 (2014)
48. X. Wu, D. Theodoratos, Homomorphic pattern mining from a single large data tree. Data Sci. Eng. **1**(4), 203–218 (2016)
49. X. Wu, F. Xie, Y. Ming, J. Gao, Mining sequential patterns with wildcards and the one-off condition. J. Soft. **24**(8), 1804–1815 (2013)
50. X. Wu, X. Zhu, Y. He, A.N. Arslan, PMBC: pattern mining from biological sequences with wildcard constraints. Comput. Biol. Med. **43**(5), 481–492 (2013)
51. Y. Wu, J. Fan, Y. Li, L. Guo, X. Wu, NetDAP: (δ, γ)-approximate pattern matching with length constraints. Appl. Intell. **50**(11), 4094–4116 (2020). https://doi.org/10.1007/s10489-020-01778-1
52. Y. Wu, C. Shen, H. Jiang, X. Wu, Strict pattern matching under non-overlapping condition. Sci. China-Inf. Sci. **60**(1), 012101 (2017)
53. Y. Wu, Z. Tang, H. Jiang, X. Wu, Approximate pattern matching with gap constraints. J. Inf. Sci. **42**(5), 639–658 (2016)
54. Y. Wu, Y. Tong, X. Zhu, X. Wu, NOSEP: nonoverlapping sequence pattern mining with gap constraints. IEEE T. Cybern. **48**(10), 2809–2822 (2018)
55. Y. Wu, L. Wang, J. Ren, W. Ding, X. Wu, Mining sequential patterns with periodic wildcard gaps. Appl. Intell. **41**, 99–116 (2014)
56. Y. Wu, Y. Wang, J. Liu, M. Yu, Y. Li, Mining distinguishing subsequence patterns with nonoverlapping condition. Cluster Comput. **22**, 5905–5917 (2019)
57. Y. Wu, X. Wu, H. Jiang, F. Min, A heuristic algorithm for solving MPMGOOC problem. Chin. J. Comput. **34**(8), 1452–1462 (2011)
58. Y. Wu, C. Zhu, Y. Li, L. Guo, X. Wu, NetNCSP: nonoverlapping closed sequential pattern mining. Knowledge-Based Syst. **196**, 105812 (2020)
59. H. Yao, H.J. Hamilton, Butz, A foundational approach to mining itemset utilities from databases, in *Proceedings of the 2004 SIAM International Conference on Data Mining* (SIAM, 2004), pp. 482–486

60. J. Yeo, S. Hwang, S. Kim, E. Koh, N. Lipka, Conversion prediction from click stream: modeling market prediction and customer predictability. IEEE Trans. Knowl. Data Eng. **32**(2), 246–259 (2020)
61. M. Zhang, B. Kao, D.W. Cheung, K.Y. Yip, Mining periodic patterns with gap requirement from sequences. ACM Trans. Knowl. Discov. Data **1**(2), 7 (2007)
62. S. Zida, P. Fournier-Viger, J.C. Lin, C. Wu, V.S. Tseng, EFIM: a fast and memory efficient algorithm for high-utility itemset mining. Knowl. Inf. Syst. **21**(2), 599–625 (2017)

Privacy Preservation of Periodic Frequent Patterns Using Sensitive Inverse Frequency

Usman Ahmed, Jerry Chun-Wei Lin, and Philippe Fournier-Viger

Abstract Pattern mining methods help to extract valuable information from a large dataset. The extraction of knowledge might result in the risk of privacy issues. Some potential information might disclosure the insights about customers' behaviors. This leads us to the issue of privacy-preserving data mining (PPDM) that hides sensitive information as much as possible but remains valid information for the further knowledge discovery methods. In this paper, we first propose a sanitization approach for hiding the sensitive periodic frequent patterns. The proposed method utilizes the Term Frequency and Inverse Document Frequency (TF-IDF) to select the transactions and items for sanitization based on the user-defined sensitive periodic frequent patterns. The designed approach can select the victim items in the transactional database for data sanitization. Experimental results showed that the model can perform better for sparse and dense datasets under different user-defined thresholds.

1   Introduction

Privacy preservation data mining (PPDM) has become a significant research issue by considering the General Data Protection Regulation (GDPR)[1] regulation. With this invention, data privacy is one of the main concerns related to the data-driven applications. Personal information includes some confidential information, i.e., social

---

[1] https://eur-lex.europa.eu/eli/reg/2016/679/oj.

U. Ahmed · J. C.-W. Lin (✉)
Department of Computer Science, Electrical Engineering and Mathematical Sciences,
Western Norway University of Applied Sciences, 5063 Bergen, Norway
e-mail: jerrylin@ieee.org

U. Ahmed
e-mail: usman.ahmed@hvl.no

P. Fournier-Viger
School of Humanities and Social Sciences, Harbin Institute of Technology (Shenzhen),
Shenzhen, China

© The Author(s), under exclusive license to Springer Nature Singapore Pte Ltd. 2021
R. Uday Kiran et al. (eds.), *Periodic Pattern Mining*,
https://doi.org/10.1007/978-981-16-3964-7_12

security numbers, address information, credit card numbers, credit ratings, and customer purchasing behaviors), is collected and used in the analytical streams to track user interests. This information helps extract the personal traits individually and then based on the traits, micro-level targeting is done, and disinformation campaigns can be conducted. Moreover, the collected information sometimes will be sold to the organization and used for further analysis and marketing. Legislation of data protection law has been developed to protect the privacy of the sensitive information. Thus, privacy preservation has become a mandatory progress in data mining tasks.

PPDM has become the primary research area as a result of prevailing issues. The main goal is to hide sensitive information while permitting usable knowledge for further knowledge discovery. A common approach for sanitizing the database is through the addition or deletion operations to hide the sensitive information. Several models are developed to sanitize the database using the above operations [8, 18, 22, 23]. Those developed approaches tend to perform well for protecting individual data. However, they can produce numerous side effects in the sanitization progress. Another issue related to PPDM is frequent itemset mining (FIM) that is the primary usage of the sanitization approaches [2]. However, FIM is not designed to discover patterns that appear periodically in a database [11, 16]. Therefore, hiding sensitive items that have periodically occurrence is much important to user privacy. Extracting and hiding the patterns are vital for the business as they result in lucrative sale offers and help in product launches [14]. Thereby, selecting and sanitizing patterns that occurred based on the periodicity helps to improve the privacy of the user-centric data.

In 2006, Aggarwal et al. introduced the concept of PPDM [1]. Lindell et al. used the ID3 algorithm to solve the PPDM issue. Clifton et al. [5] solved the associated problems of PPDM. A multiplicative perturbation algorithm was proposed that balances the utility and privacy of the data [21]. A vertical partition-based data sanitization is also proposed by the Dwork et al. [7]. Lin et al. proposed Genetic Algorithm (GA) and Particle Swarm Optimization models for data sanitization using the evolutionary computation [24, 26]. Many FIM-based algorithms were also proposed in PPDM [22, 23, 25, 27, 34].

In-text mining, the term frequency–inverse document frequency (TF-IDF) [31] is used to extract words based on their importance in the document. The TF-IDF-based method is usually based on the statistical measure that finds the importance of its appearance number in each document. Hong et al. then presented the SIF-IDF model for data sanitization [18]. However, the SIF-IDF model does not focus on the periodic patterns but only frequent patters. This paper proposes a new TF-IDF-based model called periodic sensitive items frequency–inverse database frequency (PSIF-IDF) that is used to evaluate the transactions association with sensitive itemsets based on their periodicity. The proposed approach used the TF-IDF concept to reduce the frequency of periodic itemsets for sanitization. The proposed PSIF-IDF algorithm can easily make good trade off between periodic privacy preservation, sanitized data utilization and execution time. The key contribution of the proposed PSIF-IDF are as follows:

- This is the first paper to hide the periodic frequent patterns for data sanitization in PPDM.
- Based on the user-defined sensitive threshold value, the designed model used the periodic frequent inverse transaction frequency to select and sanitize the periodic patterns efficiently.
- Experimental results indicated that the designed approach achieves good performance.

The rest of the paper is organized as follows. Section 2 described some related works. The proposed algorithm is stated in Section . An example and methodology is mentioned in Sect. 3 and experimentation are given in Sect. 4. Finally, the conclusion is given in Sect. 5.

2   Related Work

In this section, we then discuss the works related to periodic frequent pattern mining and the privacy-preserving data mining.

2.1   Periodic Frequent Pattern Mining

In data mining, several algorithms [9, 17] have been extensively discussed to find the set of frequent itemsets or the association rules based on the minimum support and minimum confidence thresholds. However, most of the existing algorithms in association-rule mining cannot be directly applied to discover the periodic behavior of patterns. In general, periodic pattern mining algorithms ignore a pattern as being non-periodic if it has a single period greater than a maximal periodicity threshold, which is mainly defined by user's preference.

The PFP-tree algorithm was first designed to mine the periodic frequent patterns (PFPs) [32]. It uses the tree-based stricture and FP-growth mining method to mine the PFPs. Amphawan et al. [3] then presented a top-k mining algorithm to apply the depth-first search and vertical database structure to mine the PFPs without threshold constraint. A model called ITL-tree [4] was presented to mine the approximate results of PFPs. Kiran et al. [19] presented an improved model for mining the approximate PFPs efficiently. A MIS-PF-tree [20] was proposed to mine the PFPs with multiple minimum support thresholds. Several works regarding the periodic frequent pattern mining were extensively studied and most of them are presented in progress [10, 11, 16, 30].

2.2 Privacy-preserving Data Mining

PPDM is being identified as a critical issue when data sanitization and side effects are kept in mind. There is also a keen interest in hiding sensitive information while keeping the database utilization preserved. Agrawal et al. proposed a metric that helps to evaluate the sanitization utility of PPDM methods [1]. Verykios et al. addressed the PPDM classification system [33]. Shared data privacy is also addressed that hides the sensitive information when data exists in the shared datasets [29]. Evmievski et al. presented new privacy measures and the development of a group of algorithms that can be used to randomize categorical and numerical data in PPDM [8]. Clifton et al. [5] then presented a toolkit that can be applied in the distributed system and environment. Dehkordi et al. [6] then introduced the GA-based model for hiding the sensitive information in PPDM. However, the artificial and missing rules are the main side effects regarding Dehkordi's model. Several GA-based algorithms for data sanitization are proposed by Lin et al. which includes sGA2DT, pGA2DT [23, 28] and cpGA2DT [24]. In addition, Lin et al. proposed PSO-based algorithms for data sanitization [26]. In recent advancement, a machine learning-based model is deployed in real time for data sanitization [2]. Recently, the periodic stable, correlated, and top-k periodic pattern mining approaches are proposed that extract pattern based on a different measure of periodicity [13–15]. To the best of our knowledge, there is no sanitization model that can be used to disturb the original database for hiding the periodic frequent patterns, and it will be investigated in this paper.

3   Proposed Methodology

In periodic frequent mining and PPDM, we used the Periodic Frequent Pattern Miner (PFPM) [11] for the periodic pattern mining, which is based on three periodic measure for the discovery of the periodic frequent pattern, i.e., the minimum, maximum, and average periodicity. This helps to mine the flexible periodic patterns. PFPM used the Eclat algorithm [35] to scan the database based on minimum periodicity and maximum periodicity first. Then the designed algorithm defines periodic itemsets having periodicity not greater than maximum periodicity. The new itemsets are then sorted based on the support of the itemsets. Another database scan was then performed in depth-first search recursively as the exploration way to scan the remaining combinations of new itemsets. The formal definitions of the studied problems are shown as follows.

Let there be a database having itemset X i.e., $s(X) = \left\{T_{s_1}, T_{s_2} \ldots, T_{s_k}\right\}$, where $1 \leq s_1 < s_2 < \ldots < s_k \leq n$ and n transactions $D = \{T_1, T_2, \ldots, T_n\}$. In transactions $T_x \supset X$ and $T_y \supset X$, if there is no transaction between them, then we consider it as consecutive with respect to X. In consecutive transactions T_x and T_y in $s(X)$, the periodicity is the number of transactions between T_x and T_y, i.e., $pe\left(T_x, T_y\right) = (y - x)$.

For example in Table 1, consider the itemset $\{a, f\}$. This itemset appears in transactions T_1, and T_4, and there are five transactions in the database, therefore, $s(\{a, f\}) = \{T_1, T_4\}$. The periods of this itemset are $ps(\{a, f\}) = \{1,2,1\}$. Periodic frequent pattern (PFP) is thus defined if $|s(X)| \geq$ minimum support $minsup$ and maximum periodicity $max\,Per\,(X) < maxPer$, where $minsup$ and $maxPer$ are user-defined thresholds [11].

In PPDM, the users should first provide the set of sensitive itemsets that is required to be hidden, i.e., $SI = \{s_i, s_2, \ldots, s_i\}$; they are required to be removed from the database. We aim to preventing sensitive itemsets and reducing its frequency from the periodic database D and let the modified database being as D'. In contrast to hiding sensitive itemsets and avoiding those patterns to be mined in the further mining progress, other objectives such as the side effects should be minimized. For instance, the non-sensitive itemsets should not be sanitized in the database D'. In addition, the artificial information that is not presented in the database D should not be included in the database D' either. An example for the scenario is then illustrated as follows.

Algorithm 1 PSIF-IDF

Input: A transaction dataset $D = \{T_1, T_2, \ldots, T_n\}$, items $I = \{i_1, i_2, \ldots, i_p\}$, minimum support threshold s, and a set of m user-specified periodic sensitive itemsets $PS = \{psi_1, psi_2, \ldots, psi_m\}$.
Output: A sanitized database.
1: Find the set of PFPs.
2: Select PS based on user-defined threshold.
3: $PSIF_{ij} = \frac{|psi_{ij}|}{|T_i|}$
4: **for** $item = 1, k$ **do**
5:    $PMRC_k = \max_{j=1}^{m} RC_{kj}$.
6:    $PIDF_k = \log \frac{|n|}{|f_k - PMRC_k|}$.
7:    $PSIF - IDF\,(T_K) = \sum_{k=1}^{n} \left(\frac{|psi_{kj}|}{|T_k|} \times \sum_{y=1}^{p} \log \frac{|n|}{|f_y - PMRC_y|} \right)$.
8: **for** $j = 1, p$ **do**
9:    Sort TID in descending order of PSIF-IDF values.
10:    Select i_j with the highest frequency.
11:    Delete the selected item i_j in TID.
12:    Update items' frequency.
13:    Repeat steps 9–13 until all sensitive itemsets are hidden.
14: **Return** A sanitized database.

Let the user define the threshold value for support. After that, the PFPM [11] is first executed to extract the periodic frequent patterns. Let us assume the threshold value is set as 40%, and we extracted the sensitive itemsets $S = \{cf, af\}$. The support represents that the minimum count is calculated as: 0.4×5, which is 2. After that, the sanitization steps are performed as mentioned in Table 1 and Algorithm 1 (Lines 1 and 2). We calculate periodic sensitive items frequency (SIF) in each transaction. For instance, $\{cf, af\}$. The number of sensitive itemsets for $\{cf, af\}$ is 2/5 and 2/5 as $\{cf\}$ and $\{af\}$ are two items, and the total items in that transaction is five (Table 2).

Table 1 A database with five transactions

TID	Item
T_1	a, b, c, d, f
T_2	a, b, d, e
T_3	b, c, d, f
T_4	a, b, c, f
T_5	c, d, e

Table 2 The PSIF of items in the transactions

TID	Item	$PSIF_{cf}$	$PSIF_{af}$
T_1	a, b, c, d, f	2/5	2/5
T_2	a, b, d, e	0/4	1/4
T_3	b, c, d, f	2/4	0/4
T_4	a, b, c, f	2/4	2/4
T_5	c, d, e	0/3	0/3

Table 3 The MRC value of transactions

Item	RC_{cf}	RC_{af}	MRC
a	–	1	1
b	–	–	0
c	2	–	2
d	–	–	0
e	–	–	0
f	2	1	2

Periodic Inverse Database Frequency (PIDF) is calculated for the transaction items. We used the reduced count (PRC) for each item and then calculated the maximum value among the sensitive items as mentioned in Table 3 that is shown in Algorithm 1 (Line 5). For example, for the item c, we calculate it as $max\{3 - 0.4 \times 5 + 1, 0\}$, where 3 is frequency of the item c with respect to $\{cf\}$, and 0.4 is support threshold. The MRC value, therefore, is calculated as 2. The result is then shown in Table 3.

The PIDF item value is then calculated. We take the item c as an example to illustrate the steps. The occurrence count of the item c is calculated as 4 and the $PMRC$ value is 2. Therefore, we calculate its PIDF value as $\log_{10}(5/(4-2))$, which is 0.39. In this way, we then calculate the values as mentioned in Table 4 and Algorithm 1 (Line 6).

After that, we calculate the Periodic Sensitive Items Frequency (PSIF) value of each sensitive itemset in the transaction. For instance, the itemset $\{cf\}$, the value for first transaction with respect to itemset $\{cf\}$, are 0.39 and 0.69. As results of 0.39

Table 4 The PIDF value of each item

Item	Count	PIDF	PMRC
a	3	0.39	1
b	4	0.09	0
c	4	0.39	2
d	4	0.09	0
e	2	0.39	0
f	3	0.69	2

Table 5 The PSIF value of each sensitive itemset in each transaction

TID	$PSIF_{cf}$	$PSIF_{af}$
T_1	1.10	1.10
T_2	0.0	0.40
T_3	1.10	0.00
T_4	1.10	1.10
T_5	0.40	0.00

+ 0.69, it is calculated as 1.10 for T_1. In this same way, we calculated the sensitive itemsets $\{cf, af\}$, as mentioned in Table 5 and Algorithm 1 (Line 7).

The PSIF-IDF value is then calculated for the sensitive itemsets. It multiples the PSIF value by its $PIDF$ value in the transaction. For instance, in the first transaction, the PSIF value of the sensitive itemset $\{cf\}$ in the first transaction is 2/5, as shown in Table 6 and its PIDF value is 1.10 as shown in Table 6 . Then PSIF-IDF value is calculated by multiplying the as $\frac{2}{5} \times 1.10$, which is 0.43. The other PSIF-IDF values for the sensitive itemsets $\{af\}$ is calculated as 0.43, and then the first transaction is summed as 0.43 + 0.43, which is 0.88. In this way, all other transactions are processed as mentioned in Table 6 and Algorithm 1 (Line 7). After that, we sorted the transactions based on the PSIF-IDF values as mentioned in Table 6, 7 and Algorithm 1 (Line 10). Based on the item frequency as mentioned in Table 4, we delete the item c in transaction 4. In this way, the occurrence frequency of the $\{cf:3, af:2\}$ has become as $\{cf:2, af:1\}$. We then update the item frequency and repeat the same steps as mentioned in Algorithm 1 (Lines 10–14) until the support value of the sensitive information is lower than the given support threshold value.

4   Experimental Results

The experiments were carried out on a Linux system with `Core I7` processor and 16 GB of RAM. We used two real-world datasets, i.e., mushrooms and foodmart. Both of the datasets are available on the SPMF data mining library [12]. For the exper-

Table 6 The PSIF-IDF values of all the transactions

TID	Item	$PSIF_{cf}$	IDF_{cf}	$PSIF_{af}$	IDF_{af}	PSIF-IDF
T_1	a, b, c, d, f	2/5	1.10	2/5	1.10	0.88
T_2	a, b, d, e	0/4	1.10	1/4	0.40	0.10
T_3	b, c, d, f	2/4	1.10	0/4	0.00	0.55
T_4	a, b, c, f	2/4	1.10	2/4	1.10	1.10
T_5	c, d, e	0/3	0.40	0/3	0.00	0.00

Table 7 The sorted transaction according PSIF-IDF values

TID	Item	PSIF-IDF
T_4	a, b, c, f	1.10
T_1	a, b, c, d, f	0.88
T_3	b, c, d, f	0.55
T_2	a, b, d, e	0.10
T_5	c, d, e	0.00

Table 8 The used datasets in the experiments

Dataset	$|I|$	$|D|$	$AvgLen.$	Type
Mushroom	8,124	119	23.0	Dense, long transactions
Foodmart	4,141	1,559	4.4	Sparse, short transactions

imental purposes, we selected the periodic patterns randomly between the threshold percentage of 1% and 2%. Also, we set the periodicity values, i.e., *minimum periodicity*, *maximum periodicity*, and *minimum average* based on the characteristics of the dataset mentioned in Table 8. Figure 1 compares the side effects of the designed algorithm under various parameter values.

4.1 Side Effect Analysis

In Fig. 1, three side effects such as α, β, and γ represented the numbers of hiding failures (ratio of periodic patterns before and after sanitization), missing rules (number of missing rules after sanitization), and artificial rules (number of pattern of artificial rules created after sanitization), respectively. Results showed that the designed model successfully hides the information and does not add any impurity into the database. This means that all the sensitive itemsets are removed from both datasets. Also, no side effects of artificial rules are generated. The number of missing rules is larger

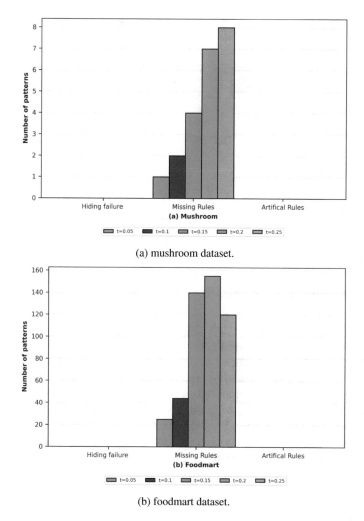

(a) mushroom dataset.

(b) foodmart dataset.

Fig. 1 Side effect analysis in terms of hiding factor α, missing rules β, and artificial rules γ

because some itemsets are removed from the transactions. As results, it causes the removal of the patterns. The lower periodicity values tend to extract patterns fast, and the reason is that model reduces the search space. With a lower number of rare sensitive items, the frequency calculation performed fast due to the less search space. The foodmart data is sparse and has high cost to find the periodicity measures. In general, the user can control the hiding factor parameters and utilization of the periodic patterns. However, reducing or hiding the patterns is good for privacy preservation, but the data has become no usage as it will reduce the number of discovered patterns or even produce *null* patterns in the knowledge discovery phase. This describes that PFPM helps to extract the periodic patterns and non-periodic patterns effectively.

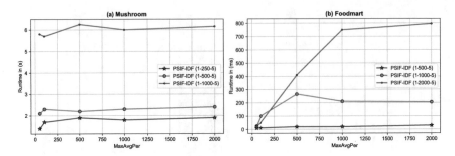

Fig. 2 Execution time analysis

As results, PSIF-IDF is able to hide the sensitive information under user-defined threshold value.

4.2   Execution Time Analysis

For each dataset, periodicity threshold values are empirically selected based on two datasets that can be observed in Fig. 2. Note that the parameters of the designed algorithm are different periodicity values, which are *minimum periodicity, maximum periodicity*, and *minimum average* consequentially. In general, the sanitization progress is much faster as the proposed model can prune search space for finding the periodic patterns and applying the sanitization model. From the results, we can see that the sparse nature of the foodmart dataset is relatively slow as calculating periodicity measures required to prune more items. In addition, the restricted threshold values help to extract and sanitize itemsets.

4.3   Pattern Analysis

In Fig. 3a, we analyze the total number of the discovered periodic patterns extracted under varied user-defined thresholds. In Fig. 3b, the number of the deleted patterns regarding the developed model for data sanitization is then discussed. First, the PFPM model helps to extract the periodic patterns effectively. As results, PSIF-IDF is able to hide the sensitive information under varied user-defined thresholds. The results showed that the lower threshold values result in producing more periodic frequent patterns. If the percentage of the sensitive item is less, then the less number of patterns is successfully deleted. When the higher percentage is set, the more sensitive itemsets are needed to be processed. For example, when the maximum periodicity is set to 1,000 under the 0.25 threshold for the mushroom dataset, 160 patterns are extracted, and 18 patterns are deleted. However, in the case of the foodmart dataset, when

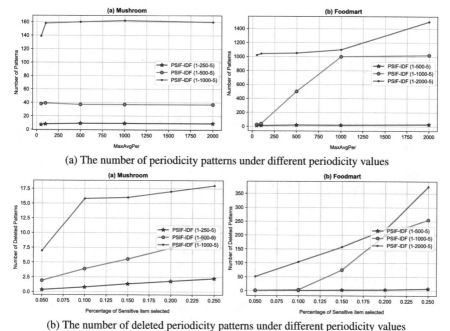

(a) The number of periodicity patterns under different periodicity values

(b) The number of deleted periodicity patterns under different periodicity values

Fig. 3 Pattern analysis results

maximum periodicity is set as 2,000, then 1,432 patterns are extracted, and 365 patterns are deleted based on a user-defined 0.25 threshold value. This due to the number of itemset variations, i.e., mushroom dataset has 119 items and foodmart has 1,559 items in the database. The deletion is highly dependent on the defined periodic threshold values, user sensitivity rate and internal dataset characteristics, i.e., the number of itemsets and transactions.

5 Conclusion and Future Work

In this paper, we proposed the privacy preservation-based periodic frequent pattern sanitization approach that uses the PFPM method to extract periodic patterns and then use the sensitive frequency–inverse document method to sanitize the database by removing the sensitive itemsets from the database. The model used the user-specified sensitive itemsets in the experiments to show the performance of the designed model regarding runtime, number of side effects, and number of the deleted patterns. In the future, more factors regarding the periodicity calculation will be discussed and the deep learning methods can be thus used to select the optimized threshold without users input. Also, the evolutionary-based models can be discussed and incorporated with the periodic preservation data mining mechanism.

References

1. C.C. Aggarwal, J. Pei, B. Zhang, On privacy preservation against adversarial data mining, in *ACM SIGKDD International Conference on Knowledge Discovery and Data Mining*, pp. 510–516 (2006)
2. U. Ahmed, G. Srivastava, J.C.W. Lin, A machine learning model for data sanitization. Comput. Netw. **189**, 107914 (2021)
3. K. Amphawan, P. Lenca, A. Surarerks, Mining top-k periodic-frequent pattern from transactional databases without support threshold, in *Advances in Information Technology*, pp. 18–29 (2009)
4. K. Amphawan, A. Surarerks, P. Lenca, P, Mining periodic-frequent itemsets with approximate periodicity using interval transaction-ids list tree, in *The International Conference on Knowledge Discovery and Data Mining*, pp. 245–248 (2010)
5. C. Clifton, M. Kantarcioglu, J. Vaidya, X. Lin, M.Y. Zhu, Tools for privacy preserving distributed data mining. ACM SIGKDD Explor. Newsl. **4**(2), 28–34 (2002)
6. M.N. Dehkordi, K. Badie, A.K. Zadeh, A novel method for privacy preserving in association rule mining based on genetic algorithms. J. Soft. **4**(6), 555–562 (2009)
7. C. Dwork, F. McSherry, K. Nissim, A. Smith, Calibrating noise to sensitivity in private data analysis, in *Theory of Cryptography Conference*, pp. 265–284 (2006)
8. A. Evfimievski, R. Srikant, R. Agrawal, J. Gehrke, Privacy preserving mining of association rules. Inf. Syst. **29**(4), 343–364 (2004)
9. P. Fournier-Viger, J.W. Lin, B. Vo, T. Truong, J. Zhang, H. Le, A survey of itemset mining. Wiley Interdiscip. Rev.: Data Min. Knowl. Discov. **7**(4), e1207 (2017)
10. P. Fournier-Viger, J.C.W. Lin, Q.H. Dong, D.T. Lan, Phm: mining periodic high-utility itemsets, in *Industrial Conference on Data Mining*, pp. 64–79 (2016)
11. P. Fournier-Viger, C.W. Lin, Q.H. Duong, T.L. Dam, L. Ševčík, D. Uhrin, M. Voznak, Pfpm: discovering periodic frequent patterns with novel periodicity measures, in *The Czech-China Scientific Conference* (2017)
12. P. Fournier-Viger, J.C.W. Lin, A. Gomariz, T. Gueniche, A. Soltani, Z. Deng, H.T. Lam, The spmf open-source data mining library version 2, pp. 36–40 (2016)
13. P. Fournier-Viger, Y. Wang, P. Yang, J.C.W. Lin, U. Yun, R.U. Kiran, Tspin: mining top-k stable periodic patterns. Appl. Intell., 1–22 (2021)
14. P. Fournier-Viger, P. Yang, R.U. Kiran, S. Ventura, J.M. Luna, Mining local periodic patterns in a discrete sequence. Inf. Sci. **544**, 519–548 (2021)
15. P. Fournier-Viger, P. Yang, Z. Li, J.C.W. Lin, R.U. Kiran, Discovering rare correlated periodic patterns in multiple sequences. Data Knowl. Eng. **126**, 101733 (2020)
16. P. Fournier-Viger, P. Yang, J.C.W. Lin, R.U. Kiran, Discovering stable periodic-frequent patterns in transactional data, in *The International Conference on Industrial, Engineering and Other Applications of Applied Intelligent Systems*, pp. 230–244 (2019)
17. J. Han, J. Pei, Y. Yin, Mining frequent patterns without candidate generation, in *ACM SIDMOD International Conference on Management of Data*, pp. 1–12 (2000)
18. T.P. Hong, C.W.Y. Lin, K. Tung, S.L. Wang, Using TF-IDF to hide sensitive itemsets. Appl. Intell. **38**(4), 502–510 (2012)
19. R.U. Kiran, M. Kitsuregawa, P.K. Reddy, Efficient discovery of periodic-frequent patterns in very large databases. J. Syst. Softw. **112**, 110–121 (2016)
20. R.U. Kiran, P.K. Reddy, Mining rare periodic-frequent patterns using multiple minimum supports, in *ACM Bangalore Conference*, pp. 1–8 (2010)
21. B.K. Pandya, U.K. Singh, K. Dixit, K. Bunkar, Effectiveness of multiplicative data perturbation for privacy preserving data mining. Int. J. Adv. Res. Comput. Sci. **5**(6) (2014)
22. C.W. Lin, T.P. Hong, C.C. Chang, S.L. Wang, A greedy-based approach for hiding sensitive itemsets by transaction insertion. J. Inf. Hiding Multimed. Signal Process. **4**(4), 201–227 (2013)
23. C.W. Lin, T.P. Hong, H.C. Hsu, Reducing side effects of hiding sensitive itemsets in privacy preserving data mining. Sci. World J. **2014** (2014)

24. C.W. Lin, T.P. Hong, K.T. Yang, S.L. Wang, The ga-based algorithms for optimizing hiding sensitive itemsets through transaction deletion. Appl. Intell. **42**(2), 210–230 (2015)
25. C.W. Lin, Y. Zhang, B. Zhang, P. Fournier-Viger, Y. Djenouri, Hiding sensitive itemsets with multiple objective optimization. Soft. Comput. **23**(4), 12779–12797 (2019)
26. J.C.W. Lin, Q. Liu, P. Fournier Viger, T.P. Hong, M. Voznak, J. Zhan, A sanitization approach for hiding sensitive itemsets based on particle swarm optimization. Eng. Appl. Artif. Intell. **53**, 1–18 (2016)
27. J.C.W. Lin, G. Srivastava, Y. Zhang, Y. Djenouri, M. Aloqaily, Privacy-preserving multiobjective sanitization model in 6g iot environments. IEEE Internet Things J. **8**(7), 5340–5349 (2021)
28. J.C.W. Lin, T.Y. Wu, P. Fournier-Viger, G. Lin, J. Zhan, M. Voznak, Fast algorithms for hiding sensitive high-utility itemsets in privacy-preserving utility mining. Eng. Appl. Artif. Intell. **55**, 269–284 (2016)
29. Y. Lindell, B. Pinkas, Privacy preserving data mining, in *Annual International Cryptology Conference*, pp. 36–54 (2000)
30. P. Fournier-Viger, Z. Li, J.C.W. Lin, R.U. Kiran, H. Fujita, Efficient algorithms to identify periodic patterns in multiple sequences. Inform. Sci. **489**, 205–226 (2019)
31. G. Salton, E.A. Fox, H. Wu, Extended boolean information retrieval. Commun. ACM **26**, 1022–1036 (1983)
32. S.K. Tanbeer, C.F. Ahmed, B. Jeong, Y.K. Lee, Discovering periodic-frequent patterns in transactional databases, in *Advances in Knowledge Discovery and Data Mining*, pp. 242–253 (2009)
33. V.S. Verykios, E. Bertino, I.N. Fovino, L.P. Provenza, Y. Saygin, Y. Theodoridis, State-of-the-art in privacy preserving data mining. ACM SIGMOD Record **33**(1), 50–57 (2004)
34. T.Y. Wu, J.C.W. Lin, Y. Zhang, C.H. Chen, The grid-based swarm intelligence algorithm for privacy-preserving data mining. Appl. Sci. **9**(4), 774 (2019)
35. M.J. Zaki, Scalable algorithms for association mining. IEEE Trans. Knowl. Data Eng. **12**(3), 372–389 (2000)

Real-World Applications of Periodic Patterns

R. Uday Kiran, Masashi Toyoda, and Koji Zettsu

Abstract Previous chapters of this textbook have mainly focused on introducing different types of periodic patterns and their mining algorithms. Some chapters have also focused on evaluating the algorithms. In this chapter, we will present three real-world applications of periodic patterns. The first case study is traffic congestion analytics, where periodic-frequent pattern mining was employed to identify the road segments in which users have regularly encountered traffic congestion in the transportation network. The second case study is flight incidents data analytics, where partial periodic pattern mining was employed to identify factors that are regularly causing flight incidents in the data. The third case study is air pollution analytics, where fuzzy periodic pattern mining was employed to identify the geographical regions where people were exposed to harmful levels of air pollution.

1 Introduction

The data generated by many real-world applications naturally exist as a temporal database. Useful information that can empower the users with the competitive knowledge to achieve socio-economic development lies in this data. Periodic pattern mining is one of the best available techniques to discover competitive information in a temporal database. It is because a periodic pattern indicates something predictable within the data. This chapter presents three use cases to demonstrate the usefulness of periodic patterns in real-world applications. The first case study is traffic congestion

R. Uday Kiran (✉)
The University of Aizu, Aizu-Wakamatsu, Fukushima, Japan
e-mail: udayrage@u-aizu.ac.jp; uday.rage@gmail.com

M. Toyoda
The University of Tokyo, Tokyo, Japan
e-mail: toyoda@tkl.iis.u-tokyo.ac.jp

K. Zettsu
National Institute of Information and Communications Technology, Tokyo, Japan
e-mail: zettsu@nict.go.jp

© The Author(s), under exclusive license to Springer Nature Singapore Pte Ltd. 2021   229
R. Uday Kiran et al. (eds.), *Periodic Pattern Mining*,
https://doi.org/10.1007/978-981-16-3964-7_13

data analytics. In this analytics, users employ periodic-frequent pattern mining to identify the road segments in which regular traffic congestion was observed in the network. In the second case study, we discuss incident data analytics, where partial periodic pattern mining was employed to discover useful information regarding the air crafts' incidents reported in Federal Aviation Authority (FAA) database. In the third case study, we present air pollution data analytics in which fuzzy periodic-frequent pattern mining was employed to identify the geographical locations in which people were regularly exposed to harmful levels of an air pollutant.

2 Traffic Congestion Data Analysis

Improving public transport in smart cities is a challenging problem of great concern in Intelligent Transportation Systems. An efficient transportation system can save the life of several thousands of people during disasters. Many countries across the globe are giving considerable research and development attention to improving their transportation networks. In particular, ASEAN countries, which are often prone to natural disasters (e.g., earthquakes, tsunamis, and typhoons), prioritize their national budgets to improve their transportation networks. Consequently, analyzing transportation data receives significant attention in many disciplines such as data mining, machine learning, and statistics.

Japan is particularly vulnerable to natural disasters because of its topography and climate. It has experienced countless earthquakes, tsunamis, typhoons, and other types of disasters. The Government of Japan has set up road traffic information center [5] to monitor traffic congestion and safely transport people from disaster-affected places to safe places. This information center has deployed a nationwide sensor network to monitor traffic congestion. Figure 1a shows the road network covered by the traffic congestion measuring sensors in Kobe, Japan. Figure 1b shows the hypothetical raw data generated by this network. Figure 1c shows the temporal database generated from the raw data. Periodic-frequent pattern mining on this database provides the information regarding the road segments in which people have regularly

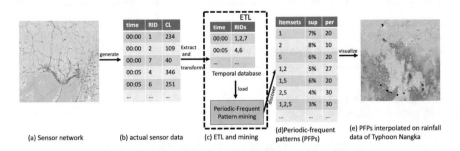

(a) Sensor network   (b) actual sensor data   (c) ETL and mining   (d)Periodic-frequent patterns (PFPs)   (e) PFPs interpolated on rainfall data of Typhoon Nangka

Fig. 1 Traffic congestion analytics using periodic patterns

Table 1 Some of the interesting periodic-frequent patterns generated in Congestion database. The terms *"sup"* and *"per"* represent *"support"* and *"periodicity,"* respectively

S. No.	Patterns	*sup*	*per*
1	{137,1487, 1473, 1471, 1442, 140, 759}	1116	225
2	{556,325, 1442, 2234, 759}	2312	216
3	{243, 325, 168, 390, 2234, 1442, 759}	2504	237
4	{1502, 2274, 168, 1442, 2234, 759}	1837	222

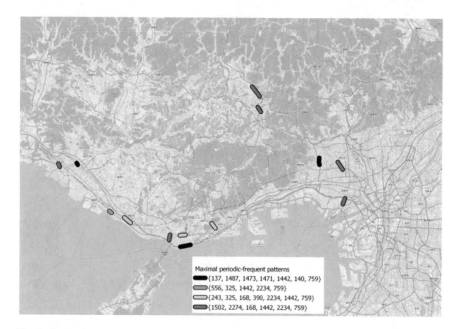

Fig. 2 Spatial locations of periodic-frequent patterns in Table 1

faced congestion in the network (see Fig. 1d). A hypothetical periodic-frequent pattern discovered from this database is as follows:

$$\{1, 2, 5\} \quad [support = 3\%, \ periodicity = 30\,\text{min}].$$

The above pattern indicates that 3% of the congestions were regularly observed (i.e., at least once in every 30 min) on the set of road segments whose identifiers were 1, 2, and 5. When such an information is visualized along with other data sources, say rainfall data of a typhoon as shown in Fig. 1e, the produced information may found to be extremely useful to the users for various purposes, such as monitoring the traffic during disastrous and suggesting police patrol routes to reduce accidents.

Kobe is the seventh-largest city in Japan. It is also the capital city of Hyogo Prefecture. On July 17, 2015, this city was stuck with a heavy typhoon called Nangka.

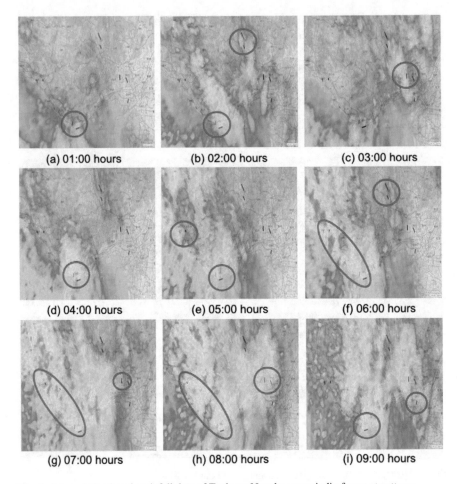

(a) 01:00 hours   (b) 02:00 hours   (c) 03:00 hours

(d) 04:00 hours   (e) 05:00 hours   (f) 06:00 hours

(g) 07:00 hours   (h) 08:00 hours   (i) 09:00 hours

Fig. 3 Interpolating hourly rainfall data of Typhoon Nangka on periodic-frequent patterns

This typhoon resulted in 29 inches of rainfall, thereby displacing 550,000 people. Uday et al. [2] collected that traffic congestion data of Kobe of this day and applied periodic-frequent pattern mining to identify the sets of regularly congested road segments. Some of the interesting patterns found in this data are shown in Table 1. Figure 2 shows the spatial locations of these patterns. When we overlay hourly rainfall data as shown in Fig. 3, the produced information can be found extremely beneficial to the users to monitor the traffic and suggest police patrol routes. The red circles in each of these figures represent the road segments that face regular congestion and encounter heavy rainfall in the respective time duration.

3 Incidents Data Analysis

Federal Aviation Administration (FAA) keeps track of various incidents that happened to air crafts in the United States of America. This incidents database, called Federal Aviation Administration-incidents (FAA-incident) database, naturally exists as a temporal database with many incidents reporting on the same day. Venkatesh et al. [6] constructed FAA-incidents database by gathering the incidents recorded by the Federal Aviation Authority (FAA) from January 1, 1978 to December 31, 2014. Partial periodic pattern mining on this database identified useful patterns that have regularly appeared in the database. Table 2 shows some of the interesting partial periodic patterns discovered in FAA-accidents at $per = 2$ years and $minPS = 100$. The first pattern reveals useful information that 546 personal flights driven by private pilots have gone through substantial damages while carrying out general operating rules. The second pattern provides the information that 110 flights were completely destroyed while following the general operating rules. The third pattern provides the information that 205 personal flights have witnessed substantial damages when the flight phase is a level-off touchdown. The final pattern provides the information that 388 Cessna flights driven by private pilots have gone through minor damages while following instrument flight rules. Such a piece of information may be handy to the users in the Federal Aviation Administration in coming up with appropriate training practices to reduce human and machine losses.

4 Air Pollution Data Analysis

Air pollution is the major cause of many cardio-respiratory problems reported in Japan. On average, 42.6 thousand people are dying every year in Japan due to pollution [3]. In this context, a nationwide sensor network called *Atmospheric Environmental Regional Observation System* (AEROS) [4] was set up by the Ministry

Table 2 Some of the interesting partial periodic patterns discovered in FAA-accidents database

S. No.	Patterns	*Periodic-support*
1	{GENERAL-OPERATING-RULES, PERSONAL, PRIVATE-PILOT, SUBSTANTIAL}	546
2	{GENERAL-OPERATING-RULES, INCIDENT, DESTROYED}	110
3	{PERSONAL, LEVEL-OFF-TOUCHDOWN, SUBSTANTIAL}	205
4	{INSTRUMENT-FLIGHT-RULES, PRIVATE-PILOT, CESSNA, MINOR}	388

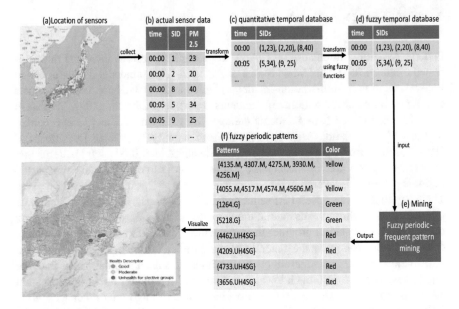

Fig. 4 Air pollution analytics

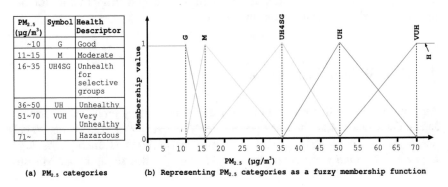

Fig. 5 Fuzzy membership functions

of Environment, Japan, to monitor pollution. Figure 4a shows the spatial locations of these sensors in Japan. This network produces raw data at hourly intervals as shown in Fig. 4b. This data can be modeled as a quantitative temporal database as shown in Fig. 4c. Using the fuzzy membership functions as shown in Fig. 5, the quantitative temporal database can be transformed into a fuzzy temporal database as shown in Fig. 4d. Fuzzy periodic-frequent pattern mining (see Fig. 5e) on this database identifies the useful patterns as shown in Fig. 4f. These patterns provide the environmentalists and policymakers with useful information regarding the areas (see Fig. 4g) in which people were regularly exposed to harmful levels of air pollution. The discovered information may benefit the users for various purposes, such as alerting local authorities and introducing new pollution control policies. More information on fuzzy periodic-frequent patterns and the related application is available at [1].

References

1. R.U. Kiran, C. Saideep, P. Ravikumar, K. Zettsu, M. Toyoda, M. Kitsuregawa, P.K. Reddy, Discovering fuzzy periodic-frequent patterns in quantitative temporal databases, in *IEEE FUZZY*, pp. 1–8 (2020)
2. R.U. Kiran, Y. Watanobe, B. Chaudhury, K. Zettsu, M. Toyoda, M. Kitsuregawa, Discovering maximal periodic-frequent patterns in very large temporal databases, in *DSAA*, pp. 11–20 (2020)
3. C. Klein, Number of deaths attributable to air pollution in japan between 2010 and 2019. https://www.statista.com/statistics/935022/number-deaths-air-pollution-japan/ (2020). [Online; Accessed 1-June-2021]
4. Japan Ministry of Environment, Atmospheric environmental regional observation system. http://soramame.taiki.go.jp/. [Online; Accessed 1-June-2021]
5. The Government of Japan, Japan road traffic information center. https://www.jartic.or.jp/. [Online; Accessed 1-June-2021]
6. J.N. Venkatesh, R.U. Kiran, P.K. Reddy, M. Kitsuregawa, Discovering periodic-correlated patterns in temporal databases. Trans. Large Scale Data Knowl. Centered Syst. **38**, 146–172 (2018)

Insights for Urban Road Safety: A New Fusion-3DCNN-PFP Model to Anticipate Future Congestion from Urban Sensing Data

Minh-Son Dao⑩, R. Uday Kiran, and Koji Zettsu

1 Introduction

Traveling is an essential demand of humankind as we all need to commute between our home and workplace by using different means of transport daily. Because of the rapid growth in population and urbanization, modeling traffic data is becoming more complicated. Various factors could, directly and indirectly, influence traffic flows, such as weather, public infrastructure, human activities, and natural disasters. Owing to IoT and AI's exponential development, those factors have been measured and utilized to model urban traffic applications in smart cities like [1, 7, 27, 35] in recent years.

Predicting future traffic data is the most prominent study topic of urban traffic modeling. In the early days, researchers employed various statistical and machine learning methods [18, 19]. As the collected data becomes more massive over time, deep neural networks have gained popularity rapidly [30]. The most significant problem of those contemporary studies is that their output results lack interpretability because of the nature of the utilized methods. Even though methods like decision trees can explain how the results are generated but are not as good as deep neural networks on multiple factors like predictive accuracy and scalability (e.g., working with multi-modal data) [21]. Due to the interpret-ability-related problem mentioned

M.-S. Dao (✉) · R. Uday Kiran · K. Zettsu
National Institute of Information and Communications Technology, Tokyo, Japan
e-mail: dao@nict.go.jp

R. Uday Kiran
e-mail: udayrage@u-aizu.ac.jp

K. Zettsu
e-mail: zettsu@nict.go.jp

R. Uday Kiran
University of Aizu, Fukushima, Japan

© The Author(s), under exclusive license to Springer Nature Singapore Pte Ltd. 2021
R. Uday Kiran et al. (eds.), *Periodic Pattern Mining*,
https://doi.org/10.1007/978-981-16-3964-7_14

above, it is challenging to explain to relevant stakeholders how traffic congestion appears. Questions like "Are there any traffic bottlenecks in the city?," "How long heavy rain (measured in what millimeters) will lead to heavy traffic congestion?," and "How long will an accident affect congestion?" remain unanswered. If they are found out, authorities can fix potential flaws in the design of traffic networks and plan resources to deal with traffic congestion based on real-life events more effectively. It is worth mentioning that traffic pattern discovery [32, 33, 37, 39], another study branch of urban traffic modeling, can somewhat answer such questions. The problem is, however, they use observed data to discover traffic patterns, so they cannot tell us if (1) whether such patterns will happen in the future, (2) when they will occur, and (3) how long they will last.

In [5], the authors proposed a system of systems concept that, in our opinion, can be utilized to address the problems mentioned above. It consists of two systems: System 1 takes care of producing fast, and intuitive answers for the problem and System 2 handles more complex problems like reasoning and explaining how System 1 reaches the answers. The authors proposed to use machine learning methods to build System 1 while left methods behind System 2 unanswered. They suggested that System 2 is built on top of System 1 and uses the data generated by System 1.

In light of the above discussions, in this work, we propose a dynamic system consisting of two systems: (1) long-term traffic congestion prediction system using multimodal urban sensing data and (2) predicted congestion patterns discovery system. To build System 1, we propose a novel deep learning model called Fusion-3DCNN. About System 2, we evaluated different periodic-frequent pattern discovery algorithms and utilized the most efficient one called maxPFP-growth proposed in [16]. The system aims to tackle the following problem: "*Predicting future long-term traffic congestion using multi-modal urban sensing data and discovering future high traffic demand regions dynamically.*" We consider future high traffic demand regions equivalent to sets of Earth portions predicted to have heavy-periodic-frequent congestion in this work. From the system's target, it is clear that the first component to predict long-term traffic congestion is straightforward in the spectrum of machine learning, while the second to discover predicted congestion patterns requires complex calculations. The system's output is interactively dynamic maps showing the people information about future traffic congestion and regions predicted to be in high traffic demand according to periods. They provide a significant value proposition to both municipal authorities and citizens. As we know, many cities worldwide have a shortage of traffic police officers, so they need to choose which regions to monitor traffic flows wisely. Without a doubt, high traffic demand regions should be prioritized. Besides, citizens could refer to the generated dynamic maps to reroute their itinerary to avoid getting stuck in traffic jams.

Our contributions are summarized as follows:

- We propose a novel deep learning model called Fusion-3DCNN, which is an enhanced version of the 3DCNN multi-source deep learning model proposed in [8]. The model aims to anticipate long-term traffic congestion using multi-modal urban sensing data. It uses multiple data sources in different formats like numeric data (e.g., congestion, rainfall, accident) and text data (e.g., Tweets) with the help of two data fusion functions (one was introduced in [8]).

- We propose a novel system of systems to integrate the traffic congestion prediction model introduced above and a periodic-frequent pattern mining algorithm to discover future high traffic demand regions that are predicted to have heavy-periodic-frequent traffic congestion. Herein, the frequency and periodicity of the predicted congestion information stored in temporal databases are fully explored.
- Experimental results on real data collected in the city of Kobe, Japan from 2014 to 2015 shows that the system is not only highly accurate in predicting traffic congestion but also time efficient in discovering predicted congestion patterns.
- Analyses of prevalent patterns, a case study, and two interesting use cases are also presented to demonstrate the usefulness of the system.

The remainder of this chapter is organized as follows. Section 2 compiles related works in the field of urban traffic modeling and introduces related concepts and algorithms to discover periodic-frequent patterns. Section 3 introduces our Fusion-3DCNN deep learning model. Section 4 explains how we integrate the traffic congestion prediction model, and a periodic-frequent pattern discovery algorithm to realize the dynamic system. Section 5 presents experiments, shows a case study of high traffic demand regions, and proposes some use cases to deal with future traffic congestion. Finally, Sect. 7 concludes the study.

2 Background

This section reviews related works and introduces the periodic-frequent pattern mining model and the PFP-growth++ and maxPFP-growth algorithms.

2.1 Related Works

In this section, we review related works in the area of urban traffic modeling. They can be broadly classified into three types: (*i*) traffic prediction [1, 2, 7, 11, 12, 18, 19, 21, 27, 30, 35], (*ii*) traffic pattern discovery [3, 22, 28, 32–34, 37, 39], and (*iii*) hybrid where a pattern discovery model are used to generate a training data for predictive models [10, 38, 40].

Studies on traffic prediction have primarily focused on predicting future traffic conditions over different geographical sizes in short-term time horizons limited under 60 min [18, 21]. In 2018, [18, 19] compiled many studies that modeled prediction systems based on statistical and machine learning techniques. In those works, only traffic data was utilized, and complexity levels of the data were low. Recent studies started to use multi-modal data to model their systems as listed following. Chou et al. used traffic and weather data [7]. Fan et al. used traffic data and sensing information collected around buildings [11]. Alkouz et al. used traffic, and social networking data [1]. Their ultimate purpose is to enhance the predictive accuracy of the model. Those contemporary works proposed prediction systems built on top deep neural networks to utilize their computational powers [30]. The mentioned studies encounter three problems as follows. First, short predictive time windows make the proposed systems not very practical in real life as traffic police forces must prepare action plans multiple times per day (24 times per day if the predictive time window is 60 min). Second, those predictive systems can predict future traffic congestion situations at a particular time instance but cannot determine if such a condition will happen periodically/frequently in the future requiring special care from traffic police officers. Third, although deep neural networks generate highly accurate results, they work as a black-box model. It means that people cannot explain how the predicted information is generated by the system (i.e., they lack interpretability). If authorities know how traffic congestion prediction systems work, they may allocate resources to tackle traffic congestion more effectively.

Studies on traffic pattern discovery aim to identify factors that may cause traffic congestion by using observed traffic data [39]. They can be classified into two minor groups. First, discovering internal road factors (e.g., designs of traffic networks and public infrastructure) and external factors (e.g., traffic accidents, natural disasters, and timings of the day) causing traffic congestion. Second, interpreting how congestion propagates from one area to others and discovering which paths will lead to large-scale traffic congestion. The studies belonging to the first group could be found in [3, 28, 37, 39]. Publications [22, 32, 33] introduced different methods belonging to the second group. Those studies' ultimate purposes are: (1) detecting potential traffic network design flaws and (2) alarming people about factors leading to traffic congestion. The methods of this topic have two main limitations as follows. First, they can discover traffic patterns from historical data, but they cannot decide whether those patterns will happen in the future and how long they will remain when occurring, so traffic congestion reaction plans cannot be made in advance for a particular time instance. Second, to discover traffic pattern, they consider only frequency and disregarded the temporal occurrence information of the events, so they are inadequate to discover periodic regularities which are an essential characteristic of traffic data [29].

Traffic prediction and traffic pattern discovery models can be used together to form hybrid methods [10, 38, 40]. They applied the same integration scheme as follows. A pattern discovery model is used to generate training data for a predictive model. The purpose is to enhance the predictive accuracy of the system. The methods of this research group encounter the same problems as the traffic prediction studies analyzed above.

To the best of our knowledge, most previous works are not capable of discovering future periodic-frequent traffic congestion patterns. The dynamic system proposed in this study will address this important issue.

2.2 Periodic-Frequent Pattern Mining Model

Periodic-frequent pattern mining is an important model in data mining, having many real-world applications. The basic model to discover periodic-frequent patterns is as follows [17]:

Let $I = \{i_1, i_2, \ldots, i_n\}$, $n \geq 1$, be the set of items. Let $X \subseteq I$ be an itemset (or a pattern). If an pattern contains k number of items, then it is called a k-pattern. A transaction $t = (ts, Y)$ is a tuple, where ts represents the timestamp and Y is a pattern. A transactional database (TDB) over I is a set of transactions, i.e., $TDB = \{t_1, t_2, \ldots, t_m\}$, $m = |TDB|$. If $X \subseteq Y$, it is said that t contains X and such timestamp is denoted as ts_j^X. Let $TS^X = \{tid_j^X, \ldots, tid_k^X\}$, $1 \leq j \leq k \leq n$, be the set of all timestamps where X occurs in TDB. The **support** of a pattern X is the number of transactions containing X in TDB, denoted as $Sup(X)$. Therefore, $Sup(X) = |TS^X|$. Let ts_i^X and ts_j^X be two consecutive timestamps where X appeared in TDB. A **period** of a pattern X is the number of transactions between ts_i^X and ts_j^X. Let $P^X = \{p_1^X, p_2^X, \ldots, p_r^X\}$, $r = Sup(X) + 1$, be the complete set of periods of X in TDB. The **periodicity** of a pattern X is the maximum difference between any two adjacent occurrences of X, denoted as $Per(X) = max(p_1^X, p_2^X, \ldots, p_r^X)$. A pattern X is said to be a periodic-frequent pattern if $Sup(X) \geq minSup$ and $Per(X) \leq maxPer$, where $minSup$ and $maxPer$ represent the user-specified threshold on *minimum support* and *maximum periodicity*, respectively.

Example 1 Let $I = \{a, b, c, d, e, f, g\}$ be the set of items (or mesh codes). Each mesh code uniquely identifies a portion of area on the earth surface. Let Table 1 be a hypothetical (predicted) temporal database generated by the Fusion-3DCNN model described in the previous subsection. This database contains 10 transactions. Therefore, the database size, i.e., $TDB = 10$. The first transaction, "1 : abcg," provides the information that traffic congestion was observed in the mesh codes of a, b, c and g at the timestamp of 1. Similar statements can be made on the remaining transactions in the database. The set of items a and b, i.e., $\{a, b\}$ (or ab, in short) is a pattern. This pattern contains two items. Therefore, it is a 2-pattern. The pattern ab appears in the transactions whose timestamps are 1, 3, 5, 7, and 9. Therefore, $TS^{ab} = \{1, 3, 5, 7, 9\}$. The *support* of ab, i.e., $Sup(ab) = |TS^{ab}| = 5$. The periods for this pattern are: $p_1^{ab} = 1$ $(= 1 - ts_{initial})$, $p_2^{ab} = 2$ $(= 3 - 1)$, $p_3^{ab} = 2$ $(= 5 - 3)$, $p_4^{ab} = 2$ $(= 7 - 5)$, $p_5^{ab} = 2$ $(= 9 - 7)$, and $p_6^{ab} = 1$ $(= ts_{final} - 9)$, where $ts_{initial} = 0$ represents the timestamp of initial transaction and $ts_{final} = |TDB| = 10$ represents the timestamp of final transaction in the database. The *periodicity* of ab, i.e., $Per(ab) = maximum(1, 2, 2, 2, 2, 1) = 2$. If the user-specified $minSup = 4$ and $maxPer = 3$, then ab is a periodic-frequent pattern. This periodic-frequent pattern is expressed as follows:

$$ab \ [support = 5 \ (= 50\%), \ periodicity = 2]. \tag{1}$$

The above pattern provides useful information that 50% of the *regular conjunctions* happen on the mesh codes a and b. It can be observed that such a discovered infor-

Table 1 Temporal database

ts	Items	ts	Items
1	abcg	6	cdef
2	bcde	7	abcd
3	abcd	8	aef
4	acdf	9	abcd
5	abcdg	10	bcde

mation may found to be extremely useful to the users for various purposes, such as suggesting police patrol routes and diverting the traffic.

Several algorithms have been described to find periodic-frequent patterns like Periodic-Frequent Pattern-growth++ (PFP-growth++) [17]. It has a very efficient runtime, but it generates too many patterns causing difficulties in visualization. To address that problem, [16] introduced the maximal periodic-frequent pattern mining model using Maximum Periodic-Frequent Pattern-growth (maxPFP-growth) algorithm. In the next two sections, we briefly introduce them. We will also compare the interestingness about patterns generated by those algorithms and their execution time in experiments.

2.3 Periodic-Frequent Pattern-growth++ (PFP-growth++) Algorithm

The PFP-growth++ algorithm is an enhanced version of PFP-growth. Different from the original, it employs greedy techniques during the process of pruning candidate itemsets. It has a very efficient runtime and works as follows:

1. **(Construction of PF-tree.)** Compress the given temporal database into a tree called Periodic-Frequent tree (PF-tree). The $maxPer$ (maximum periodicity) parameter is employed in this stage to determine aperiodic items that have periodicity larger than $maxPer$. They are removed from $tree$. At the end of this process, we have $\forall x \in tree, Per(x) \leq maxPer$.
2. **(Mining PF-tree.)** Perform recursively until $PF\text{-}tree$ is empty as follows:

   - Step 1: Choose a leaf node i of $PF\text{-}tree$, build a Prefix Tree i called PT_i. PT_i contains all branches of $PF\text{-}tree$ having i as a leaf.
   - Step 2: For each node j or a group of nodes belonging to PT_i, we combine it/them with node i to form a k-pattern. It means that $k = 1 + N$ where N is the number of nodes taken from PT_i and combined with i. We have $2 \leq k \leq |L_j|$. Assume that we combine only node j with i, we have a 2-pattern called ij. We calculate its support Sup and periodicity Per. If $Sup(ij) \geq maxSup$ and $Per(ij) \leq maxPer$, then ij is a periodic-frequent pattern.

- Step 3: Remove i out of PFT and restart from step 1.

Please refer to [17] for more information about this algorithm.

2.4 Maximal Periodic-Frequent Pattern Growth (maxPFP-Growth) Algorithm

The maxPFP-growth algorithm still employs the operations to build and mine periodic-frequent patterns on PFP trees like PFP-growth++. However, the significant difference is that this algorithm only discovers maximal periodic-frequent patterns rather than finding all periodic-frequent ones. As a result, the number of generated patterns reduces significantly. It works as follows:

1. **(Construction of PFP-tree.)** Compress the given database into a *tree*, called Periodic-Frequent Pattern tree (PFP-tree).
2. **(Initialization of maxPFP-tree.)** Initialize another *tree*, called maximum Periodic-Frequent Pattern tree (maxPFP-tree), by setting its *root* node equal to *null*.
3. **(Constructing maxPFP-tree by recursively mining PFP-tree.)** For each item i in the PFP-tree, construct i's *conditional pattern base* (CPB_i), and i's *conditional PFP-tree* ($cPFP\text{-}tree_i$). Determine whether $i \cup cPFP\text{-}tree_i$ is a maximal periodic-frequent pattern by performing *subset_checking* function of maxPFP-tree. If $i \cup cPFP\text{-}tree_i$ is a maximal periodic-frequent pattern, insert $i \cup cPFP\text{-}tree_i$ into maxPFP-tree. Else, reject $i \cup cPFP\text{-}tree_i$. **Please note that we do not generate all periodic-frequent patterns from $cPFP\text{-}tree_i$ as in traditional periodic-frequent pattern mining algorithms. Instead we stop recursive mining of $cPFP\text{-}tree_i$ once the resultant *tree* contains only one branch.**
4. **(Reducing the size of PFP-tree.)** Prune the item i in the PFP-tree by pushing its node information (i.e., the list of timestamps) to its parent nodes. Keep repeating the steps 3 and 4 until the PFP-tree is empty.

Please refer to [16] for more information about this algorithm.

3 Fusion-3DCNN Deep Learning Model

This section explains the process of building the Fusion-3DCNN deep learning model. It is used for long-term traffic congestion using multi-modal urban sensing data. There are two main stages involved: (1) converting multiple sources of urban sensing data to 3D multi-layer raster images, and (2) fusing and feeding the data sources wrapped in those images to the deep learning model used for long-term traffic congestion prediction. The first step utilizes the method proposed in [9], while the second leverages two fusion functions and a deep neural network, which will be described in later sections.

3.1 3D Multi-layer Raster Image Creation

This section explains the process to create a 3D multi-layer raster image called $3DMRI$ from multi-modal urban sensing data. The process is illustrated in Fig. 1. It assumes that we have three urban sensing data sources: congestion, rainfall, and Tweets having content related to heavy traffic congestion, heavy rain, and traffic accidents. Firstly, we introduce related concepts of $3DMRI$. Each image consists of many layers. If we separate $3DMRI$ images by layers, we will obtain 3D single-layer raster images called $3DSRI$. Each $3DSRI$ image contains the evolution of an urban sensing data type over a particular period. Each pixel on the images represents a rectangular geographic portion on the Earth denoted by mesh code. The value of each pixel is an accumulation of related data appearing in the related mesh code. Below we explain the workflow in more detail:

1. For each urban sensing data type j, at time T_i, we create the representative layer RI_{ij} as follows:

   a. We divide the examined area into a grid of mesh codes and create the associated 2D single-channel raster image called RI_{ij} whose pixel values are initialized by zero. To do so, we reuse the Quarter Grid Square standard defined by the Statistics Bureau of Japan under Announcement No. 143 issued by the Administrative Management Agency on July 12, 1973.[1] The area size denoted by each mesh code is equivalent to a 250m×250m portion on the Earth.

   b. We accumulate the value of data that appears in each mesh code and assign it to the associated pixel. For example, assume that mesh code M has three road segments whose average congested length measured at T_i are 200m, 150m, and 180m. It means that the value of the pixel associated with mesh code M for the traffic congestion data is $200 + 150 + 180 = 530$.

   c. We normalize the pixels' value into [0, 255] range and have the final data of the RI_{ij} image.

2. We superimpose all 2D single-channel images RI_{ij} generated from multiple urban sensing data at time T_i to create a 2D multi-channel image MRI_i.

3. We arrange all MRI_i images along the time dimension over a predefined period to get the 3D multi-channel raster image $3DMRI$, which becomes the input of our Fusion-3DCNN model.

For more information about this process, please refer to [9].

---

[1] http://www.stat.go.jp/english/data/mesh/02.html.

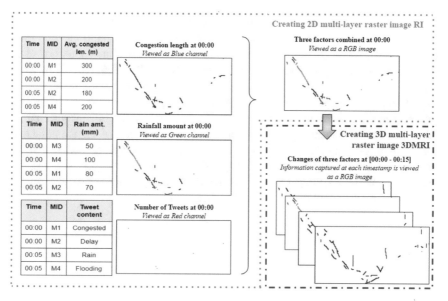

Fig. 1 Creating 3D multi-layer raster image from multi-modal urban sensing data

3.2 Architecture of the Fusion-3DCNN Deep Learning Model

3.2.1 Data Fusion Functions

This section explains how we feed multiple urban sensing data sources into the Fusion-3DCNN deep learning model. It is done with the help of two data fusion functions. We observe that the traffic is affected by two types of sensing data: social networking data and environmental factors.

The social networking data (the first group) concerns users' posts on social networking platforms like Twitter or Facebook or others consisting of keywords about the bad surrounding environment. We consider such posts are online warnings. As we know, if a person sees an online warning about bad environmental conditions like heavy rain, earthquakes at a location in his commuting route, he would avoid reaching it. Therefore, the adverse effects of those events on traffic congestion would be alleviated. Thus, the severity of traffic congestion is reduced. In [24], the authors attempted to use locations where posts on Twitter intending to warn about bad environmental conditions were collected. The data was used alongside traffic data to anticipate future congestion. The aim was to enhance the predictive accuracy of the model but failed as the improvement was limited. We think the reason for that problem is they treated the social networking data as independent learnable factors. It means that features related to social networking data were treated on par with traffic congestion data. This work makes social networking data dependent on environmental factors (e.g., traffic congestion, rainfall amount, traffic accident-will be explained later). We integrate the former into the latter by using a weight multiplication function. The integration is done as follows. Firstly, we categorize the content of social networking posts by using a set of specific keywords into three different online warning groups: (1) heavy traffic congestion/traffic delay, (2) rain/natural disasters, and (3) traffic accidents. We divide the posts into three groups because we have three different environmental factors, namely, traffic congestion, rainfall, and accident, in our experimental dataset. For other studies that use different environmental factors, the number of online warning types should be grouped accordingly. After that, we integrate online warnings into the corresponding environment factor using Eq. 2. It means that online warnings related to heavy traffic congestion/traffic delay will be integrated into observed traffic congestion data, and so on.

$$
x_{(i,j)}^f = \begin{cases} (1-p) \times x_{(i,j)}^f, & \text{if } y_{(i,j)}^f = 1 \\ x_{(i,j)}^f, & \text{otherwise,} \end{cases} \tag{2}
$$

where

- $x_{(i,j)}^f$ is the normalized value of the environmental factor f collected at location (i, j) on raster images at time t;
- $y_{(i,j)}^f$ is a binary value indicating an existence of online warnings relating to factor f at time t and location (i, j). If $y_{(i,j)}^f = 1$, there is an online warning collected. Otherwise, $y_{(i,j)}^f = 0$; and
- p is a hyperparameter defining the integrating weight.

The second group of data contains traffic congestion, precipitation, and vehicle collisions. They are called "environmental factors." Those data sources are firstly integrated with online warnings explained above. After that, the integrated data sources are treated as learnable factors and supplied directly to the Fusion-3DCNN deep learning model as follows:

$$
X = \sum_{i=1}^{N} W_{f_i} \times x_{f_i}, \tag{3}
$$

where

- f_i is individual learnable factors;
- x_{f_i} is integrated data of factor f_i;
- W_{f_i} is weights of the neural network branch relating to factor f_i; and
- X is learned features.

3.3 Disclosure

The Fusion-3DCNN deep learning model proposed in this study is an enhanced version of the 3DCNN multi-source deep learning model proposed in [8] which is our previous work. The significant difference between this work and [8]'s is that we newly introduce a new data fusion to integrate social networking data into environmental factors. It is related to Eq. 2. In [8], we treated all data sources as independent learnable factors. It means that we did not separate social network posts into online warnings. We treated it in a way that if a social network post related to (1) heavy traffic congestion/traffic delay, (2) rain/natural disasters, and (3) traffic accidents is detected, we set 1 to the corresponding locations on raster images. The data was supplied directly to the model. However, in this work, we treat social networking data as a dependent source, which has been explained in detail in Sect. 3.2.1.

3.3.1 Predictive Model

This section discusses the Fusion-3DCNN deep learning model's architecture. It is illustrated in Fig. 2. The model receives 3D multi-layer raster images prepared in Sect. 3.1. It then separates the images into 3D single-layer raster images containing spatiotemporal information of a specific data source over a particular period. Each image contains k ($k \in \mathbf{N}^+$) 2D single-layer raster image, which comprises spatial information of a specific data source at k particular time instances. The data contained in the 3D single-layer raster images are fused by two data fusion functions explained in Sect. 3.2. The model's output is 3D single-layer raster images containing predicted traffic congestion of the whole examined geographic region over a predefined period. Each image contains m ($m \in \mathbf{N}^+$) 2D single-layer raster image showing predicted congestion information at m time instances. In Fig. 2, please note that the "Decision making block" is used only when $m = \frac{1}{2}k$.

The models are implemented in Keras[2] with Tensorflow backend.[3] All 3D-CNN layers used in the model have kernel size equal to (3, 3, 3) and are set to the same padding, one-step striding. In Feature learning and Fused learning blocks, each 3DCNN layer is followed by a Batch Normalization layer. The purpose is to speed up the learning process of the model [4]. The models are optimized with Adam optimizer and the Mean Square Error (MSE) loss function.

---

[2] https://keras.io/.

[3] https://www.tensorflow.org/.

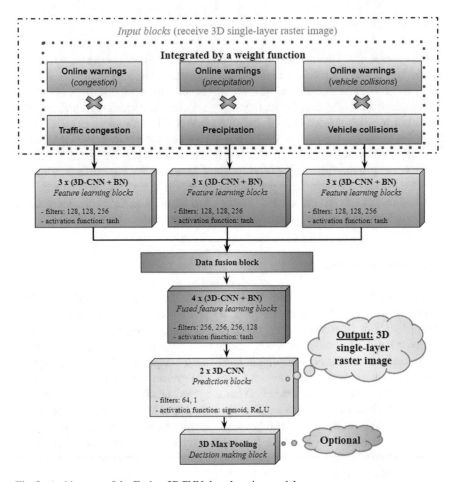

Fig. 2 Architecture of the Fusion-3DCNN deep learning model

4 Dynamic System to Discover Future Traffic Congestion Patterns

In this section, we introduce the dynamic system to discover predicted traffic congestion patterns dynamically. The system aims to generate dynamic maps containing future high traffic demand regions over different periods. It provides holistic information to authorities about future traffic congestion situations and high traffic demand regions according to different periods, helping people make effective traffic management strategies. The system is developed in three steps as follow:

1. Predicting long-term traffic congestion using multi-modal spatiotemporal urban sensing data;
2. Extracting future heavily congested mesh codes; and

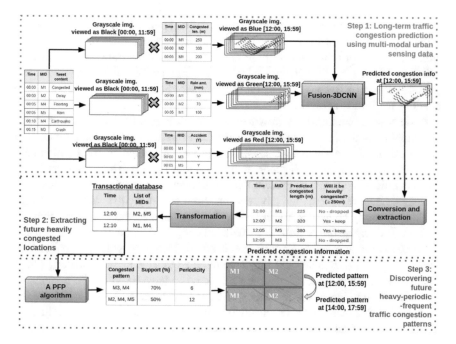

Fig. 3 Development process of the dynamic system (*"MID" stands for "Mesh code IDentifier"*)

3. Mining future high traffic demand regions.

The workflow of the system is briefly illustrated in Fig. 3.

4.1 Predicting Future Long-Term Traffic Congestion

The first step is to build a traffic congestion prediction system, which is viewed as System 1 of the dynamic system. It aims to anticipate the long-term congestion length of citywide mesh codes in the examined region. It utilizes the Fusion-3DCNN deep learning model introduced in Sect. 3. The model uses multi-modal spatiotemporal urban sensing data for congestion prediction.

4.2 Extracting Future Heavily Congested Mesh Codes

The main target of the second step is to extract future heavily congested mesh codes. As we all know, there are many locations (i.e., mesh codes) having traffic congestion at a time instance in real life. It is also true for predicted congestion data. The majority of mesh codes would be predicted to have light and normal traffic congestion, which will disappear in normal circumstances without police intervention. Currently, traffic police forces in many cities worldwide are running short of resources, so authorities must prioritize some mesh codes to monitor the traffic at certain time instances. Obviously, the mesh codes included in high traffic demand should be paid special attention to. Because of that, this step aims to choose future heavily congested locations for further analysis.

Besides, since both the PFP-growth++ and maxPFP-growth algorithms only work on temporal databases, we need to convert the extracted data to that format. It is performed as follows. Firstly, we convert the predicted congestion information generated by the Fusion-3DCNN model to a time series dataset. There are three fields per row: timestamp, mesh code identifier and predicted congestion length measured in meters. Secondly, we define a binary threshold called *heavyCongestion* to filter future heavily congested mesh codes. For example, in Fig. 3, assume that we set *heavilyCongestion* = 250. The predicted congested length in M1 at 02:00 and M2 at 02:05 are dropped since their predicted congestion lengths are smaller than *heavyCongestion*. It is because they are predicted to have light traffic congestion. Finally, we group mesh codes that are predicted to have heavy traffic congestion according to time instances. In other words, at time instance T_i, we will have the list L_i containing mesh codes predicted to be heavily congested. Please note that the database does not store the predicted congestion length of mesh codes.

4.3   Discovering Future High Traffic Demand Regions

The purpose of Step 3 is to discover future high traffic demand regions. They contain sets of mesh codes predicted to have heavy-periodic-frequent traffic congestion. It is done by using either the PFP-growth++ or maxPFP-growth algorithms based on the temporal database prepared in Step 2. The operations behind those algorithms have been explained in Sect. 2.2. This component is viewed as System 2 of the dynamic system.

5   Experiments and Promising Applications

This section explains the dataset used for experiments and followed by the results of our proposed methods. The section finishes with some compelling use cases that we have discovered which could be used to tackle traffic congestion in the real world.

Fig. 4 The geographical region of collected data

5.1 Experimental Setup

For the experimental purposes, we use the datasets about traffic congestion, rainfall amount, traffic accidents, and Twitter posts collected in Osaka Bay, Kobe City, Japan. That region is illustrated in Fig. 4 and covers a portion of 20 km × 15 km on the Earth. The data is collected in two periods: 01/05/2014 to 31/10/2014 and 01/05/2015 to 31/10/2015 (format: day/month/year).

- The **traffic congestion** data [13] contains 658 items (equivalent to mesh codes) and 8,707,516 transactions. The format of the data is "timestamp: mesh code identifier: average congestion length of road segments running through the mesh code (m)."
- The **precipitation** (or rainfall) data [14] contained 1,327 items and 747,398 transactions. The format of the data is "timestamp: mesh code identifier: rainfall amount (mm)."
- The **traffic accidents** data [25] contains 283 items and 1,100 transactions. The data format is "timestamp: mesh code identifier."
- The **social networking** data [31] contains 1,562 items and 15,627 transactions. The format of the data is "timestamp: mesh code identifier: Tweet's content."

All the related algorithms, namely, (1) training learning models for traffic congestion prediction presented in Sect. 5.2, (2) extracting future heavily congested locations and storing in a temporal database, and (3) PFP-growth++, maxPFP-growth were written in Python 3.6.9. They are executed on a machine with an Intel i5-3470 CPU running at 3.20GHz and 12GB of RAM. The graphics card of this machine is Nvidia GeForce GTX 750 Ti with 2GB of VRAM. The operating system of the machine is Ubuntu 18.04LTS.

5.2 Evaluation of the Fusion-3DCNN Deep Learning Model

In this section, we evaluate the predictive performance of the Fusion-3DCNN deep learning models. In this context, we compare the predictive accuracy of Fusion-3DCNN against the 3DCNN multi-source deep learning model [8] and some other baselines listed below. The models' predictive accuracies are measured by the Mean Absolute Error (MAE) metric. The lower the predictive error of the model is, the higher the predictive accuracy it obtains.

In the evaluation, all models will predict traffic congestion in the next 6 h with three immediate timestamps (the interval between two consecutive timestamps is 2 h). To do so, the models look back 12 h of collected urban sensing data sources. In a nutshell, we prepare and compare the following state-of-the-art and highly influential models' predictive accuracy. Models 1–3 use single-source data that is traffic congestion information. Models 4–6 utilize multi-source data that consists of traffic congestion, rainfall amount, number of accidents, and Tweets.

1. **Historical Average:** this model is used in many real-world traffic congestion prediction systems [21] and has a very efficient runtime. Many traffic congestion prediction studies also employed this model as a baseline like [6, 12, 23, 35, 36]. We use the average traffic congestion lengths of observed data as the predicted result for all three immediate timestamps.
2. **Vector Autoregression:** this is a traditional statistical model and used widely in traffic congestion prediction studies like [26, 29, 41].
3. **3DCNN:** this deep learning model uses only one data source that is traffic congestion. We self-develop this model and make its architecture similar to the 3DCNN multi-source and Fusion-3DCNN models. The difference between this model and the other two is it does not have two data fusion functions explained in Sect. 3.2.1 and learning branches related to exogenous data factors shown in Fig. 2. The use of this model aims to evaluate the effectiveness of our approach in using multi-modal urban sensing data.
4. **Seq2Seq:** this model was proposed in [20]. It employed the Sequence to Sequence (Seq2Seq) architecture based on LSTM. We implement Liao et al.'s proposed Seq2Seq + AT (stands for Attributes) + NB (stands for Neighboring–Spatial Relation) model that uses auxiliary information and neighboring traffic data. Specifically, the AT component contains exogenous data like weather, accidents, and online warnings. It is placed at the Decoder part of the model. For the NB module, we use traffic data of two nearby locations. It is placed in the Encoder part of the network.

Table 2 Experimental results of long-term traffic congestion prediction models (measured by MAE, lower is better)

Model	MAE
Historical average	10.24
Vector autoregression	9.44
3DCNN	8.82
Seq2Seq AT+NB [20]	8.75
3DCNN multi-source [8]	8.67
Fusion-3DCNN (this work)	**8.13**

5. **3DCNN multi-source**: this model was proposed in [8] and has been briefly explained in Sect. 3.3. The model uses all four data sources introduced in Sect. 5.1 and treats them as independent learnable factors.
6. **Fusion-3DCNN** ($p = 0.5$): this model uses all four data sources and treats the social networking data dependent on environmental factors. $p = 0.5$ indicates the integrating weight used in Eq. 2. It means that the impacts of the environmental factors on traffic congestion are reduced by 50%.

We use the data collected from 01/05/2014 to 31/10/2014 to train models. The data got from 01/05/2015 to 31/10/2015 is reserve to test/evaluate the models. Table 2 presents predictive error values of the models.

As shown clearly from Table 2, the Fusion-3DCNN model is the best performer among participants. Paying attention to detail, models 4–6 have higher predictive accuracy than models 1–3. The results indicate that using multi-source data helps improve the predictive performance of the models. Comparing predictive performance between models 1–3, we can observe that the 3DCNN model is the best performer. It shows that reserving spatiotemporal correlations of data done by 3DCNN layers enhances predictive accuracy. Therefore, our choice in utilizing such layers in building our models is correct. Moving models 4–6, we can see that Seq2Seq [20] has lower predictive accuracy than our proposed models even though all three models use the same set of data sources. We can conclude that LSTM components used inside Seq2Seq are not as good as 3DCNN in capturing spatiotemporal dependencies. Again, it rectifies our choice in using 3DCNN. Finally, the Fusion-3DCNN model has predictive accuracy higher than the 3DCNN multi-source model [8] by about 6%. It indicates that making online warnings dependent on environmental factors will significantly enhance predictive accuracy rather than treating all data sources as independent factors.

Since the Fusion-3DCNN model proposed in this work is the best performer, using the predicted data generated by it will ensure the output's quality of the subsequent system. Therefore, we will perform experiments on the PFP-growth++ and maxPFP-growth algorithms on the data generated by it. They will be presented in later sections.

5.3   Execution and Evaluation of the PFP-growth++ and MaxPFP-Growth Algorithms

This section explains how we execute and evaluate the PFP-growth++ and maxPFP-growth algorithms. The two algorithms' input data is the predicted congestion information generated by the Fusion-3DCNN model proposed in this work. Please note that the data has been post-processed, as explained in Sect. 4.2.

As explained in Sect. 5.2, the Fusion-3DCNN model predicts future traffic congestion information in 6-hour intervals. During the predicted periods, three immediate timestamps are evenly spaced by two hours. As we all know, two-hour is usually the duration of traffic rush hour in most cities worldwide. For example, the rush hour in Japan is 07:00 to 09:00 in the mornings and 17:00 to 19:00 in the evenings. Therefore, we will discover future high traffic demand regions in two-hour periods as they will give authorities more useful information to deal with traffic congestion. We also slide discovery periods by one hour in order to detect interesting patterns during the transitional time. As a result, we will have 24 transactional databases containing predicted congestion information per day. The first one is called $data$-1 contains predicted congestion data from 00:00 to 01:59. The second one $data$-2 concerns data from 01:00 to 03:59, and so on. Each transactional database contains many transactions. Each consists of "timestamp" and "the list of mesh code identifiers predicted to have heavy traffic congestion at $timestamp$." In other words, the future heavily congested mesh code is predicted to have an average congestion length of at least $heavyCongestion$ meters. The described transactional databases are used by the PFP-growth++ and maxPFP-growth algorithms, which will generate k−patterns ($k > 1$). Each consists of k mesh codes that are predicted to have heavy-periodic-frequent traffic congestion. In other words, each k−pattern is a high traffic demand region concerning k mesh codes.

Next, we evaluate the PFP-growth++ and maxPFP-growth algorithms by two criteria: (1) the number of generated patterns and (2) execution time. We fix the value of $heavyCongestion$ to 350. It means that future heavy congestion will have an average traffic congestion length of at least 350 m. That value is defined by expert knowledge. We vary the values of $minSup$ by {30%, 50%, 70%, 90%} and $periodity$ by {15 min, 30 min, 45 min, 1 h}. For example, any k traffic congestion patterns discovered on the combination of $(heavyCongestion, minSup, periodicity) = (350, 50\%, 30$ minutes) will have the following characteristics: "all k involved mesh codes are predicted to have congestion length larger than or equal to 350m at least 50% and repeat per 30 min."

Firstly, we fix the value of $minSup$ to 30%. We illustrate the statistics according to the two criteria mentioned above in Fig. 5. For each parameter's combination shown in the chart, the bars show the number of generated patterns by the algorithms, and the lines reveal their execution time. We also write the algorithms' statistical values on the corresponding setting to show differences more clearly. As can be seen, the numbers of patterns discovered by the PFP-growth++ algorithm are significantly higher than that of maxPFP-growth. At the same time, the execution time of the former is also

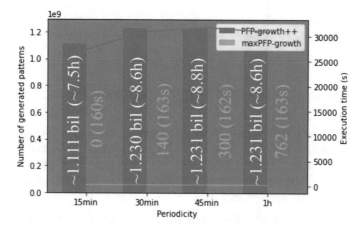

Fig. 5 Comparison of PFP-growth++ and maxPFP-growth at fixed $minSup = 30\%$

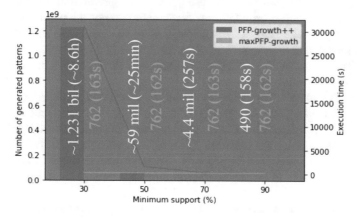

Fig. 6 Comparison of PFP-growth++ and maxPFP-growth at fixed $periodicity = 1h$

longer than the latter. Digging into patterns generated by PFP-growth++, we see that they are mostly uninteresting as the patterns are redundant. Taking the following 4-pattern "89114;98143;98144;98145" as an example. The PFP-growth++ algorithm discovers in total $2^4 - 1 = 15$ patterns containing items in that 4-pattern, which leads to difficulty in analyzing the data. In contrast, the output of the maxPFP-growth algorithm contains only one pattern, which is the longest. This finding also confirms the observation discussed in [16].

Next, we fix the value of $periodicity$ to 1h and vary the values of $minSup$. We show the statistics according to the two criteria mentioned above in Fig. 6. We can also observe the same patterns discussed in Fig. 5. The number of generated patterns of the PFP-growth++ algorithm is higher than maxPFP-growth. Besides, the former's execution time is also longer than the latter.

Obviously, the maxPFP-growth algorithm is better than PFP-growth++ at discovering periodic-frequent patterns, so we will utilize it later to analyze the data.

5.4 Analyses of Discovered Traffic Congestion Patterns

We found that the maxPFP-growth algorithm produces the most adequate patterns on combination $(heavyCongestion, minSup, periodicity) = (350, 30\%, 1hour)$. Therefore, in this section, we analyze future traffic congestion patterns discovered by executing the algorithm on that combination. Firstly, we present some interesting patterns according to the time below:

1. The periods from 02:00 to 05:00 and 07:00 to 09:00 per day are the most anticipated to have high traffic demand regions. For the periods from 07:00 to 09:00, they are coincident with Japan's rush hour. We suggest authorities allocate more resources to monitor traffic during that timeframe. For the periods from 02:00 to 05:00, please see the analysis below. It is worth mentioning that few patterns are discovered in afternoons and evenings (from 12:00 onwards). Except for the timeframe from 16:00 to 18:00, many patterns were discovered but not as many as of morning's rush hour.
2. Saturdays and Sundays are the days that are most often predicted to have high traffic demand regions. Interestingly, 02:00 to 04:00 is the most crowded time on Saturdays. In indicates the locals' intent to have late parties on Friday nights and come back home around that time. The most congested period on Sunday is from 08:00 to 13:00, explained by people's weekend trips. We suggest the authorities allocate resources to monitor traffic on Sunday mornings to prevent long-term congestion.
3. June and August are the two months with the highest traffic demand regions discovered, collectively accounting for 50%. June is the month with the highest average rainfall amount annually in Kobe City, Japan, indicating that more rain leads to more severe traffic congestion. The same observation is explained for August as this month witnesses many thunderstorms.

Next, we describe some aspiring patterns discovered according to mesh codes below:

1. The region covering Route 428 connecting two big residential areas of Kobe City is mostly predicted to be in high traffic demand and has the biggest number of high traffic demand mesh codes. Route 428 is the only and nearest way to connect those areas. The most crowded period of this region is from 06:00 to 09:00.
2. The second crowded region involves a working complex (water treatment plant) and airport. The region is located right at the center of Osaka Bay, Kobe City, and on the city's main coastal road where people have to travel through on their paths. The most crowded time of this region is from 08:00 to 09:00, being coincident with Japan's rush hour.

5.5   *A Case Study of Future High Traffic Demand Regions*

In this section, we present a case study, apart from patterns presented in the previous section, to explain the factors behind forming a traffic demand region. As we are all aware, regions that usually witness heavy traffic congestion are likely to involve high public infrastructure. In Fig. 7, we illustrate such an example. The shown region is usually predicted to be in high traffic demand. Looking at the figure, we can see it contains many public facilities like three hospitals and two elementary schools. It is straightforward for authorities to determine which geographical regions have complex public infrastructure. However, it is challenging for them to know when those regions will experience high traffic demand. With the help of the proposed dynamic system, they can get such insights with ease, thus, enhancing their efficacy in managing traffic flows and dealing with traffic congestion.

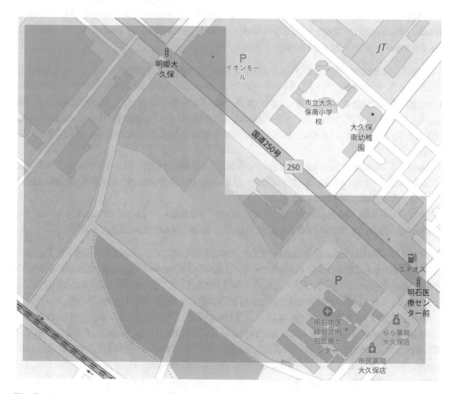

Fig. 7   A case study explaining future high traffic demand regions

5.6   Proposals of Use Cases to Tackle Future Traffic Congestion

This section proposes two use cases that the authorities can build to tackle future traffic congestion. They are discovered by using the output of the dynamic system.

5.6.1   Building a Database of Future Traffic Management Strategies

In this section, we present the first use case. It is done solely by using the output of the dynamic system. As we know, in cities, some traffic hotspots usually encounter high traffic demand. The traffic flows at those locations require special care from traffic police officers. However, the effectiveness of dealing with traffic congestion varies among officers. Some may employ very effective strategies, while some may not. Therefore, if the effective plans are shared among officers, a high level of efficacy in dealing with traffic congestion could be ensured. Motivated by that goal, we propose the following procedure to achieve that ambition.

- Step 1: Determining the duration of traffic congestion patterns. As explained in Sect. 5.3, if a k-pattern is discovered in a transactional database $data\text{-}x$, the corresponding region of k mesh codes is predicted to have high traffic demand for 2 h covered by $data\text{-}x$. In this context, when observing one or many consecutive $data\text{-}x_i$ transactional databases that the pattern is discovered, we will know its duration.
- Step 2: Grouping similar patterns and accumulating them in a database for future analyses. We propose to gather patterns with similar duration and time instances of starting time in the same group.
- Step 3: When the number of accumulated patterns is large enough, choose the most popular patterns and prepare a traffic management strategy for them.

Figure 8 illustrates an example of the presented use case. Herein, using the Fusion-3DCNN model allows us to know how traffic congestion will happen in the future with exact time instances ($T1$, $T5$, ...). We will then discover traffic congestion patterns employing a periodic-frequent pattern (PFP) mining algorithm (preferably, maxPFP-growth). Suppose during periods [$T3$, $T5$] and [$T7$, $T9$], the dynamic system predicts a set of patterns that have been anticipated to happen many times. In this case, the officers just need to take the predefined strategies stored in the database to manage traffic flows in those time instances. For other milestones, they may apply flexible plans according to real-life situations. If they happen more frequently in the future, a strategy to tackle congestion in such regions could be defined.

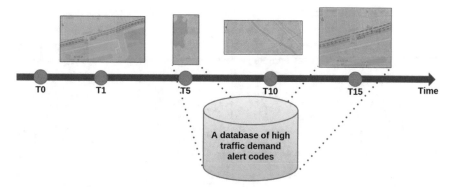

Fig. 8 Making strategies to deal with future traffic congestion

5.6.2 Determining the Right Time to Tackle Traffic Congestion

This section presents another use case to mitigate traffic congestion by combining the Fusion-3DCNN model's output information and the dynamic system. The Fusion-3DCNN model aims to predict the average traffic congestion length of citywide mesh codes at different time instances. The dynamic system's output is a set of traffic congestion patterns that concern regions predicted to be heavily-periodically-frequently congested. Combining such information will give officers how the traffic demand of a region transform from low to high. Therefore, looking at that, authorities can decide the right timing to start monitoring traffic to prevent the traffic demand of that region from becoming high.

Figure 9 illustrates an example and related concepts. The figure shows the predicted congestion information of a location starting from 14/08/2015 06:00:00 GMT+09 (shown as $(T + 1)$ on the chart). The height of bars indicates the average congestion length of the corresponding time instances predicted by the Fusion-3DCNN model. The dynamic system discovered that this location would be in high traffic demand from 09:00:00 to 14:00:00 (totally, 5 h). The bars concerning the time instances during which the location will be in high traffic demand are shown in red. In contrast, blue bars reveal time instances that the location will be in low traffic demand. It can be seen from the chart, the average traffic congestion length of this location is predicted to increase from $(T + 1)$ to $(T + 3)$, resulting in the traffic demand of the location become high from $(T + 4)$ to $(T + 8)$. If the traffic policemen could manage traffic flows effectively to reduce traffic congestion severity during the period from $(T + 1)$ to $(T + 3)$, congestion length in later time instances could be alleviated. Residents could also be notified about that potentially dangerous region so that they may avoid traveling. As a result, the duration of high traffic demand in this region might be shortened (i.e., from 12 h to 4–8 h). Please note that although the congestion lengths from $(T + 9)$ to $(T + 11)$ are higher than $[(T + 4), (T + 8)$, that period will not be considered as a high traffic demand period. It is because this region is predicted to be not heavily-periodically-frequently congested.

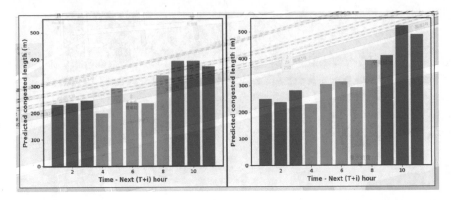

Fig. 9 Example of a predicted congestion situation of a high traffic demand location

6 Discussions

In this section, we discuss open topics relating to what people might investigate to improve our system.

6.1 Abnormal and Sudden Factors

Abnormal factors also play a significant role in traffic congestion management. For example, social events (e.g., Olympic, Woodstock, expo fair, film festival) regularly happen at a particular time that might increase traffic congestion. Nevertheless, upon completion of these events, traffic should come back to normal. Another example of abnormal factors is Covid times. During this time, traffic congestion may decrease in many parts of the world. The common point of these factors is that they can be known or aware before. Therefore, to cope with these factors, we can utilize the fixed weight to consider their impacts, as described in Equation (2). The adaptive model, specially designed for these events, has one more input channel with a related fixed weight.

Sudden factors are another concern in traffic congestion management. These factors do not give us time to prepare. When they happen, the chaos will immediately burst out. For example, natural disasters such as sudden snowfall, earthquake, and flood can cause chaos in traffic congestion management. Thus, we can consider Equation (3) with learnable weight representing these factors. The learnable model is upgraded from the original model with learnable weights to have the ability to predict traffic patterns in this case.

Another solution that might work for both cases mentions above is the conditional-Fusion-3D-CNN, when one layer of <spation, temporal> condition is added to specify when and where abnormal or sudden events happened (or will happen). The training can be done with historical data.

6.2 Visualization Dashboard

The strengths of our system can be maximized if there is a visualization dashboard that would cater to different types of stakeholders. Different stakeholders need to see different information. For example, traffic police need to see causality of congestion such as accidents and road obstacles or the priority of each congestion zone to coordinate workforce. Fire brigade authorities need to see the optimal route, especially the possibility of passing through, to reach the target (because fire trucks are large and need more space to travel on the road). Town planners need additional information about the neighborhood (e.g., shops, facilities) on the roads to cope with heavy traffic infrastructure zones management.

We will integrate our system into the cross-data collaboration platform introduced in [15]. Using the smart navigation design tools of this platform, we expect to have a visualization dashboard that satisfies the criteria mentioned above.

7 Conclusion

This study introduces a novel dynamic system to discover future heavy-periodic-frequent traffic congestion (high traffic demand) regions consisting of two systems: (1) traffic congestion prediction system and (2) periodic-frequent pattern discovery model. The system works as follows. Firstly, the traffic congestion prediction system predicts citywide traffic congestion at multiple time instances. After that, a periodic-frequent pattern discovery algorithm is utilized to extract useful traffic congestion patterns. The output of the system is (1) predicted congestion situations at multiple time instances, (2) high traffic demand regions, and (3) duration that regions are predicted to experience high traffic demand.

Our work also proposes the Fusion-3DCNN deep learning model which is an enhanced version of the 3DCNN multi-source deep learning model [8] for long-term traffic congestion by proposing a new data fusion function. It makes social networking data dependent on environmental information (congestion, rainfall, accident). The experimental results show that the predictive accuracy of the enhanced model improved significantly compared to the original.

Finally, we propose two use cases that authorities can use to deal with future traffic congestion more effectively.

In the future, we will present some new use cases of the system and propose new integration schemes of models.

Acknowledgements We appreciate the contribution of Ngoc-Thanh Nguyen, currently Ph.D. students of Western Norway University of Applied Sciences, to this chapter.

References

1. B. Alkouz, Z. Al Aghbari, SNSJam: road traffic analysis and prediction by fusing data from multiple social networks. Inf. Process. Manag. **57**(1), 102139 (2020)
2. B. Alsolami, R. Mehmood, A. Albeshri, *Hybrid Statistical and Machine Learning Methods for Road Traffic Prediction: A Review and Tutorial* (Springer International Publishing, Cham, 2020), pp. 115–133
3. S. An, H. Yang, J. Wang, N. Cui, J. Cui, Mining urban recurrent congestion evolution patterns from GPS-equipped vehicle mobility data. Inf. Sci. **373**, 515–526 (2016)
4. N. Bjorck, C.P. Gomes, B. Selman, K.Q. Weinberger, Understanding batch normalization, in *Advances in Neural Information Processing Systems*, pp. 7694–7705 (2018)
5. G. Booch, F. Fabiano, L. Horesh, K. Kate, J. Lenchner, N. Linck, A. Loreggia, K. Murugesan, N. Mattei, F. Rossi et al., Thinking fast and slow in ai (2020). arXiv:2010.06002
6. M. Chen, X. Yu, Y. Liu, PCNN: deep convolutional networks for short-term traffic congestion prediction. IEEE Trans. Intell. Transp. Syst. **19**(11), 3550–3559 (2018)
7. C. Chou, Y. Huang, C. Huang, V. Tseng, Long-term traffic time prediction using deep learning with integration of weather effect, in *Advances in Knowledge Discovery and Data Mining* (Springer International Publishing, Cham, 2019), pp. 123–135
8. M. Dao, N. Nguyen, K. Zettsu, Multi-time-horizon traffic risk prediction using spatio-temporal urban sensing data fusion, in *2019 IEEE International Conference on Big Data (Big Data)*, pp. 2205–2214 (2019)
9. M.-S. Dao, K. Zettsu, Complex event analysis of urban environmental data based on deep CNN of spatiotemporal raster images, in *2018 IEEE International Conference on Big Data (Big Data)* (IEEE, 2018), pp. 2160–2169
10. X. Di, Y. Xiao, C. Zhu, Y. Deng, Q. Zhao, W. Rao, Traffic congestion prediction by spatiotemporal propagation patterns, in *2019 20th IEEE International Conference on Mobile Data Management (MDM)*, pages 298–303 (2019)
11. X. Fan, C. Xiang, C. Chen, P. Yang, L. Gong, X. Song, P. Nanda, X. He, BuildSenSys: reusing building sensing data for traffic prediction with cross-domain learning. IEEE Trans. Mob. Comput., pp. 1–1 (2020)
12. S. Guo, Y. Lin, S. Li, Z. Chen, H. Wan, Deep spatial-temporal 3d convolutional neural networks for traffic data forecasting. IEEE Trans. Intell. Transp. Syst. **20**(10), 3913–3926 (2019)
13. Japan Road Traffic InformationCenter. Traffic congestion data. www.jartic.or.jp
14. Data Integration and Analysis System Program. Precipitation data. www.diasjp.net
15. S. Ito, Z. Koji, Assessing a risk-avoidance navigation system based on localized torrential rain data. MATEC Web Conf. **308**, 03006 (2020)
16. R.U. Kiran, Y. Watanobe, B. Chaudhury, K. Zettsu, M. Toyoda, M. Kitsuregawa, Discovering maximal periodic-frequent patterns in very large temporal databases, in *2020 IEEE 7th International Conference on Data Science and Advanced Analytics (DSAA)*, pp. 11–20 (2020)
17. R.U. Kiran, M. Kitsuregawa, P.K. Reddy, Efficient discovery of periodic-frequent patterns in very large databases. J. Syst. Softw. **112**, 110 – 121 (2016)
18. I. Lana, J. Del Ser, M. Velez, E.I. Vlahogianni, Road traffic forecasting: recent advances and new challenges. IEEE Intell. Transp. Syst. Mag. **10**(2), 93–109 (2018)
19. Y. Li, C. Shahabi, A brief overview of machine learning methods for short-term traffic forecasting and future directions. SIGSPATIAL Spec. **10**(1), 3–9 (2018)
20. B. Liao, J. Zhang, C. Wu, D. McIlwraith, T. Chen, S. Yang, Y. Guo, F. Wu, Deep sequence learning with auxiliary information for traffic prediction, in *Proceedings of the 24th ACM SIGKDD International Conference on Knowledge Discovery & Data Mining*, KDD '18 (ACM, New York, 2018), pp. 537–546

21. A.M. Nagy, V. Simon, Survey on traffic prediction in smart cities. Pervasive Mob. Comput. **50**, 148–163 (2018)
22. H. Nguyen, W. Liu, F. Chen, Discovering congestion propagation patterns in spatio-temporal traffic data. IEEE Trans. Big Data **3**(2), 169–180 (2016)
23. Z. Pan, Y. Liang, W. Wang, Y. Yu, Y. Zheng, J. Zhang, Urban traffic prediction from spatio-temporal data using deep meta learning, in *Proceedings of the 25th ACM SIGKDD International Conference on Knowledge Discovery & Data Mining*, KDD '19 (ACM, New York, 2019), pp. 1720–1730
24. N. Pourebrahim, S. Sultana, J. Thill, S. Mohanty, Enhancing trip distribution prediction with twitter data: comparison of neural network and gravity models, in *Proceedings of the 2Nd ACM SIGSPATIAL International Workshop on AI for Geographic Knowledge Discovery*, GeoAI'18 (ACM, New York, 2018), pp. 5–8
25. Traffic Accident Research and Analysis. Traffic accident data. www.itarda.or.jp
26. F. Schimbinschi, L. Moreira-Matias, V.X. Nguyen, J. Bailey, Topology-regularized universal vector autoregression for traffic forecasting in large urban areas. Expert. Syst. Appl. **82**, 301–316 (2017)
27. X. Shi, H. Qi, Y. Shen, G. Wu, B. Yin, A spatial-temporal attention approach for traffic prediction. IEEE Trans. Intell. Transp. Syst., pp. 1–10 (2020)
28. J. Song, C. Zhao, S. Zhong, T.A.S. Nielsen, A.V. Prishchepov, Mapping spatio-temporal patterns and detecting the factors of traffic congestion with multi-source data fusion and mining techniques. Comput., Environ. Urban Syst. **77**, 101364 (2019)
29. J. Tang, F. Liu, Y. Zou, W. Zhang, Y. Wang, An improved fuzzy neural network for traffic speed prediction considering periodic characteristic. IEEE Trans. Intell. Transp. Syst. **18**(9), 2340–2350 (2017)
30. D.A. Tedjopurnomo, Z. Bao, B. Zheng, F. Choudhury, A.K. Qin, A survey on modern deep neural network for traffic prediction: trends, methods and challenges. IEEE Trans. Knowl. Data Eng., pp. 1–1 (2020)
31. Twitter. Social networks data. www.twitter.com
32. D. Xie, M. Wang, X. Zhao, A spatiotemporal apriori approach to capture dynamic associations of regional traffic congestion. IEEE Access **8**, 3695–3709 (2020)
33. L. Yang, L. Wang, Mining traffic congestion propagation patterns based on spatio-temporal co-location patterns. *Evolutionary Intelligence* (2019)
34. L. Yang, L. Wang, Mining traffic congestion propagation patterns based on spatio-temporal co-location patterns. *Evolutionary Intelligence*, pp. 1–13 (2019)
35. R. Yu, Y. Li, C. Shahabi, U. Demiryurek, Y. Liu, Deep learning: a generic approach for extreme condition traffic forecasting, in *Proceedings of the 2017 SIAM international Conference on Data Mining* (SIAM, 2017), pp. 777–785
36. X. Yuan, Z. Zhou, T. Yang, Hetero-ConvLSTM: a deep learning approach to traffic accident prediction on heterogeneous spatio-temporal data, in *Proceedings of the 24th ACM SIGKDD International Conference on Knowledge Discovery & Data Mining*, KDD '18 (ACM, New York, 2018), pp. 984–992
37. P. Zhao, H. Haoyu, Geographical patterns of traffic congestion in growing megacities: big data analytics from Beijing. Cities **92**, 164–174 (2019)
38. Y. Zheng, L. Liao, F. Zou, M. Xu, Z. Chen, PLSTM: long short-term memory neural networks for propagatable traffic congested states prediction, in *International Conference on Genetic and Evolutionary Computing* (Springer, Berlin, 2019), pp. 399–406
39. H. Zhou, K. Hirasawa, Spatiotemporal traffic network analysis: technology and applications. Knowl. Inf. Syst., pp. 1–37 (2018)
40. J. Zhu, C. Huang, M. Yang, G. Pui, C. Fung, Context-based prediction for road traffic state using trajectory pattern mining and recurrent convolutional neural networks. Inform. Sci. **473**, 190–201 (2019)
41. Y. Zou, X. Hua, Y. Zhang, Y. Wang, Hybrid short-term freeway speed prediction methods based on periodic analysis. Can. J. Civ. Eng. **42**(8), 570–582 (2015)

Printed in the United States
by Baker & Taylor Publisher Services